Fatos Xhafa and Ajith Abraham (Eds.)

Metaheuristics for Scheduling in Distributed Computing Environments

Studies in Computational Intelligence, Volume 146

Editor-in-Chief
Prof. Janusz Kacprzyk
Systems Research Institute
Polish Academy of Sciences
ul. Newelska 6
01-447 Warsaw
Poland
E-mail: kacprzyk@ibspan.waw.pl

Further volumes of this series can be found on our homepage:
springer.com

Vol. 124. Ricardo Zavala Yoe
Modelling and Control of Dynamical Systems: Numerical Implementation in a Behavioral Framework, 2008
ISBN 978-3-540-78734-1

Vol. 125. Larry Bull, Bernadó-Mansilla Ester and John Holmes (Eds.)
Learning Classifier Systems in Data Mining, 2008
ISBN 978-3-540-78978-9

Vol. 126. Oleg Okun and Giorgio Valentini (Eds.)
Supervised and Unsupervised Ensemble Methods and their Applications, 2008
ISBN 978-3-540-78980-2

Vol. 127. Régie Gras, Einoshin Suzuki, Fabrice Guillet and Filippo Spagnolo (Eds.)
Statistical Implicative Analysis, 2008
ISBN 978-3-540-78982-6

Vol. 128. Fatos Xhafa and Ajith Abraham (Eds.)
Metaheuristics for Scheduling in Industrial and Manufacturing Applications, 2008
ISBN 978-3-540-78984-0

Vol. 129. Natalio Krasnogor, Giuseppe Nicosia, Mario Pavone and David Pelta (Eds.)
Nature Inspired Cooperative Strategies for Optimization (NICSO 2007), 2008
ISBN 978-3-540-78986-4

Vol. 130. Richi Nayak, Nikhil Ichalkaranje and Lakhmi C. Jain (Eds.)
Evolution of the Web in Artificial Intelligence Environments, 2008
ISBN 978-3-540-79139-3

Vol. 131. Roger Lee and Haeng-Kon Kim (Eds.)
Computer and Information Science, 2008
ISBN 978-3-540-79186-7

Vol. 132. Danil Prokhorov (Ed.)
Computational Intelligence in Automotive Applications, 2008
ISBN 978-3-540-79256-7

Vol. 133. Manuel Graña and Richard J. Duro (Eds.)
Computational Intelligence for Remote Sensing, 2008
ISBN 978-3-540-79352-6

Vol. 134. Ngoc Thanh Nguyen and Radoslaw Katarzyniak (Eds.)
New Challenges in Applied Intelligence Technologies, 2008
ISBN 978-3-540-79354-0

Vol. 135. Hsinchun Chen and Christopher C. Yang (Eds.)
Intelligence and Security Informatics, 2008
ISBN 978-3-540-69207-2

Vol. 136. Carlos Cotta, Marc Sevaux and Kenneth Sörensen (Eds.)
Adaptive and Multilevel Metaheuristics, 2008
ISBN 978-3-540-79437-0

Vol. 137. Lakhmi C. Jain, Mika Sato-Ilic, Maria Virvou, George A. Tsihrintzis, Valentina Emilia Balas and Canicious Abeynayake (Eds.)
Computational Intelligence Paradigms, 2008
ISBN 978-3-540-79473-8

Vol. 138. Bruno Apolloni, Witold Pedrycz, Simone Bassis and Dario Malchiodi
The Puzzle of Granular Computing, 2008
ISBN 978-3-540-79863-7

Vol. 139. Jan Drugowitsch
Design and Analysis of Learning Classifier Systems, 2008
ISBN 978-3-540-79865-1

Vol. 140. Nadia Magnenat-Thalmann, Lakhmi C. Jain and N. Ichalkaranje (Eds.)
New Advances in Virtual Humans, 2008
ISBN 978-3-540-79867-5

Vol. 141. Christa Sommerer, Lakhmi C. Jain and Laurent Mignonneau (Eds.)
The Art and Science of Interface and Interaction Design (Vol. 1), 2008
ISBN 978-3-540-79869-9

Vol. 142. George A. Tsihrintzis, Maria Virvou, Robert J. Howlett and Lakhmi C. Jain (Eds.)
New Directions in Intelligent Interactive Multimedia, 2008
ISBN 978-3-540-68126-7

Vol. 143. Uday K. Chakraborty (Ed.)
Advances in Differential Evolution, 2008
ISBN 978-3-540-68827-3

Vol. 144. Andreas Fink and Franz Rothlauf (Eds.)
Advances in Computational Intelligence in Transport, Logistics, and Supply Chain Management, 2008
ISBN 978-3-540-69024-5

Vol. 145. Mikhail Ju. Moshkov, Marcin Piliszczuk and Beata Zielosko
Partial Covers, Reducts and Decision Rules in Rough Sets, 2008
ISBN 978-3-540-69027-6

Vol. 146. Fatos Xhafa and Ajith Abraham (Eds.)
Metaheuristics for Scheduling in Distributed Computing Environments, 2008
ISBN 978-3-540-69260-7

Fatos Xhafa
Ajith Abraham
(Eds.)

Metaheuristics for Scheduling in Distributed Computing Environments

Springer

Dr. Fatos Xhafa
Departament de Llenguatges i Sistemes Informàtics
Universitat Politècnica de Catalunya
Campus Nord, Ed. Omega, C/Jordi Girona 1-3
08034 Barcelona
Spain
Email: fatos@lsi.upc.edu

Dr. Ajith Abraham
Center for Quantifiable Quality of
Service in Communication Systems
Faculty of Information Technology,
Mathematics and Electrical Engineering
Norwegian University of Science and Technology
O.S. Bragstads plass 2E
7491 Trondheim
Norway
Email: ajith.abraham@ieee.org

ISBN 978-3-540-69260-7   e-ISBN 978-3-540-69277-5

DOI 10.1007/978-3-540-69277-5

Studies in Computational Intelligence   ISSN 1860949X

Library of Congress Control Number: 2008928448

© 2008 Springer-Verlag Berlin Heidelberg

This work is subject to copyright. All rights are reserved, whether the whole or part of the material is concerned, specifically the rights of translation, reprinting, reuse of illustrations, recitation, broadcasting, reproduction on microfilm or in any other way, and storage in data banks. Duplication of this publication or parts thereof is permitted only under the provisions of the German Copyright Law of September 9, 1965, in its current version, and permission for use must always be obtained from Springer. Violations are liable to prosecution under the German Copyright Law.

The use of general descriptive names, registered names, trademarks, etc. in this publication does not imply, even in the absence of a specific statement, that such names are exempt from the relevant protective laws and regulations and therefore free for general use.

Typeset & Cover Design: Scientific Publishing Services Pvt. Ltd., Chennai, India.

Printed in acid-free paper

9 8 7 6 5 4 3 2 1

springer.com

To our families, for their love and gratitude!

Preface

Grid computing has emerged as one of the most promising computing paradigms of the new millennium! This paradigm can be seen as a main facet of Sun's lemma "The internet is the computer": Grid computing systems are about sharing computational resources, software and data at a large scale. Grid computing, although recent, is attracting each time more large masses of researchers, projects, applications and investment from academia and industry. We are witnessing thus an explosion in Grid research projects (Google web search returns about 2,810,000 entries for "Grid project"!) To make the Grid computing fully beneficial to researchers, practitioners, academia and industry, there are still plenty of issues to deal with and currently researchers are very actively investigating. One such issue is the performance requirement on the resulting Grid system or the Virtual Grid-enabled Supercomputer. Achieving high performance Grid computing requires techniques to efficiently and adaptively allocate jobs and applications to available resources in a large scale, highly heterogenous and dynamic environment.

This volume presents meta-heuristics approaches for Grid scheduling problems. Due to the complex nature of the problem, meta-heuristics are primary techniques for the design and implementation of efficient Grid schedulers. The volume brings new ideas, analysis, implementations and evaluation of meta-heuristic techniques for Grid scheduling, which make this volume novel in several aspects. First, Grid scheduling is tackled as a family of problems, it takes different forms depending on system requirement, application requirements, user requirements, etc. The chapters of this volume have identified several important formulations of the problem, which we believe will serve as a reference for the researchers in the Grid computing community. Second, the selected chapters for this volume comprise a variety of successful meta-heuristic approaches including: (a) Local Search based meta-heuristics (Local search, Simulated Annealing, Variable Neighborhood Search, ...); (b) Population-based approaches (Genetic Algorithms, Memetic Algorithms, Ant Colony Optimization, Particle Swarm Optimization, ...); (c) Fuzzy, QoS, dynamic programming and optimization approaches; and, (d) Hybridization of meta-heuristics among them as well as with other approaches. All these approaches aim to explore the capabilities of the meta-heuristics in dealing with many facets of the Grid scheduling. This is actually the best way to deal with the complexity of the problem, in particular with its multi-objective nature. Third, the contributed chapters in the book include formal definitions and theoretical results, implementation and experimental studies as well as practical insights on how to approach Grid scheduling.

All in all, Grid scheduling and novel meta-heuristics approaches for its resolution are presented in a comprehensive way, which we believe, makes this volume an important contribution to Grid computing, meta-heuristics and optimization research areas.

Chapters were selected after a careful review process by at least three reviewers on the basis of the originality, soundness and their contribution to both meta-heuristics and Grid scheduling. The volume consists of 13 chapters, which are organized as follows.

In Chapter 1, *Xhafa and Abraham* present Grid scheduling problems by first introducing different types of current Grid systems. Several computational models for the problem and multi-objective optimization criteria that arise in Grid scheduling are presented. An in depth analysis in the chapter shows why meta-heuristics are a defacto approach for this problem.

Montana and Zinky in Chapter 2 address the problem of optimizing the flow of compute jobs in a distributed system of compute servers through a hybrid approach of dynamic programming and a Genetic Algorithm.

In Chapter 3, *Gu and Welch* study task allocation and scheduling approach for dynamic, distributed real-time systems. The authors present an approach that offers systems explicit real-time guarantees as well as maximized robustness of unpredictable changes in computing environment.

LaTorre et al. in the fourth Chapter propose a theoretical framework to combine multi evolutionary algorithms and use it to combine multiple codings and genetic operators for Supercomputer scheduling.

In the fifth Chapter *Kaya et al.* consider the problem of scheduling an application on a computing system consisting of heterogeneous processors and one or more file repositories. The authors present iterative-improve-based heuristics by exploring complex neighborhood structures for the considered scheduling problem.

Byun et al. in the sixth Chapter report an advanced job scheduler based on Markov model in desktop Grid computing environment. The authors propose and analyze several advanced resource selection schemes in order to satisfy time requirements to complete job allocation and adapt to the needs of the user and the application on the fly.

In the seventh Chapter *Yu et al.* present workflow scheduling algorithms for Grid computing. Several heuristic methods and meta-heuristics including Simulated Annealing and Genetic Algorithms for Grid workflow scheduling are considered. Examples of experimental comparisons for workflow scheduling algorithms are also given.

Iordache et al. in the eighth Chapter address a Genetic Algorithms approach for decentralized Grid scheduling. GAs are combined with lookup services for obtaining a scalable and highly reliable Grid scheduler. The authors experimentally analyze their approach and compare it with other scheduling approaches using a monitoring environment.

Abraham et al. in the ninth Chapter introduce several nature inspired metaheuristics for Grid scheduling including Simulated Annealing, Genetic Algorithms, Ant Colony optimization and Particle Swarm Optimization. Also, the authors illustrate the usage of Multi-objective Evolutionary Algorithm for two scheduling problems.

In the tenth Chapter *Xhafa et al.* exploit the capabilities of a new class of population-based meta-heuristics, namely the Cellular Memetic Algorithms aiming, at minimizing the makespan and flowtime simultaneously using a weighted sum method. The approach is analyzed under a simulation model and showed to be effective for batch scheduling problem in Grids.

Bendjoudi et al. in the eleventh Chapter present a P2P hybrid approach that combines B&B and GA for the Flow-Shop Scheduling Problem. The authors aim at distributing at large scale the computation, using Peer-to-Peer computing to reach high computing performance. To this end, the authors propose P2P-based parallelization of the B&B and GA algorithms for the computational Grid.

In the eleventh Chapter, *Liu et al.* introduce the Peer-to-Peer neighbor selection problem for which single and multi-objective population-based metaheuristics are presented. Specifically, the authors address the Particle Swarm Optimization and Genetic Algorithms for the problem. The performance and effectiveness of the proposed approach is also illustrated with computational examples.

Khoo and Veeravalli in the last Chapter propose an approach for resource-scheduling strategy capable of handling multiple resource requirements for jobs that arrive in a Grid Computing Environment. The authors include in their method the resource availabilities in the Grid environment. The performance of the proposed approach is experimentally analyzed.

We are very much grateful to the authors of this volume and to the reviewers for their great efforts by reviewing and providing interesting feedback to authors of the chapter. The editors would like to thank Dr. Thomas Ditzinger (Springer Engineering Inhouse Editor, Studies in Computational Intelligence Series), Professor Janusz Kacprzyk (Editor-in-Chief, Springer Studies in Computational Intelligence Series) and Ms. Heather King (Editorial Assistant, Springer Verlag, Heidelberg) for the editorial assistance and excellent cooperative collaboration to produce this important scientific work. We hope that the reader will share our joy and will find it useful!

Fatos Xhafa acknowledges partial support by Projects ASCE TIN2005-09198-C02-02, FP6-2004-ISO-FETPI (AEOLUS) and MEC TIN2005-25859-E and FORMALISM TIN2007-66523. *Ajith Abraham* acknowledges the support by the

Centre for Quantifiable Quality of Service in Communication Systems, Norwegian Centre of Excellence, appointed by The Research Council of Norway, and funded by the Research Council, Norwegian University of Science and Technology and UNINETT.

Fatos Xhafa
Departament de Llenguatges i Sistemes Informàtics
Universitat Politècnica de Catalunya
Barcelona (Spain)

Ajith Abraham
Centre for Quantifiable Quality of Service in Communication Systems
Norwegian University of Science and Technology
Trondheim (Norway)

April 2008   Barcelona (Spain), Trondheim (Norway)

Contents

1   Meta-heuristics for Grid Scheduling Problems
Fatos Xhafa, Ajith Abraham 1

2   Optimizing Routing and Backlogs for Job Flows in a Distributed Computing Environment
David Montana, John Zinky 39

3   Robust Allocation and Scheduling Heuristics for Dynamic, Distributed Real-Time Systems
Dazhang Gu, Lonnie Welch 61

4   Supercomputer Scheduling with Combined Evolutionary Techniques
A. LaTorre, J.M. Peña, V. Robles, P. de Miguel 95

5   Adapting Iterative-Improvement Heuristics for Scheduling File-Sharing Tasks on Heterogeneous Platforms
Kamer Kaya, Bora Uçar, Cevdet Aykanat 121

6   Advanced Job Scheduler Based on Markov Availability Model and Resource Selection in Desktop Grid Computing Environment
EunJoung Byun, SungJin Choi, HongSoo Kim, ChongSun Hwang, SangKeun Lee 153

7   Workflow Scheduling Algorithms for Grid Computing
Jia Yu, Rajkumar Buyya, Kotagiri Ramamohanarao 173

8   Decentralized Grid Scheduling Using Genetic Algorithms
George Iordache, Marcela Boboila, Florin Pop, Corina Stratan, Valentin Cristea 215

9 Nature Inspired Meta-heuristics for Grid Scheduling: Single and Multi-objective Optimization Approaches
Ajith Abraham, Hongbo Liu, Crina Grosan, Fatos Xhafa 247

10 Efficient Batch Job Scheduling in Grids Using Cellular Memetic Algorithms
Fatos Xhafa, Enrique Alba, Bernabé Dorronsoro, Bernat Duran, Ajith Abraham 273

11 P2P B&B and GA for the Flow-Shop Scheduling Problem
A. Bendjoudi, S. Guerdah, M. Mansoura, N. Melab, E.-G. Talbi 301

12 Peer-to-Peer Neighbor Selection Using Single and Multi-objective Population-Based Meta-heuristics
Hongbo Liu, Ajith Abraham, Fatos Xhafa 323

13 An Adaptive Co-ordinate Based Scheduling Mechanism for Grid Resource Management with Resource Availabilities
B.T. Benjamin Khoo, Bharadwaj Veeravalli 341

Index 361

Author Index 365

List of Contributors

Ajith Abraham Center for
Quantifiable Quality of Service
in Communication Systems, Faculty
of Information Technology,
Mathematics and Electrical
Engineering, Norwegian
University of
Science and Technology, Norway
ajith.abraham@ieee.org

Enrique Alba Dpto. de Lenguajes y
Ciencias de la
Computación, E.T.S.I. Informática,
Campus de Teatinos, 29071
Málaga, Spain
eat@lcc.uma.es

Cevdet Aykanat Department of
Computer Engineering,
Bilkent University, Turkey
aykanat@cs.bilkent.edu.tr

Ahcene Bendjoudi Université
A/Mira de Béjaia,
Département d'Informatique CEntre
de Recherche sur l'Information
Scientique et Technique (CERIST),
Laboratoire d'Ingénierie et
Théories des Systèmes Informatiques,
Algiers, Algeria
ahcene.bendjoudi@gmail.com

Marcela Boboila Stony Brook
University, USA
mboboila@cs.sunysb.edu

Rajkumar Buyya Grid Computing
and Distributed
Systems (GRIDS) Laboratory,
Department of Computer Science and
Software Engineering, The University
of Melbourne, Austraila
raj@csse.unimelb.edu.au

EunJoung Byun Department of
Computer Science &
Engineering, Korea University, Korea
fvision@disys.korea.ac.kr

SungJin Choi Department of
Computer Science &
Engineering, Korea University, Korea
lotieye@disys.korea.ac.kr

Pedro de Miguel CeSViMa,
Department of Computer
Architecture and Technology,
Universidad Politéecnica de Madrid,
Spain
pmiguel@fi.upm.es

Bernabé Dorronsoro Faculty of
Science, Technology

and Communication University of
Luxembourg 6, rue Richard
Coudenhove-Kalergi L-1359
Luxembourg
bernabe.dorronsoro@uni.lu

Bernat Duran Department of
Languages and Informatics
Systems, Polytechnic University of
Catalonia, Barcelona, Spain
bduran@lsi.upc.edu

Dazhang Gu Center for Intelligent,
Distributed, and
Dependable Systems, School of
Electrical Engineering and Computer
Science, Ohio University, USA
gud@ohio.edu

Crina Grosan Department of
Computer Science, Faculty
of Mathematics and Computer
Science, Babeş Bolyai University,
Romania
cgrosan@cs.ubbcluj.ro

Samir Guerdah Université Mouloud
Mammeri de Tizi
Ouzou, Département Informatique,
Tizi Ouzou, Algeria
samir.guerdah@gmail.com

ChongSun Hwang Department of
Computer Science &
Engineering, Korea University, Korea
hwangg@disys.korea.ac.kr

George Iordache Stony Brook
University, USA
georgei@cs.sunysb.edu

Kamer Kaya Department of
Computer Engineering,
Bilkent University, Ankara, Turkey
kamer@cs.bilkent.edu.tr

Benjamin Khoo B.T. Department
of ECE, Singapore

HongSoo Kim Department of
Computer Science &
Engineering, Korea University, Korea
hera@disys.korea.ac.kr

Antonio LaTorre CeSViMa,
Department of Computer
Architecture and Technology,
Universidad Politécnica de Madrid,
Spain
atorre@fi.upm.es

Hongbo Liu School of Computer
Science and
Engineering, Dalian Maritime
University and Department of
Computer and Dalian University of
Technology, Dalian, China
lhb@dlut.edu.cn

SangKeun Lee Department of
Computer Science &
Engineering, Korea University, Korea
yalphy@korea.ac.kr

Madjid Mansoura Université
Mouloud Mammeri de Tizi
Ouzou, Département Informatique,
Tizi Ouzou, Algeria
madjid.mansoura@gmail.com

Nouredine Melab Universitée des
Sciences et
Technologies de Lille, France
melab@lifl.fr

David Montana BBN Technologies,
Cambridge, MA, USA
dmontana@bbn.com

José María Peña CeSViMa,
Department of Computer

Architecture and Technology,
Universidad Politécnica de Madrid,
Spain
jmpena@fi.upm.es

Florin Pop University "Politehnica"
of Bucharest,
Romania
florinpop@cs.pub.ro

Kotagiri Ramamohanarao Grid
Computing and
Distributed Systems (GRIDS)
Laboratory, Department of Computer
Science and Software Engineering,
The University of Melbourne,
Austraila
rao@csse.unimelb.edu.au

Víctor Robles CeSViMa,
Department of Computer
Architecture and Technology,
Universidad Politécnica de Madrid,
Spain
vrobles@fi.upm.es

Corina Stratan University
"Politehnica" of
Bucharest, Romania
corina@cs.pub.ro

El-Ghazali Talbi Universitée des
Sciences et
Technologies de Lille, France
talbi@lifl.fr

Bora Uçar CERFACS, France
ubora@cerfacs.fr

Bharadwaj Veeravalli Department
of ECE, Singapore
elebv@nus.edu.sg

Fatos Xhafa Department of
Languages and Informatics
Systems, Polytechnic University of
Catalonia, Barcelona, Spain
fatos@lsi.upc.edu

Jia Yu Grid Computing and
Distributed Systems
(GRIDS) Laboratory, Department of
Computer Science and Software
Engineering, The University of
Melbourne, Australia
jiayu@csse.unimelb.edu.au

John Zinky BBN Technologies,
Cambridge, MA, USA
jzinky@bbn.com

Lonnie Welch Center for Intelligent,
Distributed,
and Dependable Systems, School of
Electrical Engineering and
Computer Science,
Ohio University, USA
welch@ohio.edu

1
Meta-heuristics for Grid Scheduling Problems

Fatos Xhafa[1] and Ajith Abraham[2]

[1] Departament de Llenguatges i Sistemes Informàtics, Universitat Politècnica de Catalunya Barcelona, Spain
`fatos@lsi.upc.edu`
[2] Center of Excellence for Quantifiable Quality of Service, Norwegian University of Science and Technology, Trondheim, Norway
`ajith.abraham@ieee.org`

Summary. In this chapter, we review a few important concepts from Grid computing related to scheduling problems and their resolution using heuristic and meta-heuristic approaches. Scheduling problems are at the heart of any Grid-like computational system. Different types of scheduling based on different criteria, such as static vs. dynamic environment, multi-objectivity, adaptivity, etc., are identified. Then, heuristics and meta-heuristics methods for scheduling in Grids are presented. The chapter reveals the complexity of the scheduling problem in Computational Grids when compared to scheduling in classical parallel and distributed systems and shows the usefulness of heuristics and meta-heuristics approaches for the design of efficient Grid schedulers.

Keywords: Grid Computing, Scheduling, Independent Scheduling, Grid workflow, Multi-objective Optimization, Heuristics, Meta-heuristics.

1.1 Introduction

Grid Computing is a powerful computing paradigm penetrating each time more in every activity of our lives! Grid computing is about benefiting from large computing power never known before, is about scientific progress, business and much more! What would it mean if:

- A researcher from Computer Science could solve to optimality his favorite NP-hard problem within a few hours?
- A researcher from Chemistry could obtain a new drug design not known before?
- A researcher from Biomedicine could discover the DNA sequencing and use it for investigating diseases?
- A climate forecast center could predict in advance a possible tsunami?
- A medical team could remotely run a complex surgery operation using virtual laboratories?
- An economist could analyze almost on real time portfolio values?
- A student of online distance university could contribute his computer to a computational infrastructure and work online with his team for achieving the academic goals?

- An enterprise could never run short of computational resources?
- ...an many more real life scenarios?

It would really mean increasing our knowledge on complex problems, improving our lives, improving our productivity and achieving ambitious goals not possible before! Today all these are possible thanks to advances in Grid Computing!

Grid computing and Grid technologies have primarily emerged for scientific and technical work, where geographically distributed computers, linked through Internet, are used to create virtual supercomputers of vast amount of computing capacity able to solve complex problems from eScience in less time than known before. Thus, within the last years we have witnessed how Grid Computing has helped to achieve breakthroughs in meteorology, physics, medicine and other computing-intensive fields. Examples of such large scale applications are known from optimization (e.g. Casanova et al. [17], Goux et al. [34], Wright [66], Wright et al. [43]), Collaborative/eScience Computing (e.g. Newman et al. [51], Paniagua et al. [54]), Data-Intensive Computing (e.g. Beynon al. [6]), to name a few.

Grid computing is still in the development stage, and most projects are still from academia and large IT enterprises; it has however developed very quickly and more and more scientists are currently engaged to solve many challenges in Grid Computing. Among these, improving its efficiency is imperative! The question is:

> "How to make use of millions of computers world-wide, ranging from simple laptops, to clusters of computers and supercomputers connected through heterogenous networks in an efficient, secure and reliable manner?"

The above question is a real challenge for Grid computing community. The good news are the reported advances in both scientific research in Grid Computing and technological achievements and software development for enabling Grid computing systems. Software packages exist and have been successfully deployed and it is now possible to build Grid systems joining together both single computers and clusters of computers yet, *the challenging problem of dynamically and adaptively allocating resources in response to demanding application requests remains unsolved.* For the majority of grid systems, scheduling is a very important mechanism. In the simplest of cases, scheduling of jobs can be done in a blind way by simply assigning the incoming tasks to the compatible resources according to their availability. Nevertheless, it is a lot more profitable to use more advanced and sophisticated schedulers. Moreover, the schedulers would generally be expected to react to the dynamics of the grid system, typically by evaluating the present load of the resources, and notifying when new resources join or drop from the system. Additionally, schedulers can be organized in a hierarchical form or can be distributed in order to deal with the large scale of the grid.

In this chapter, we focus on the design of efficient Grid schedulers using heuristics and meta-heuristics methods. Heuristic and meta-heuristics methods have proven to be efficient in solving many computationally hard problems. They are showing their usefulness also in the Grid Computing domain, especially for scheduling and resource allocation. We analyze why heuristics and meta-heuristics methods are good alternatives to more traditional scheduling techniques and what make them appropriate for Grid scheduling. An important issue here is how to formally define the Grid scheduling problem. We have presented the most important and useful computational models for this purpose.

The rest of the chapter is organized as follows. We present in Section 1.2 a few important concepts from Grid computing, introduce a few types of Grids in view of needs for different types of scheduling and resource allocation. Then, in Section 1.3 we identify different types of scheduling problems arising in Grid systems. In Section 1.4, we focus in the current state of using heuristic and meta-heuristic methods for solving scheduling problems in Grid systems, as *de facto* approaches for dealing with the complexity of the problem. A few other issues such as security and grid services scheduling are discussed in Section 1.5. We end the chapter in Section 1.6 with a few conclusions.

1.2 The *many* Grids

The present state of the computation systems is, in some aspects, analogous to that of the electricity systems at the beginning of the 20th century. At that time, the generation of electricity was possible, but still it was necessary to have available generators of electricity. The true revolution that permitted its establishment was the discovery of new technologies, namely the networks of distribution and broadcast of electricity. These discoveries made possible to provide a reliable, low price service and thus the electricity became universally accessible.

By analogy, the term *grid* is adopted to designate a computational infrastructure of distributed resources, highly heterogeneous (as regards their computing power and architecture), interconnected by heterogeneous communication networks and by a middleware that offers reliable, simple, transparent, efficient and global access to their potential of computation.

The Grid Computing domain has witnesses a fast development over a relatively short time period, pushed by important technology advancements and interest of large IT companies such as IBM, Sun Microsystems, Oracle and HP. The roots of Grid Computing can be traced back to the late 1980s and the first concept that laid the basis of today's Grid systems were developed by researchers from distributed super-computing for numerical or optimization with particular emphasis on scheduling algorithms to achieve high performance computing (e.g. Condor-G). By the late 1990s, the term of Computational Grids and Grid Computing were popularized by Foster et al. [26, 27] who developed the Globus toolkit as a general middleware for Grid Systems. Since then, Grid Computing, Grid systems and Grid technology are advancing in unstoppable way! In the following subsections we briefly review most important types of Grids pushing Grid technology, actually, it is by large impossible to review all existing types of Grids and Grid projects running world-wide!

1.2.1 Computational Grids

One of the first questions raised by this emerging technology is its utility or the need of disposing computational grids. On the one hand, the computational proposals have usually shown to have a great success in any field of the human activity. Guided by the increase of the complexity of the real life problems, and prompted by the increment of the capacity of the technology, the human activity (whether scientific, engineering, business, personal, etc.) is highly based on computation. Computers are very often used to model and to simulate complex problems, for diagnoses, plant control, weather forecast, and many other fields of interest. Even so, there exist many problems that challenge or exceed our ability to solve them, typically because they require processing a large quantity of operations or data. In spite of the fact that the capacity of the computers continues improving, the computational resources do not respond to the continuous demand for more computational power.

On the other hand, statistical data show that computers are usually infrautilized. Most of computers from companies, administration, etc. are most of the time idle or are used for basic tasks that do not require the whole computation power. It is pointed out however by several statistic studies that a considerable amount of money is spent for the acquisition of these resources. One of the main objectives of the grid technology is, therefore, to benefit from the existence of many computation resources through the sharing. As pointed out by Foster & Kesselman *"the sharing that we are concerned with is not primarily file exchange but rather direct access to computers, software, data, and other resources..."*

1.2.2 Scavenging Grids

In a simple Computational Grid, such as united devices, the politics of *"scavenging"* is applied. This means, each time a machine remains idle, it reports its state to the grid node responsible for the management and planning of the resources. Then, this node usually assigns to the idle machine the next pending task that can be executed in that machine. Scavenging normally hinders the owner of the application, since in the event that the idle machine changes its state to be busy with tasks not coming from the grid system, the application is suspended or delayed. This situation would create completion times not predictable for grid-based applications. With the objective of having a predictable behavior, the resources participating in the grid often are dedicated resources (exclusive use in the grid), and they do not suffer from preemptions caused by external works. Moreover, this permits the tools associated to the schedulers (generally known as *profilers*) to compute the approximate completion time for an assembly of tasks, when their characteristics are known in advance. Sethi@home project is an example of scavenging Grids.

1.2.3 eScience Grids

Under the name of eScience Grids are known types of Grids that are primarily devoted to the solution of problems from science and engineering. Such Grids give support to the computational infra-structure (access to computational and data resources) needed to solve many complex problems arising in areas of science and engineering. Representative examples are UK eScience Grid, German D-Grid, BIG GRID (the Dutch e-Science Grid) and French Grid'5000, to name a few.

1.2.4 Data Grids

Data grids are Grid computing systems that primarily deal with data repositories, sharing, access and management of large amounts of distributed data. Many scientific and engineering applications require access to large amounts of distributed data, however, different data could have their own format. An application that needs access to data in different source data needs transparent and secure access to the data. In such Grid systems many types of algorithms, such as replication, are important to increase the performance of Grid enabled applications that use large amount of data. Also, data movement is an issue here in order to achieve high throughput.

1.2.5 Enterprise Grids

Although Grid technologies were motivated by High Performance Computing and have been used for several years now in scientific labs, nowadays Grid computing is becoming a significant component of business. Indeed, today's e-business must be able to respond to increasing costumer demands and adjust dynamically and efficiently to marketplace shifts and customer demands. Enterprise Grids make possible to run several projects within one large enterprise or many departments to share resources (computational and/or data) in a transparent way. It should be noted that in such Grids the security and resource policy management issues are not of first concern. Enterprise Grids are thus showing great and innovative changes on how computing is used. Indeed, Grid Computing is envisaged as a significant factor for increasing the productivity and efficiency to the world-wide business. The Grid offers a large potential to solving business problems by facilitating global access to enterprise computing services and data. Examples of enterprise grids are "Sun Grid Engine", "IBM Grid", "Oracle Grid" and "HP Grid".

A new form of enterprise grids is also emerging in institutions, the so called desktop grids, which use the idle cycles of mainly desktop PC's. Small enterprises and institutions usually are equipped with hundreds or thousands of desktops mainly used for office tasks. This amount of PCs is thus a good source for setting up a Grid system for the institution. In this case, the particularity of the Grid system is its unique administrative domain, which makes it easier to manage due to low heterogeneity and volatility of resources (for instance, all PC's could be running under the same OS). Of course, the desktop Grid can cross many administrative domains and in this case the heterogeneity and volatility of the resources is an issue as in a general Grid system setting.

1.3 Scheduling Problems in Computational Grids

Rather than a problem, scheduling in Grid systems can be viewed as a whole family of problems. This is due to the many parameters that intervene scheduling as well as to the different needs of Grid-enabled applications. In the following, we give some basic concepts of scheduling in Grid systems and identify most common scheduling types. Needless to say, job scheduling in its different forms is computationally hard; it has been shown that the problem of finding optimum scheduling in heterogeneous systems is in general NP-hard [30].

1.3.1 Basic Concepts and Terminology

Although many types of resources can be shared and used in a Computational Grid, normally they are accessed through an *application* running in the grid. Normally, an application is used to define the piece of work of higher level in the Grid. A typical grid scenario is as follows: an application can generate several jobs, which in turn can be composed of sub-tasks, in order to be solved; the grid system is responsible for sending each sub-task to a resource to be solved. In a simpler grid scenario, it is the user who selects the most adequate machine to execute its sub-tasks. However, in general, grid systems must dispose of *schedulers* that automatically and efficiently find the most appropriate machines to execute an assembly of tasks.

New characteristics of Scheduling in Grids

The scheduling problem in distributed systems is not new at all; as a matter of fact it is one of the most studied problems in the optimization research community. However, in the grid setting there are several characteristics that make the problem different from its traditional version of conventional distributed systems. Some of these characteristics are the following:

- *The dynamic structure* of the Computational Grid. Unlike traditional distributed systems such as clusters, resources in a Grid system can join or leave the Grid in an unpredictable way. It could be simply due to loosing connection to the system or because their owners switch off the machine or change the operating system, etc. Given that the resources cross different administrative domains, there is no control over the resources.
- The *high heterogeneity of resources*. Grid systems act as large virtual supercomputers, yet the computational resources could be very disparate, ranging from laptops, desktops, clusters, supercomputers and even small devices of limited computational resources. Current Grid infrastructures are not yet much versatile but heterogeneity is among most important features to take into account in any Grid system.
- The *high heterogeneity of jobs*. Jobs arriving to any Grid system are diverse and heterogenous in terms of their computational needs. For instance, they could be computing intensive or could be data intensive; some jobs could be

full applications having a whole range of specifications other could be just atomic tasks. Importantly, Grid system could not be aware of the type of tasks, jobs or applications arriving in the system.
- The *high heterogeneity of interconnection networks*. Grid resources will be connected through Internet using different interconnection networks. Transmission costs will often be very important in the overall Grid performance and hence smart ways to cope with the heterogeneity of interconnection networks is necessary.
- The *existence of local schedulers* in different organizations or resources. Grids are expected to be constructed by the "contribution" of computational resources across institutions, universities, enterprises and individuals. Most of these resources could eventually be running *local* applications and use their local schedulers, say, a Condor batch system. In such cases, one possible requirement could be to use the local scheduler of the domain rather than an external one.
- The *existence of local policies on resources*. Again, due to the different ownership of the resources, one cannot assume full control over the Grid resources. Companies might have unexpected computational needs and may decide to reduce their contribution to the Grid. Other policies such as rights access, available storage, *pay-per-use*, etc. are also to be taken into account.
- *Job-resource requirements*. Current Grid schedulers assume full availability and compatibility of resources when scheduling. In real situations, however, many restrictions and/or incompatibilities could be derived from job and resource specifications.
- *Large scale of the Grid system*. Grid systems are expected to be large scale, joining hundreds or thousands of computational nodes world-wide. Moreover, the jobs, tasks or applications submitted to the Grid could be large in number since different independent users and/or applications will send their jobs to the Grid without knowing previous workload of the system. Therefore, the efficient management of resources and planning of jobs will require the use of different types of scheduling (super-schedulers, meta-schedulers, decentralized schedulers, local schedulers, resource brokers, etc.) and their possible hierarchical combinations.
- *Security*. This characteristic, which is inexisting in classical scheduling, is an important issue in Grid scheduling. Here the security can be seen as a two-fold objective: on the one hand, a task, job or application could have a security requirement to be allocated in a secure node, that is, the node will not "watch" or access the processing and data used by the task, job or application. On the other hand, the node could have a security requirement, that is, the task, job or application running in the resource will not "watch" or access other data in the node.

A general definition and terminology

A precise definition of a Grid scheduler will much depend on the way the scheduler is organized (whether it is a super-scheduler, meta-scheduler, decentralized scheduler or a local scheduler) and the characteristics of the environment such as dynamics of the system. In a general setting, however, a Grid scheduler will be permanently running as follows: receive new incoming jobs, check for available resources, select the appropriate resources according to feasibility (job requirements to resources) and performance criteria and produce a planning of jobs (making decision about job ordering and priorities) to selected resources.

Usually the following terminology is employed for scheduling in Grids:

Task: represents a computational unit (typically a program and possibly associated data) to run on a Grid node. Although in the literature there is no unique definition of task concept, usually a task is considered as an indivisible schedulable unit. Tasks could be independent (or loosely coupled) among them or there could have dependencies, as it is the case of Grid workflows.

Job: A job is a computational activity made up of several tasks that could require different processing capabilities and could have different resource requirements (CPU, number of nodes, memory, software libraries, etc.) and constraints, usually expressed within job description. In the simplest case, a job could have just one task.

Application: An application is a software for solving a (large) problem in a computational infrastructure; it may require splitting the computation into many jobs or it could be a "monolithic" application. In the later case, the whole application is allocated in a computational node and is usually referred to as application deployment. As in the case of jobs, applications could have different resource requirements (CPU, number of nodes, memory, software libraries, etc.) and constraints, usually expressed within application description.

Resource: A resource is a basic computational entity (computational device or service) where tasks, jobs and applications are scheduled, allocated and processed accordingly. Resources have their own characteristics such as CPU characteristics, memory, software, etc. Several parameters are usually associated with a resource, among them, the processing speed and workload, which change over time. As in the case of jobs and applications, resource characteristics are usually given by the resource description. It should be noted that in a Grid computing environment resources are geographically distributed and may belong to different administrative domains implying different usage policies and access rights.

Specifications: Task, job and application requirements are usually specified using high level specification languages (meta-languages). Similarly, the resource characteristics are expressed using specification languages. One such language is the ClassAds language [56].

Resource pre-reservation: The pre-reservation is needed either when tasks, jobs or applications have requirements on the finishing time or when there are dependencies/precedence constraints that require advance resource reservation to assure the correct execution of the workflow. The advance reservation goes through negotiation and agreement protocols between resource providers and consumers.

Planning: A planning is the mapping of tasks, jobs and applications to computational resources.

Grid Scheduler: Software components in charge of computing a mapping of tasks, jobs or applications to Grid resources under multiple criteria and Grid environment configurations. Different levels within a Grid scheduler have been identified in the Grid computing literature comprising: super-schedulers, meta-scheduler, local/cluster scheduler and enterprise scheduler. As a main component of any Grid system, Grid scheduler interacts with other components of the Grid system: Grid information system, local resource management systems and network management systems. It should be noted that in Grid environments, all these kinds of schedulers must co-exists, and they could in general pursue conflicting goals, thus, there is need for interaction between the different schedulers in order to execute the tasks.

Super-scheduler: This kind of schedulers corresponds to a centralized scheduling approach in which local schedulers are used to reserve and allocate resources in the Grid. The local schedulers manage their job queue processing. The super-scheduler is in charge of managing the advance reservation, negotiation and service level agreement. Notice that tasks, jobs or applications are entirely completed in unique resource.

Meta-scheduler: This kind of schedulers (also known as Meta-broker in the literature) arise when a single job or application is allocated in more than one resource across different systems. As in the case of super-schedulers, a meta-scheduler uses local schedulers of the particular systems. Thus, meta-schedulers coordinate local schedulers to compute an overall schedule. Performing load balancing across multiple systems is a main objective of such schedulers.

Local/Cluster Scheduler: This kind of scheduler is in charge of assigning tasks, jobs or applications to resources in the same local area network. The scheduler manages the local resources and the local job queuing system and is this a "close to resource" scheduler type.

Enterprise Scheduler: This type of scheduler arises in large enterprises having computational resources distributed in many enterprise departments. The enterprise scheduler uses the different local schedulers belonging to the same enterprise.

Immediate mode scheduling: In the immediate mode scheduling, tasks, jobs or applications are scheduled as soon as they enter the system.

Batch model scheduling: In the batch mode scheduling, tasks, jobs or applications are grouped into *batches* which are allocated to the resources by the scheduler. The results of processing are usually obtained at a later time.

Non-preemptive/preemptive scheduling: This classification of scheduling establishes whether a task, job or application can be interrupted or not, once allocated to the resource. In the non-preemptive mode, a task, job or application should entirely be completed in the resource (the resource cannot be taken away from the task, job or application). In the preemptive mode, the preemption is allowed, that is, the current execution of the job can be interrupted and the job is migrated to another resource. Preemption can be useful if job priority is to be considered as one of the constraints.

High-throughput schedulers: The objective of this kind of scheduler is to maximize the throughput (average number of tasks or jobs processed per unit of time) in the system. These schedulers are thus task-oriented schedulers, that is, the focus is in task performance criteria.

Resource-oriented schedulers: The objective of this kind of scheduler is to maximize resource utilization. These schedulers are thus resource-oriented schedulers, that is, the focus is in resource performance criteria.

Application-oriented schedulers: This kind of schedulers are concerned with scheduling applications in order to meet user's performance criteria. To this end, the scheduler have to take into account the application specific as well as system information to achieve the best performance of the application. The interaction with the user could also be considered.

Phases of scheduling in Grids

In order to perform the scheduling process, the Grid scheduler has to follow a series of steps which could be classified into five blocks: (1) Preparation and information gathering on tasks, jobs or applications submitted to the Grid; (2) Resource selection; (3) Computation of the planning of tasks (jobs or applications) to selected resources; (4) Task (job or application) allocation according to the planning (the mapping of tasks, jobs or applications to selected resources); and, (5) Monitoring of task, job or application completion (the user is referred to [61] for a detailed description).

Preparation and information gathering: The Grid scheduler will have access to the Grid information on available resources and tasks, jobs or applications (usually known as "Grid Information Service" in the Grid literature). Moreover, the scheduler will be informed about updated information (according to the scheduling mode). This information is crucial for the scheduler in order to compute the planning of tasks, jobs or applications to the resources.

Resource selection: Not all resources could be candidates for allocation of task, jobs or applications. Therefore, the selection process is carried out based on job requirements and resource characteristics. The selection process, again, will depend on the scheduling mode. For instance, if tasks were to be allocated in a batch mode, a pool of as many as possible candidate resources will be identified out of the set of all available resources. The selected resources are then used to compute the mapping that meets the optimization criteria.

As part of resource selection, there is also the advanced reservation of resources. Information about future execution of tasks is crucial in this case. Although the queue status could be useful in this case, it is not accurate, especially if priority is one of the task requirements. Another alternative is using prediction methods based on historical data or users specifications of job requirements.

Computation of the planning of tasks: In this phase the planning is computed.

Task allocation: In this phase the planning is made effective: tasks (jobs or applications) are allocated to the selected resources according to the planning.

Task execution monitoring: Once the allocation is done, the monitoring will inform about the execution progress as well as possible failures of jobs, which depending on the scheduling policy will be rescheduled again (or migrated to another resource).

1.3.2 Types of Scheduling in Grids

As mentioned above, scheduling is a family of problems: on the one hand, different applications could have different scheduling needs such as batch or immediate mode, task independent or dependent; on the other hand, the Grid environment characteristics itself imposes restrictions such as dynamics, use of local schedulers, centralized or decentralized view of the system, etc. It is clear that in order to achieve a good performance of the scheduler, both problem specifics and Grid environment information should be "embedded" in the scheduler. In the following, we describe the main types of scheduling arising in Grid environments.

Independent Scheduling

Computational Grids are parallel in nature. The potential of a massive capacity of parallel computation is one of the most attractive characteristics of the computational grids. Aside from the purely scientific needs, the computational power is causing changes in important industries such as biomedical one, oil exploration, digital animation, aviation, in financial field, and many others. They also appear in intensive computing applications and data intensive computing, data mining and massive processing of data, etc. The common characteristic in these uses is that the applications are written to be able to be partitioned into almost independent parts (or loosely coupled). For instance, an application of intensive use of CPUs can be thought of as an application composed by subtasks (also known as bags-of-tasks applications in Grid computing literature), each one capable to be executed in a different machine of the Computational Grid. This kind of applications require independent scheduling, according to the following scenario: the tasks being submitted to the grid are independent.

Grid workflows

Solving many complex problems in Grids require the combination and orchestration of several processes (actors, services, etc.). This arises due to the dependencies in the solution flow (determined by control and data dependencies). This class of applications are know as Grid workflows, which can take advantage of the power of Grid computing, however, the characteristics of the Grid environment make the coordination of its execution very complex [15, 76]. As in other types of scheduling, performance is an important issue in order to enable high performance Grid applications. Unlike independent scheduling, it is more difficult to achieve efficient allocation of workflow tasks to the appropriate Grid resources, which largely depends on data movement between tasks and services as well as interaction with different data sources.

Besides the efficiency, Grid workflows should deal with robustness. Certainly, on the one hand, a Grid workflow could run for a long period, which in a dynamic setting increases the possibility of process failure, which could cause failure of the whole workflow if failure mechanisms are not used.

Centralized, hierarchical and decentralized scheduling

Both centralized and decentralized scheduling are useful in Grid computing, showing advantages and limitations. Essentially, they differ in the control of the resources as well as knowledge of the overall Grid system. In the case of centralized scheduling, there is more control on resources, the scheduler has knowledge of the system by monitoring of the resources state and therefore, it's easier to obtain efficient schedulers. This type of scheduling, however, suffers from limited scalability. Therefore such type of scheduling are not appropriate for large scale Grids.

Centralized schedulers have a single point of failure. Another way to organize different Grid schedulers is in a hierarchic way, which allows to coordinate scheduler at a certain level. In this case, schedulers at the lowest level in the hierarchy has knowledge of the resources. This scheduler type still suffers from lack of scalability and fault-tolerance, yet it scales better and is more fault-tolerant than centralized schedulers.

In the decentralized or distributed scheduling there is no central entity controlling the resources. The autonomous Grid sites makes it more challenging to obtain efficient schedulers. In decentralized schedulers, the local (site) schedulers play an important role. The scheduling requests, either by local users or other Grid schedulers, are sent to local schedulers, which manage and maintain the state of the queue job. This type of scheduling is more realistic for real Grid systems of large scale although decentralized schedulers could be less efficient than centralized schedulers.

Static vs. dynamic scheduling

There are essentially two main aspects that determine the dynamics of the Grid scheduling, namely:

- *The dynamics of job execution*: This refers to the situation when job execution could fail or, in the preemptive mode, job execution is stopped due to the arrival in the system of high priority jobs.
- *The dynamics of resources*: Resources can join or leave the Grid in an unpredictable way, their workload can significantly vary over time, the local policies on usage of resources could change over time, etc.

The above factors decide the behavior of the Grid scheduler, ranging from static to highly dynamic scheduling. For instance, in the static case, there is no job failure and resources are assumed available all the time and fluctuations on computing capacity and workload are not considered. Although this is unrealistic for real Grids, it could be useful to consider for batch mode scheduling: the number of jobs and resources is considered fixed during short intervals of time (time interval between two successive activations of the scheduler) and the computing capacity is also considered unchangeable. Other variations are possible to consider, for instance, just the dynamics of resources but not that of jobs.

Immediate vs. batch mode scheduling

Immediate and batch scheduling are well-known methods, largely explored for many computing environments and different types of applications. They are also useful for Grid scheduling. In immediate mode, jobs are scheduled as soon as they enter the system, without waiting for the next time interval when the scheduler will get activated or the job arrival rate is small having thus available resources to execute jobs immediately.

In batch mode, tasks jobs or applications are grouped in batches and scheduled as a group. Batch mode scheduling methods are simple and yet powerful heuristics that are distinguished for their efficiency. In contrast to immediate scheduling, batch scheduling could take better advantage of job and resource characteristics in deciding which job to allocate to which resource since they dispose of the time interval between two successive activations of the batch scheduler. Immediate scheduling methods include Opportunistic Load Balancing, Minimum Completion Time, Minimum Execution Time, Switching Algorithm and k-Percent Best and among batch mode methods there are Min-Min, Max-Min, Sufferage, Relative Cost and Longest Job to Fastest Resource - Shortest Job to Fastest Resource [1, 9, 44, 67].

Adaptive Scheduling

The changeability over time of the Grid computing environment requires adaptive scheduling techniques [42] which will take into account both current status of the resources and predictions for their future status with the aim of detecting and avoiding performance deterioration. Rescheduling can also be seen as a form of adaptive scheduling in which running jobs are migrate to more suitable resources.

Casanova *et al.* [18] considered a class of Grid applications with large numbers of independent tasks (Monte Carlo simulations, parameter-space searches, etc.), also known as task farming applications. For these applications with loosely coupled tasks, the authors developed a general adaptive scheduling algorithm. The authors used NetSolve [17] as a testbed for evaluating the proposed algorithm.

Othman *et al.* [52] stress the need for the Grid system's ability to recognize the state of the resources. The authors presented an approach for system adaptation, in which Grid jobs are maintained, using an adaptable Resource Broker.

Huedo et al. [38] reported a scheduling algorithm built on top of the GridWay framework, which uses internally adaptive scheduling to reflect the dynamic Grid characteristics.

Scheduling in Data Grids

Grid computing environments are making possible applications that work on distributed data and even across different data centers. In such applications, it is not only important to allocate tasks, jobs or application to fastest and reliable nodes but also to minimize data movement and ensure fast access to data. In other terms, data location is important in such type of scheduling. In fact, the usefulness of large computing capacity of the Grid could be compromised by slow data transmission, which could be affected by both network bandwidth and available storage resources. Therefore, data should be "close" to tasks, jobs or applications to achieve efficient access.

1.3.3 Computation Models for Formalizing Grid Scheduling

Given the versatility of scheduling in Grid environments, one needs to consider different computation models for Grid scheduling that would allow to formalize, implement and evaluate –either in real Grid or through simulation– different scheduling algorithms. Following we present some important computation models for Grid scheduling. It should be noted that such models have much in common with computation models for scheduling in distributed computing environments. We notice that in all the models described below, tasks, jobs or applications are submitted for completion to a single resource.

Expected Time To Compute model

In the model proposed by Ali et al. [5], it is assumed that we dispose of estimation or prediction of the computational load of each task (e.g. in millions of instructions), the computing capacity of each resource (e.g. in millions of instructions per second, MIPS), and an estimation of the prior load of each one of the resources. Moreover, the Expected Time to Compute matrix ETC of size number of tasks by number of machines, where each position $ETC[t][m]$ indicates the expected time to compute task t in resource m, is assumed to be known or computable in this model. In the simplest of cases, the entries $ETC[t][m]$ could be computed by dividing the workload of task t by the computing capacity of resource m. This formulation is usually feasible, since it is possible to know the computing capacity of resources while the computation need of the tasks (task workload) can be known from specifications provided by the user, from historic data or from predictions [36, 37].

Modelling heterogeneity and consistency of computing

The ETC matrix model is able to describe different degrees of heterogeneity in distributed computing environment through consistency of computing. The consistency of computing refers to the coherence among execution times obtained by a machine with those obtained by the rest of machines for a set of tasks. This feature is particularly interesting for Grid systems whose objective is to join in a single large virtual computer different resources ranging from laptops and PCs to clusters and supercomputers. Thus, three types of consistency of computing environment can be defined using the properties of the ETC matrix: consistent, inconsistent and semi-consistent.

An ETC matrix is said to be consistent if for every pair of machines m_i and m_j, if m_i executes a job faster than m_j then m_i executes all the jobs faster than m_j. On the contrary, in an inconsistent ETC matrix, a machine m_i may execute some jobs faster than another machine m_j and some jobs slower than the same machine m_j. Partially-consistent ETC matrices are inconsistent matrices having a consistent sub-matrix of a predefined size. Further, the ETC matrices are classified according to the degree of job heterogeneity, machine heterogeneity and consistency of computing. Job heterogeneity expresses the degree of variance of execution times for all jobs in a given machine. Machine heterogeneity indicates the variance of the execution times of all machines for a given job.

Problem instance

From the above description, it can be seen that formalizing the problem instance is easy under the ETC model; it consists of: a vector of tasks workloads, a vector of computing capacity of machines and the matrix ETC. As we will see in next subsection, it is almost straightforward to define several optimization criteria in this model to measure the quality of a feasible schedule. It is worth noting that incompatibilities among tasks and resources can also be expressed in ETC model, for instance, a value of $+\infty$ to $ETC[t][m]$ would indicate that task t is incompatible with resource m. Other restrictions of running a job on a machine can be simulated using penalties to ETC values. It is, however, more complicated to simulate communication and data transmission costs.

Total Processor Cycle Consumption model

Despite of its interesting properties, the ETC model has an important limitation, namely, the computing capacity of resources remains unchanged during task computation. This limitation becomes more evident when we consider Grid systems in which not only the resources have different computing capacities but also they could change over time due to Grid system's computing overload. The computing speed of resources could be assumed constant only for short or very short periods of time. In order to remedy this, Fujimoto and Hagihara [28] introduced the Total Processor Cycle Consumption (TPCC) model. The total processor cycle consumption is defined as the total number of instructions the Grid resources could complete from the starting time of executing the schedule to the completion time. As in ETC model, task workload is expressed in number of instructions and the computing capacity of resources in number of instructions computed per unit time. However, now is measured the total consumption of computing power due to Grid application completion. Clearly, this model takes into account that resources could change their computing speed over time, as it happens in large-scale computing systems whose workload is in general unpredictable.

Problem instance

A problem instance in TPCC model consists of the vector of task workloads (denoted task lengths in [28]) and a matrix expressing the computing speed of resources. Since the computing speed can change over time, one should fix a short time interval in which the computing speed remains unchanged (for instance, a unit time interval could be considered). Then a matrix PS (for processor speed) is built overtime in which one dimension is processor number and the other dimension is time (discretized by unit time); the component $PS[p][t]$ represents the processor's speed during interval time $[t, t+1)$. As the availability and processing speed of a resource vary over time, the processor speed distribution is used.

This model has shown to be useful for independent and coarse-grain task scheduling, i.e., scheduling in which the computation time in Grid nodes is superior to data transmission time, such as stand-alone applications.

Grid Information System model

The computation models for Grid scheduling presented so far allow for a precise description of problem instance however they are based on predictions, distributions or simulations. Currently, other Grid scheduling models are developed from a higher level perspective. In the Grid Information System model the Grid scheduler uses task (job or application file descriptions) and resource file descriptions as well as state information of resources (CPU usage, number of running jobs per grid resource), provided by the Grid Information System. The Grid scheduler then computes the best matching of tasks to resources based on the up-to-date workload information of resources. This model is more realistic for Grid environments and is especially suited for the implementation of simple heuristics such as FCFS (First Come First Served), EDF (Earliest Deadline First), SJF (Shortest Job First), etc.

Problem instance

The problem instance in this model is constructed, at any point in time, from the information on task file descriptions, resource file descriptions and the current state information on resources.

Cluster and Multi-Cluster Grids model

Cluster and Multi-Cluster Grids refer to Grid model in which the system is made up of several clusters. For instance the Cluster Grid of an enterprise comprises different clusters located at different departments of the enterprise. One main objective of cluster grids is to provide a common computing infrastructure at enterprise or department levels in which computing services are distributed to different clusters. More generally, clusters could belong to different enterprises and institutions, that is, are autonomous sites having their local users (both local and grid jobs are run on resources) and usage policies.

The most common scheduling problem in this models is a Grid scheduler which makes use of local schedulers of the clusters. The benefit of cluster grids is to maximize the usage of resources and at the same time, increase throughput for user tasks (jobs or applications). This model has been exploited in Lee and Zomaya [47] for scheduling data-intensive bag-of-tasks applications.

Problem instance

The problem instance in this model is constructed, at any point in time, from the information on task file descriptions; again, it is assumed that the workload of each task is known *a priori*. On the other hand, the (multi-)cluster grid can be formally represented as a set of clusters, each one with the information on its resources. Note that in this model the Grid scheduler need not to know the information on resources within a cluster nor the state information or control on every Grid resource. In this way, it is possible to reduce dependencies on Grid information services and respect local policies on resource usage.

1.3.4 Grid System Performance and Scheduling Optimization Criteria

Several performance requirements and optimization criteria can be considered for Grid scheduling problem –the problem is multi-objective in its general formulation. We could distinguish proper Grid system performance criteria from scheduling optimization criteria although both performance and optimization objectives allow to establish the overall Grid system performance.

Grid system performance criteria include: CPU utilization of Grid resources, load balancing, system usage, queuing time, throughput, turnaround time, cumulative throughput (i.e. cumulative number of completed tasks) waiting time and response time. In fact other criteria could also be considered for characterizing Grid system's performance such as deadlines, missed deadlines, fairness, user priority, resource failure, etc. Scheduling optimization criteria include: makespan, flowtime, resource utilization, load balancing, matching proximity, turnaround time, total weighted completion time, lateness, weighted number of tardy jobs, weighted response time, etc. Both performance criteria and optimization criteria are desirable for any Grid system; however, their achievement depends also on the considered model (batch system, interactive system, etc.). Importantly, it should be stressed that these criteria are conflicting among them; for instance, minimizing makespan conflicts with resource usage and response time.

Among most popular and extensively studied optimization criterion is the minimization of the makespan. Makespan is an indicator of the general productivity of the grid system: small values of makespan mean that the scheduler is providing good and efficient planning of tasks to resources. Considering makespan as a stand alone criterion not necessarily implies optimization of other objectives. As mention above, its optimization could in fact go in detriment to other optimization criteria. Another important optimization criterion is that of flowtime, which refers to the response time to the user submissions of task executions. Minimizing the value of flowtime means reducing the average response time of the Grid system. Essentially, we want to maximize the productivity (*throughput*) of the grid and at the same time we want to obtain planning of tasks to resources that offer an acceptable *QoS*.

Makespan, completion time and flowtime

In Grid scheduling we aim, among other criteria, to minimize the makespan and flowtime. Makespan is the time when finishes the latest task and flowtime is the sum of finalization times of all the tasks. Formally they can defined as:

- minimization of *makespan*: $\min_{S_i \in Sched}\{\max_{j \in Jobs} F_j\}$ and,
- minimization of *flowtime*: $\min_{S_i \in Sched}\{\sum_{j \in Jobs} F_j\}$

where F_j denotes the time when the task j finalizes, *Sched* is the set of all possible schedules and *Jobs* the set of all jobs to be scheduled. Note that makespan is not affected by any particular execution order of tasks in a concrete resource, while in order to minimize flowtime of a resource, tasks that have been assigned to should be executed in a ascending order of their workload (computation time).

Completion time of a machine m is the time in which machine m will finalize the processing of the previous assigned tasks as well as of those already planned tasks for the machine. This parameter measures the previous workload of a machine. Notice that this definition requires knowing both the ready time for a machine and the expected time to complete of the tasks assigned to the machine.

The expression of makespan, flowtime and completion time depends on the computational model. For instance, in the ETC model, $completion[m]$ is calculated as follows:

$$completion[m] = ready\_times[m] + \sum_{\{j \in Tasks \mid schedule[j]=m\}} ETC[j][m].$$

where $ready\_times[m]$ is the time when machine m will have finished the previously assigned tasks.

Makespan can be expressed in terms of the completion time of a resource, as follows:

$$makespan = \max\{completion[i] \mid i \in Machines\}.$$

Similarly, for the flowtime we use the completion times of machines, but now by first sorting in ascending order according to their ETC values the tasks assigned to a machine. Thus for machine m the flowtime $flowtime[m]$ can be expressed as follows ($S[m]$ is a vector representing the schedule for machine m):

```
flowtime[m]=0;
completion = ready_times[m];
for (i = 0; i < S[m].size(); ++i) {
   completion += ETC[S[m][i]][m];
       flowtime[m] += completion;
}
```

In the case of TPCC model, for a schedule S of makespan M, the TPCC is expressed as follows:

$$\sum_{p=1}^{m}\sum_{t=0}^{\lfloor M \rfloor -1} S[p][t] + \sum_{p=1}^{m}(M - \lfloor M \rfloor)S[p][\lfloor M \rfloor],$$

where m is the total number of Grid resources used in the schedule, p denotes a processor (resource) and $S[p][t]$ is the speed of processor during time interval $[t, t+1)$. Note that there is no direct relation between TPCC value and makespan value, however the longer makespan, the larger the value of TPCC and vice-versa. In other terms it could be established that any schedule with good TPCC value is a schedule also with good makespan value. In fact it is claimed that the set of makespan optimal schedules is the same as the set of TPCC optimal schedules.

It should be noted that this model is appropriate not only for heuristic-based scheduling methods without guarantee of fitness value of the TPCC but also for *approximation*[1]-based schedulers ensuring a quality of delivered schedule.

Resource utilization

Maximizing the resource utilization of the grid system is another important objective. This criterion is gaining importance due to the economic aspects of Grid systems such as the contribution of resources by individuals or institutions in exchange for economic benefits. Achieving a high resource reutilization becomes a challenge in Grid systems given the disparity of computational resources of the Grid. Indeed, to increment the benefit of the resource owners, the scheduler should use any resource, yet this contradicts with the high performance criteria since limited resources could be the bottleneck of the system. It could then be said that from the resource owners perspective, resource utilization is a quality of service criterion.

One possible definition of this parameter is to consider the average utilization of resources. For instance, in the ETC model, for a schedule S, it can be defined as follows:

$$avg\_utilization = \frac{\sum_{\{i \in Machines\}} completion[i]}{makespan \times nb\_machines}.$$

and we aim at maximizing this value over all possible schedules.

---

[1] An approximation algorithm is one that delivers a feasible solution whose fitness value is within a certain bound of the fitness of the optimal solution. Constant factor approximation algorithms, for instance, deliver a solution whose fitness is within a constant factor of the fitness of the optimal solution.

Matching proximity

The Grid scheduler should not only map tasks onto resources according to task requirements and resource characteristics but also it aims at matching the tasks to resources that best fit them according to desired computational criteria. Matching proximity is one such facet of the Grid scheduler, which is usually implicitly pursued in Grid schedulers. Expressing this criterion explicitly is sort of more difficult, as compared to other criteria.

In the *ETC* model, matching proximity could be defined as the degree of proximity of a given schedule with regard to the schedule produced by *Minimum Execution Time* (MET) method. MET assigns a job to the machine having the smallest execution time for that job. Observe that a large value of matching proximity means that a large number of jobs is assigned to the machine that executes them faster. Formally, for a schedule S, matching proximity can be computed as follows:

$$matching\_proximity = \frac{\sum_{i \in Tasks} ETC[i][S[i]]}{\sum_{i \in Tasks} ETC[i][MET[i]]}.$$

Turnaround time

Turnaround time is a useful criterion when the (mean) elapsed time of computation, from the submission of the first task to the completion of the last submitted task, is to be measured. Dominguez et al. [21] considered this objective for scheduling bags-of-tasks applications in desktop Grids. This objective is usually more important in batch scheduling than in interactive applications. Kondo [40] and Kondo et al. [41] characterized four real desktop grid systems and designed scheduling heuristics based on resource prioritization, resource exclusion, and task replication for fast application turnaround.

Total weighted completion time

This criterion is appropriate when user's tasks, jobs or applications have priorities. As usually, this criterion is implemented through weights associated to tasks [33, 25] and thus the weighted completion time is expressed as:

$$Total\ weighted\ completion\ time = \sum_{j \in Jobs} w_j F_j,$$

where w_j is the priority (weight) of job j and F_j denotes the time when the task j finalizes (completion time of job j). As in the case of flowtime, this criterion can be seen as QoS to the Grid user.

In a similar way are defined the *total weighted tardiness* and the *weighted number of tardy jobs* for the case of jobs having due dates.

Average Weighted Response Time

In interactive Grid applications, response time is an important parameter. Let w_j be the weight associated to job j, F_j its finalization time and R_j its submission time to the Grid system. This criterion can then be expressed as follows:

$$\frac{\sum_{j \in Jobs} w_j (F_j - R_j)}{\sum_{j \in Jobs} w_j},$$

where $(F_j - R_j)$ is the response time of job j. In [24, 60], the response time of a job is weighted by its resource consumption (long jobs have larger resource consumption than short jobs) to balance the impact of short jobs vs. long jobs with a higher resource consumption.

Similarly can be defined the *average weighted wait time*, in which the wait time is defined as the difference between the starting time (when job starts execution) and submission time.

1.3.5 Multi-objective Optimization Approaches

As described in the previous subsections, Grid scheduling is multi-objective in its general formulation. Therefore, the optimization criteria, when considered together, have to be combined in a way that a good tradeoff among them is achieved. There are several approaches in multi-objective optimization theory to deal with the multi-criteria condition of the problem. Among them we could distinguish the hierarchical and the simultaneous approach.

Hierarchical approach

This approach is useful when we would like, depending on the type of the application or Grid scenario, to establish the priority among the criteria. For instance, in a high performance computing we could give more priority to the makespan and less priority to the response time; yet, if the user requirements are concerned, we could consider the reverse priority. Let c_i, $1 \leq i \leq N$ be a set of optimization criteria. In the hierarchic approach, these criteria are sorted by their priority, in a way that if a criterion c_i is of smaller importance than criterion c_j, the value for the criterion c_j cannot be varied while optimizing according to c_i. These approach has the limitations that one should a priori establish the priority among the criteria and it is not possible to optimize more than one criterion at a time. Nonetheless, its is especially useful when the criteria are measured in different units and can't be combined in a single aggregate objective function (for instance, optimizing makespan and the number of tardy jobs).

This approach has been considered in Xhafa [16, 68, 73] for the independent job scheduling under ETC model.

Simultaneous approach

In this approach, an optimal planning is that in which any improvement with respect to a criterion causes a deterioration with respect to another criterion.

Dealing with many conflicting optimization criteria at the same time has certainly a high computation cost. It should be addressed through Pareto[2] optimization theory [23, 62]. However, in the Grid scheduling problem, rather than knowing many Pareto points in solution space, we could be interested to know a schedule having a tradeoff among the considered criteria and which could be computed very fast. Therefore, we can consider a small number of objectives at the same time, which in general suffices for practical applications (usually two or three criteria at the same time would suffice for practical purposes).

In the Pareto optimization theory we could distinguish two different approaches:

(a) **Weighted sum approach:** in this case the optimization criteria are combined in a single aggregate function, which is then solved via heuristic, metaheuristic, AI and hybrid approaches for single-objective problems. There are two issues here: the first is how to combine the different objectives in a meaningful way in a single objective function –in fact this is not always possible! The other problem is that suitable values to the weights of the criteria should be found, which *per se* introduces new variables to the problem definition. For practical cases, however, one could fix a priori the weights either based on a certain (user, application, system performance) priority or conduct a tuning process to identify appropriate values.

(b) **General approach:** In the general approach the objective is to efficiently compute the Pareto optimal front [23, 62]. Many classes of meta-heuristics algorithms have been developed for multi-objective optimization, e.g., Multi-objective Genetic Algorithms (MOGA) [22].

As an example let's consider the case a) when makespan and flowtime are considered simultaneously. As mention before, the first concern is to combine them into a single meaningful objective function. Indeed, when summing them up, we have to take into account that even though makespan and flowtime are measured in the same time unit, the values they can take are in incomparable ranges, due to the fact that flowtime has a higher magnitude order over makespan, and its difference increases as more jobs and machines are considered in the Grid system. In order to deal with this we consider the normalized or mean flowtime: $flowtime/nb\_machines$. Next we have to weight both values to balance their importance:

$$fitness = \lambda \cdot makespan + (1 - \lambda) \cdot mean\_flowtime.$$

In Xhafa et al. [16, 68, 71, 72, 73] the value of λ is fixed, based on preliminary tuning, to $\lambda = 0.75$, that is, more priority is given to makespan. In many metaheuristic implementations, it was observed that this single aggregate objective function shows good performance and outperforms known approaches in the literature for the independent Grid scheduling.

---

[2] Vilfredo Pareto, 1848-1923, Italian economist. He introduced the notion of Pareto-optimality, the idea that a society is enjoying maximum ophelimity *(economic satisfaction)* when no one can be made better off without making someone else worse off.

1.4 Heuristics and Meta-heuristics Methods for Scheduling in Grids

From the exposition in the previous sections, it is clear that Grid scheduling problem is really challenging. Dealing with the many constraints and optimization criteria in a dynamic environment is very complex and computationally hard. Meta-heuristic approaches are considered undoubtedly the *de facto* approach. Why meta-heuristics are useful for scheduling in Computational Grids? Following we point out the main reasons that explain the strength of meta-heuristics approaches for designing efficient Grid schedulers:

- *Meta-heuristics are well-understood*: there is a vast body of literature for meta-heuristic approaches. Meta-heuristics have been studied for a large number of optimization problems, from theoretical, practical and experimental perspectives. Certainly, the known studies, results and experiences with meta-heuristic approaches are a good starting point for designing meta-heuristics based Grid schedulers.
- *No "need" for optimal solutions*: In Grid scheduling problem, for most practical applications, any scheduler delivering good quality planning of jobs would suffice rather than searching for optimality. In fact, in highly dynamic Grid environment, there is not possible to even define optimality of planning, as it is defined in combinatorial optimization. This is so due to the fact that Grid schedulers run as long as the Grid system exist and thus the performance is measured not only for particular applications but also in the long run. It is well-known that meta-heuristics are able to compute in short time high quality feasible solutions. Therefore, in such situation meta-heuristics are among best candidates to cope in practice with Grid scheduling.
- *Efficient solutions in short time:* the research work on meta-heuristics has by large tried to find ways to avoid getting stuck in local optima and ensure convergence to sub-optimal or optimal solutions. However, meta-heuristics dispose of mechanisms that allow to "play" with the convergence speed. For instance, in Genetic Algorithms, by choosing appropriate genetic operators one can achieve a very fast convergence of the algorithm to local optima. Similarly, in Tabu Search method, one can work with just short-term memory (recency) in combination with intensification procedure to produce high quality feasible solutions in very short time. This feature of meta-heuristics is very useful for Grid schedulers in which we might want to have a very fast reduction in makespan, flowtime and other parameters.
- *Dealing with multi-objective nature:* Meta-heuristics has proven to efficiently solve not only single objective optimization problems but also multi-objective optimization problems as is the case of Grid scheduling.
- *Appropriateness for periodic and batch scheduling:* Periodic scheduling is a particular case of Grid scheduling. It arises often when companies and users submit their applications to the Grid system periodically. For instance, a bank may wish to run once a month an application that processes the log file keeping bank's clients transaction activity with the bank online system.

In this case suitable resource provisioning can be done in the Grid infrastructures and, which is more important in our context, there are no strong time restrictions. This means that we can run meta-heuristics based schedulers for longer execution times and increase significantly the quality of planning of jobs to resources. Similarly, in batch scheduling, we could run the meta-heuristics based scheduler for the time interval comprised within two successive batches activations.

- *Appropriateness for decentralized approaches:* Since Grid systems are expected to be large or very large scale, decentralization and co-existence of many Grid schedulers in the system is desirable. We could thus have many instances of the meta-heuristics based schedulers running in the system which are coordinated by higher level schedulers.
- *Hybridization with other approaches:* Meta-heuristics can be easily hybridized with other approaches. This is useful to make Grid schedulers to better respond to concrete types of Grid infrastructures, specific types of applications etc. The hybridization has in general shown to produce better solutions than those delivered by single approaches; in fact, meta-heuristics are themselves hybrid approaches.
- *Designing robust Grid schedulers:* The changeability of the Grid environment over time is among the factors that directly influences the performance of the Grid scheduler. A robust scheduler would be one that is able to deliver high quality planning even under constant changes of the characteristics of the Grid infrastructure such as changeability in heterogeneity of resources, of the underlying interconnection networks, in heterogeneity of jobs, etc. Evidence in meta-heuristics literature exist that in general meta-heuristic approaches are robust.
- *Libraries and frameworks for meta-heuristics:* Since meta-heuristic approaches are high level approaches, many libraries and frameworks have been developed in the literature. For instance, Mallba library [1], Paradiseo [14] and EasyLocal++ [45] are such libraries. These libraries can be easily used for Grid scheduling problem; for instance, the meta-heuristic approaches in Xhafa et al. [16, 72, 73] use skeletons defined in Mallba Library. It is worth to note that libraries have been also developed for the meta-heuristics to deal with multi-objective optimization case.

In the next subsections we briefly review most important heuristic and meta-heuristic approaches and the benefits of using them for Grid scheduling problem (the reader is referred to [31, 48] for a survey on meta-heuristic approaches).

1.4.1 Local Search Based Heuristic Approaches

Local search heuristics [39] is a family of methods that explore the solution space by jumping from one solution to another one and constructing thus a path in solution space with the aim of finding the best solution for the problem. Methods in this family range from simple ones such as Hill Climbing, Simulated Annealing to more sophisticated ones such as Tabu Search method.

Simple local search methods (Hill Climbing-like) are of interest, at least, for two reasons: (1) they produce a feasible solution of certain quality within very short time; and, (2) they can be used to feed (initialize) population-based meta-heuristics with genetically diverse individuals. Such methods has been studied for the scheduling under ETC model in Ritchie and Levine [58]. Xhafa [68] used several local search methods in implementing Memetic Algorithms for the same problem.

Simulated Annealing (SA) is more powerful than simple local search by accepting also worse solutions with certain probability. This method has been proposed for Grid scheduling problem by Abraham et al. [1] and Yarkhan and Dongarra [75].

Tabu Search [32] is more sophisticated and usually requires more computation time for computing good solutions. However, its mechanisms of tabu lists, aspiration criteria, intensification and diversification make it very powerful search algorithm. Abraham et al. [1] also considered Tabu Search as candidate solution method for the problem. Ritchie [57] implemented the TS for the problem under ETC model and used it in combination with ACO approach. Recently, Xhafa et al. [74] has presented the design, implementation and evaluation of a full TS for the scheduling problem under ETC model. The proposed TS approach showed to outperform Ritchie's approach for the problem.

Following we present the design of simple local search methods for Grid scheduling problem and present some computational results for the problem under ETC model. Makespan and flowtime objectives are considered for this purpose.

Design of local search methods for Grid scheduling

Local search methods like Hill Climbing can be applied straightway to the Grid scheduling problem. However, many variations of these methods can be designed by considering different neighborhood structures and move types in order to increase their performance. Indeed, many Hill Climbing versions are obtained by defining appropriate neighborhood relationships. Moreover, different variations are due to the order and the way in which neighboring solutions are visited. For instance, if in each iteration the best neighboring solution is accepted, we have the *steepest descent* version (in minimization case) and *steepest ascent*, in maximization case.

- *Move-based local search*: In this group of methods, the neighborhood is fixed by moving a task from one resource to another one. Thus, two solutions are neighbors if they only differ in a position of their vector of assignments task-resource. The following methods are obtained: (a) *Local Move (LM)*: moves a randomly chosen task from the resource where it was assigned to, to another randomly chosen resource; (b) *Steepest Local Move (SLM)*: moves a randomly chosen task to the resource yielding the largest improvement; (c) *Local MCT Move (LMCTM)*: this method is based on the MCT (*Minimum Completion Time*) heuristic. Here, a task is moved to the resource yielding

the smallest completion time among all the resources; (d) *Local Minimum Flowtime Move (LMFTM)*: applies the movement of a randomly chosen task that yields the largest reduction in the flowtime.
- *Swap-based local search*: In this group of methods, the neighborhood is fixed by swapping two tasks of different resources. This group includes: (a) *Local Swap (LS)*: the resources of two randomly chosen tasks are swapped; (b) *Steepest Local Swap (SLS)*: the movement swap yielding the largest improvement is is applied; (c) *Local MCT Swap (LMCTS)*: in this case, a randomly chosen task t_1 is swapped with a task t_2 so that the maximum completion_time of the two implied resources is the smallest of all possible exchanges; (d) *Local MFT Swap (LMFTS)*: the exchange of the two tasks yields the largest reduction in the value of flowtime; and, (e) *Local Short Hop (LSH)*: this method is based on the the process of *Short Hop* [7]. In our case, pairs of resources are chosen one from the subset of the most loaded resources and the other from the subset of the less loaded resources together with the subset of tasks that are assigned to these resources. In each iteration (*hop*) the swap of a task of a most loaded resource with a task assigned to a less loaded resource is evaluated and accepted if the completion_time of the implied resources is reduced.
- *Rebalance-based local search*: load balancing of resources is used as a criterion for the neighborhood definition. We can thus design: (a) *Local Rebalance (LR)*: the movement from a solution to a neighboring one is done by rebalancing the most loaded resources; (b) *Deep Local Rebalance (DLR)*: applies a movement with the largest improvement in rebalancing; (d) *Local Flowtime Rebalance (LFR)*: the swap is done for a task from the most loaded resource and a task of a resource that reduces the value of the flowtime contributed by the most loaded resource; (e) *Emptiest Resource Rebalance (ERR)*: in this method the aim is to balance the workload of the resources but now the less loaded resource are used as a basis; and, (f) *Emptiest Resource Flowtime Rebalance (ERFR)*: this is similar to the previous method but now the less loaded resource is considered the one that contributes the smallest flowtime.
- *Variable Neighborhood Search (VNS)*: in this method a generalized concept of neighborhood is considered. More precisely, the neighborhood relationship is defined so that two solutions are considered neighbors if they differ in k positions of their vectors of assignments task-resource, where k is a parameter. This method in general could yield better solutions, however its computational cost is higher since the size of the neighborhood is much larger than in the case of simple neighborhood (for $k = 1$, VNS is just the Local Move).

Computational results for local search methods (ETC model)

We exemplify the usefulness of the local search methods presented above through their implementation for the independent Grid scheduling under ETC model (see subsection 1.3.3).

For the purposes of this experimental study problem instances from ETC model consisting of 512 jobs and 16 resources are used. The aim was to study

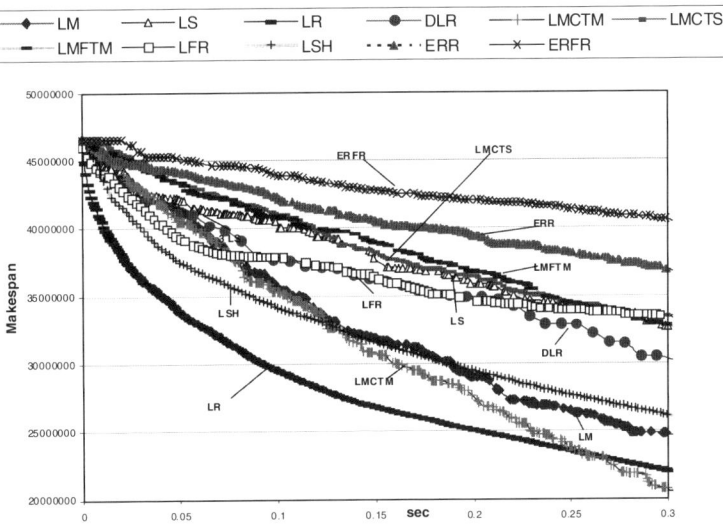

Fig. 1.1. Comparison of makespan (in arbitrary time units) reductions obtained with different local search procedures

the makespan reduction of different local search methods presented above (the initial solution –staring point– of the local search was generated randomly). Since local search methods are based on random decisions, 20 independent runs (of 500 iterations each) were performed on the same instance and the performance evaluation is done based on averaged makespan values. We show in Fig. 1.1 the graphical representation of the makespan reduction for 11 of the local search methods introduced above; the rest of them (LMFTS, SLS, SLM, ERFR) performed worse and are omitted from the graph.

From Fig. 1.1, we can see that: (a) all the considered local search methods achieve a reduction of makespan in very short time; (b) the fastest reduction in makespan is achieved by LR (Local Rebalance) although in the long run LM-CTM (Local MCT Move) obtained better makespan reduction; (c) the method ERFR (Emptiest Resource Flowtime Rebalance) based on flowtime reduction performed poorly, which is expected since it tries to minimize flowtime, not the makespan.

On the other hand we measured the makespan reduction of the VNS method for $k = 3$ and $k = 8$. We show in Fig. 1.2 the graphical representation of the makespan reduction; we have also included in the graph the LR and LMCTM methods for ease of comparison between VNS and simple local search methods presented above.

From Fig. 1.2 we can see that VNS(18) achieves the fastest reduction of the maksepan but is soon "stagnated" and VNS(3) performs better. This could be explained by the fact that doing a considerable number of movements (8 movements in this case as compared to just three movements in VNS(3)) could

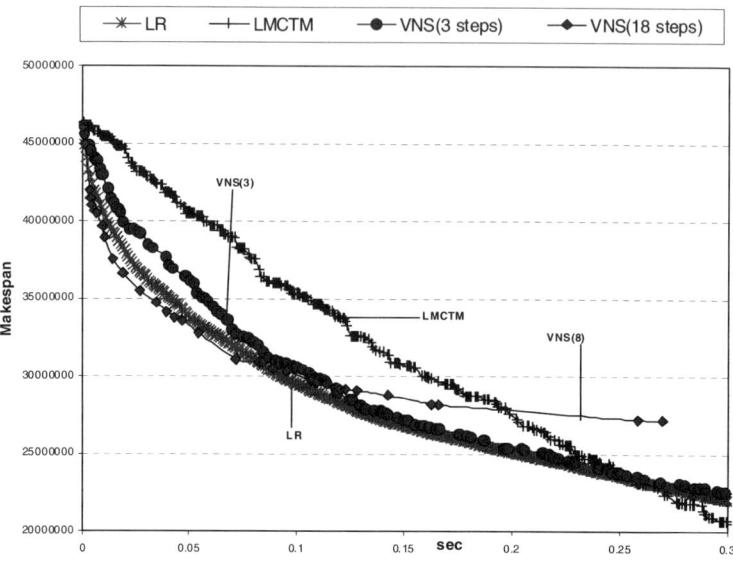

Fig. 1.2. Comparison of makespan (in arbitrary time units) reductions obtained with VNS(k) method for $k = 3$ and $k = 8$

damage the structure of the schedule. It is therefore suggestive to keep the value of k small. It should also be noted that VNS, despite of being considered more powerful method than simple local search, does not perform significantly better than; in fact, LMCTM seems to perform better than VNS in the long run.

1.4.2 Population-Based Heuristic Approaches

Population-based heuristics is a large family of methods that has shown their efficiency for solving combinatorial optimization problems. Population based methods usually require large running times if sub-optimal or optimal solutions are to be found. However, when the objective is to find feasible solutions of good quality in short execution times, as in case of Grid scheduling, we can exploit the inherent mechanisms of these methods to increase the *convergence* rapidity of the method.

We could distinguish three different categories of population-based methods: Evolutionary Algorithms (Genetic Algorithms (GAs), Memetic Algorithms (MAs) and their variations), Ant Colony Optimization (ACO) and Particle Swarm Optimization (PSO).

GAs for Grid scheduling problems have been addressed by Abraham et al. [1], Braun et al. [9], Zomaya and Teh [81], Martino and Mililotti [46], Page and Naughton [53], Carretero and Xhafa [16], Gao et al. [29], Xhafa et al. [70, 73].

MAs [49] is a relatively new class of population-based methods, which combine the concepts of evolutionary search and local search by taking advantage of good characteristics of both of them. In this sense MAs could be considered as hybrid evolutionary algorithms, in fact, MAs arose as an attempt to combine concepts and strategies of different meta-heuristics. There has been few work on MAs for Grid scheduling problem. Xhafa [68] applied unstructured MAs and Xhafa et al. [71] proposed Cellular MAs (structured MAs) for the independent scheduling problem under ETC model.

An ACO implementation for the problem under ETC model has been reported by Ritchie [57]. Abraham et al. [3] proposed an approach for scheduling jobs on Computational Grids using fuzzy PSO algorithm.

Specific methods for population initialization

In population-based methods, its is important to dispose a wide variety of initialization methods for the generation the first population. Typically, the initial solutions are generated randomly, however, introducing a few genetically good individuals would be helpful to accelerate the search. Thus, besides a random method, other specific or *ad hoc* methods can be used to generate solutions, among them, the *ad hoc* heuristics Opportunistic Load Balancing (OLB), Minimum Completion Time (MCT), Minimum Execution Time (MET), Switching Algorithm (Switch), K-percent Best (KPB), Min-min, Max-min, Sufferage, Relative-cost and Longest Job to Fastest Resource-Shortest Job to Fastest Resource (LJFR-SJFR) [1, 9, 44, 67].

In [73], the LJFR-SJFR method was used for generating one individual (the rest were generated through random perturbations, that is, by reassignment of a subset of tasks). By monitoring some of the runs, we observed that our GA spends roughly 55-70% of the total number of iterations to reach a solution of the quality of Min-Min method, which is due to the fact that Min-Min performs much better than LJFR-SJFR.

1.4.3 Hybrid Heuristics Approaches

Meta-heuristic methods are *per se* hybridized approaches. For instance, MAs combine evolutionary search with local search. However, hybridization among different meta-heuristics has shown to be effective for many problems by outperforming single methods [63]. However, hybrid meta-heuristics have been less explored for the problem. Abraham et al. [1] addressed the hybridization of GA, SA and TS heuristics; the hybridization GA+SA is expected to have a better convergence than pure GA search and GA+TS could improve the efficiency of GA. In these hybridizations a heuristic capable to deal with a population of solutions, such as GA, is combined with two other local search heuristics, such as TS and SA, that deal with only one solution at a time. Another hybrid approach for the problem is due to Ritchie and Levine [57, 59] who combine an ACO algorithm with a TS algorithm for the problem. In [68], a basic unstructured MA is combined with 16 local search algorithms in order to identify the best performance of the resulting MA.

1.4.4 Other Approaches

Many other approaches can be applied to Grid scheduling problem. We briefly present them next.

Hyper-heuristic approaches

Hyper-heuristic approaches [9] are methods that guide the search, at a higher level as compared to the meta-heuristics approaches, through other heuristic methods for the resolution of optimization problems. Hyper-heuristics have shown effective for scheduling and timetabling (Burke et al. [11]). Hyper-heuristics can also be combined to design hybrid approaches for general scheduling and timetabling problems [8, 10]. They are therefore candidate approaches also for Grid scheduling problem.

Xhafa [69] presented a simple hyper heuristic for the problem, which uses as underlying heuristics a set of *ad hoc* (immediate and batch mode) scheduling methods to provide the scheduling of jobs to Grid resources according to the Grid and job characteristics.

The hyper-heuristic is a high level algorithm, which examines the state and characteristics of the Grid system (jobs and resources), and selects and applies the *ad hoc* method that yields the best planning of jobs. The resulting hyper-heuristic based scheduler can be thus used to develop network-aware applications that need efficient planning of jobs to resources.

Reinforced learning

Some research work in the literature addressed the use of reinforced learning techniques for scheduling in Grid systems. Perez *et al.* [55], proposed to implement a Reinforcement Learning based scheduling approach for large Grid computing systems. Vengerov [64] presented a utility-based framework for making repeated scheduling decisions dynamically; the observed information about unscheduled jobs and system's resources is used for this purpose.

Fuzzy logic, neural networks and QoS approaches

Zhou et al. [80] used Fuzzy Logic techniques to design an adaptive Fuzzy Logic scheduler, which utilizes the Fuzzy Logic control technology to select the most suitable computing node in the Grid environment. A Fuzzy Neural Networks was proposed by Yu et al. [77] to develop a high performance scheduling algorithm. The algorithms uses Fuzzy Logic techniques to evaluate the Grid system load information, and adopt the Neural Networks to automatically tune the membership functions. Hao et al. [35] presented a Grid resource selection based on Neural Networks aiming at offering QoS on distributed, heterogeneous resources. To this end, the authors propose to select Grid resources constrained by QoS criteria. The resource selection problem is solved using a novel neural networks.

Chunlin and Layuan [20] proposed a joint QoS optimization approach to optimize global QoS by adopting cross-layer design and information exchange among multiple Grid layers.

Economy-based scheduling

Economy-based models are important for the design of resource management architecture for Grid systems. Several recent works [4,12,13,19,78] are addressing the resource allocation through market-oriented approaches. These approaches are suitable, on the one hand, to exploit the interaction of different scheduling layers, and on the other, different negotiation and agreement strategies can be implemented for resource allocation.

Grid services scheduling

W3C defined a service is a set of actions that form a coherent whole from the point of view of service providers and service requesters. Although this definition originated for web systems, services were defined similarly for Grid systems. There are two aspectes related to Grid scheduling and Grid services: (a) Grid services need to be discovered and scheduled to appropriate resources; for instance, scheduling a service in the Grid system to process a requested transaction; and (b) achieving Grid scheduling functionalities through services. Several recent research work [50,65,79] explore these aspects, yet there is still few research work in this direction.

1.5 Further Issues

Besides the many aspects and facets of Grid scheduling problem presented in the previous sections, there still remain other issues to be considered. We briefly mention here the Grid security as an important aspect to be considered in Grid scheduling. The security can be seen as a two-fold objective: on the one hand, a task, a job or application could have a security requirement to be allocated in a secure node, that is, the node will not "watch" or access the processing and data used by the task, job or application. On the other hand, the node could have a security requirement, that is, the task, job or application running in the resource will not "watch" or access other data in the node.

It should be noted that current security approaches are treated at different levels of Grid systems and independently of the Grid schedulers. It is challenging to incorporate the security/trust level as one of the objectives of the scheduling by using trust values that span from very trustworthy to very untrustworthy scale. Moreover, one of the aims to pursue here is to reduce the possible overhead to the Grid scheduler and to the overall system that would introduce a secure scheduling approach.

1.6 Conclusions

In this Chapter, we have reviewed the most important concepts from Grid computing related to scheduling problems and their resolution using heuristic and meta-heuristic approaches. After introducing a few important Grid types that have appeared in the Grid computing domain, we identify different types of scheduling based on different criteria, such as static vs. dynamic environment, multi-objectivity, adaptivity, etc. Our exposition aims to reveal the complexity of the scheduling problem in Computational Grids when compared to scheduling in classical parallel and distributed systems and shows the usefulness of heuristics and meta-heuristics approaches for the design of efficient Grid schedulers. We have reasoned about the importance and usefulness of meta-heuristic approaches for the design of efficient Grid schedulers when considering the scheduling as a multi-objective optimization problem. Also, a few other approaches and current research issues in the context of Grid scheduling are discussed.

Acknowledgment

The first author acknowledges partial support by Projects ASCE TIN2005-09198-C02-02, FP6-2004-ISO-FETPI (AEOLUS) and MEC TIN2005-25859-E and FORMALISM TIN2007-66523.

References

1. Alba, E., Almeida, F., Blesa, M., Cotta, C., Díaz, M., Dorta, I., Gabarró, J., León, C., Luque, G., Petit, J., Rodríguez, C., Rojas, A., Xhafa, F.: Efficient parallel LAN/WAN algorithms for optimization. The MALLBA project. Parallel Computing 32(5-6), 415–440 (2006)
2. Abraham, A., Buyya, R., Nath, B.: Nature's heuristics for scheduling jobs on computational grids. In: The 8th IEEE International Conference on Advanced Computing and Communications (ADCOM 2000), India (2000)
3. Abraham, A., Liu, H., Zhang, W., Chang, T.: Scheduling jobs on computational grids using fuzzy particle swarm algorithm. In: 10th Int. Conf. on Knowledge-Based & Intelligent Information & Engineering Systems. LNCS. Springer, Heidelberg (2006)
4. Abramson, D., Buyya, R., Giddy, J.: A computational economy for grid computing and its implementation in the Nimrod-G resource broker. Future Generation Computer Systems Journal 18(8), 1061–1074 (2002)
5. Ali, S., Siegel, H.J., Maheswaran, M., Hensgen, D.: Task execution time modeling for heterogeneous computing systems. In: Proceedings of Heterogeneous Computing Workshop (HCW 2000), pp. 185–199 (2000)
6. Beynon, M.D., Sussman, A., Catalyurek, U., Kure, T., Saltz, J.: Optimization for data intensive grid applications. In: Third Annual International Workshop on Active Middleware Services, California, pp. 97–106 (2001)
7. Braun, T.D., Siegel, H.J., Beck, N., Boloni, L.L., Maheswaran, M., Reuther, A.I., Robertson, J.P., Theys, M.D., Yao, B.: A comparison of eleven static heuristics for mapping a class of independent tasks onto heterogeneous distributed computing systems. J. of Parallel and Distributed Comp. 61(6), 810–837 (2001)
8. Burke, E., Kendall, G., Landa Silva, D., O'Brien, R., Soubeiga, E.: An ant algorithm hyperheuristic for the project presentation scheduling problem. The 2005 IEEE Congress on Evolutionary Computation 3, 2263–2270 (2005)

9. Burke, E.K., Kendall, G., Newall, J., Hart, E., Ross, P., Schulemburg, S.: Hyper-heuristics: an Emerging Direction in Modern Search Technology. In: Glover, F.W., Kochenberger, G.A. (eds.) Handbook of Meta-heuristics. Kluwer, Dordrecht (2003)
10. Burke, E.K., Kendall, G., Soubeiga, E.: A Tabu-Search Hyperheuristic for Timetabling and Rostering. J. Heuristics 9(6), 451–470 (2003)
11. Burke, E., Soubeiga, E.: Scheduling Nurses Using a Tabu-Search Hyperheuristic. In: Proceedings of the 1st Multidisciplinary International Conference on Scheduling: Theory and Applications (MISTA 2003), Nottingham, UK, pp. 180–197 (2003)
12. Buyya, R.: Economic-based Distributed Resource Management and Scheduling for Grid Computing. PhD thesis, Monash University, Australia (2002)
13. Buyya, R., Abramson, D., Giddy, J.: Nimrod/G: An architecture for a resource management and scheduling system in a global computational grid. In: The 4th Int. Conf. on High Performance Comp., Asia-Pacific, China (2000)
14. Cahon, S., Melab, N., Talbi, E.: ParadisEO: A Framework for the Reusable Design of Parallel and Distributed Meta-heuristics. Journal of Heuristics 10(3), 357–380 (2004)
15. Cao, J., Jarvis, S.A., Saini, S., Nudd, G.R.: GridFlow: Workflow Management for Grid Computing. In: Proc. of the 3rd International Symposium on Cluster Computing and the Grid (CCGrid 2003), Tokyo, Japan, May 2003, pp. 198–205 (2003)
16. Carretero, J., Xhafa, F.: Using Genetic Algorithms for Scheduling Jobs in Large Scale Grid Applications. Journal of Technological and Economic Development –A Research Journal of Vilnius Gediminas Technical University 12(1), 11–17 (2006)
17. Casanova, H., Dongarra, J.: Netsolve: Network enabled solvers. IEEE Computational Science and Engineering 5(3), 57–67 (1998)
18. Casanova, H., Kim, M., Plank, J.S., Dongarra, J.J.: Adaptive Scheduling for Task Farming with Grid Middleware. Int. J. High Perform. Comput. Appl. 13(3), 231–240 (1999)
19. Chin, S., Lee, J., Yoon, T., Yu, H.: List Scheduling Method for Service Oriented Grid Applications. In: Proceedings of the Second international Conference on Semantics, Knowledge, and Grid, p. 44. IEEE Computer Society, Los Alamitos (2006)
20. Chunlin, L., Layuan, L.: Joint QoS optimization for layered computational grid. Inf. Sci. 177(15), 3038–3059 (2007)
21. Domingues, P., Andrzejak, A., Silva, L.: Scheduling for fast touraround time on institutional desktop grid. CoreGRID TechRep No. 0027
22. Deb, K., Pratap, A., Agarwal, S., Meyarivan, T.: A Fast and Elitist Multi-objective Genetic Algorithm: NSGA-II. IEEE Transactions on Evolutionary Computation 6(2), 182–197 (2002)
23. Ehrgott, M., Gandibleux, X.: Approximative solution methods for multiobjective combinatorial optimization. TOP –Trabajos de Investigación Operativa 12(1), 1–88 (2004)
24. Ernemann, C., Hamscher, V., Yahyapour, R.: Benefits of Global Grid Computing for Job Scheduling. In: Proceedings of the Fifth IEEE/ACM International Workshop on Grid Computing. International Conference on Grid Computing, pp. 374–379. IEEE Computer Society, Washington (2004)
25. Fibich, P., Matyska, L., Rudová, H.: Model of Grid Scheduling Problem. In: Exploring Planning and Scheduling for Web Services, Grid and Autonomic Computing, pp. 17–24. AAAI Press, Menlo Park (2005)
26. Foster, I., Kesselman, C.: The Grid - Blueprint for a New Computing Infrastructure. Morgan Kaufmann, San Francisco (1998)
27. Foster, I., Kesselman, C., Tuecke, S.: The anatomy of the grid. International Journal of Supercomputer Applications 15(3) (2001)

28. Fujimoto, N., Hagihara, K.: Near-Optimal Dynamic Task Scheduling of Precedence Constrained Coarse-Grained Tasks onto a Computational Grid. In: Second International Symposium on Parallel and Distributed Computing (ISPDC 2003), pp. 80–87 (2003)
29. Gao, Y., Rong, H., Huang, J.Z.: Adaptive Grid job scheduling with genetic algorithms. Future Gener. Comput. Syst. 21(1), 151–161 (2005)
30. Garey, M.R., Johnson, D.S.: Computers and Intractability – A Guide to the Theory of NP-Completeness. W.H. Freeman and Co., New York (1979)
31. Gendreau, M., Potvin, J.-Y.: Meta-heuristics in Combinatorial Optimization. Annals of Operations Research 140(1), 189–213 (2005)
32. Glover, F.: Future Paths for Integer Programming and Links to Artificial Intelligence. Computers and Op. Res. 5, 533–549 (1986)
33. Gomoluch, J., Schroeder, M.: Market-based Resource Allocation for Grid Computing: A Model and Simulation. In: Middleware Workshops 2003, pp. 211–218 (2003)
34. Goux, J.P., Kulkarni, S., Linderoth, J., Yoder, M.: An enabling framework for master-worker applications on the computational grid. In: 9th IEEE Int. Symposium on High Performance Distributed Computing (HPDC 2000) (2000)
35. Hao, X., Dai, Y., Zhang, B., Chen, T., Yang, L.: QoS-Driven Grid Resource Selection Based on Novel Neural Networks. In: Chung, Y.-C., Moreira, J.E. (eds.) GPC 2006. LNCS, vol. 3947, pp. 456–465. Springer, Heidelberg (2006)
36. Hotovy, S.: Workload evolution on the Cornell Theory Center IBM SP2. In: Job Scheduling Strategies for Parallel Proc. Workshop, IPPS 1996, pp. 27–40 (1996)
37. The Hebrew University Parallel Systems Lab. Parallel workload archive, http://www.cs.huji.ac.il/labs/parallel/workload/
38. Huedo, E., Montero, R.S., Llorente, I.M.: Experiences on Adaptive Grid Scheduling of Parameter Sweep Applications. In: 12th Euromicro Conference on Parallel, Distributed and Network-Based Processing (PDP 2004), p. 28 (2004)
39. Hoos, H.H., Stützle, Th.: Stochastic Local Search: Foundations and Applications. Elsevier/Morgan Kaufmann (2005)
40. Kondo, D.: Scheduling Task Parallel Applications for Rapid Turnaround on Desktop Grids. Doctoral Thesis, University of California at San Diego (2005)
41. Kondo, D., Chien, A., Casanova, H.: Scheduling Task Parallel Applications for Rapid Turnaround on Enterprise Desktop Grids. Journal of Grid Computing 5(4), 379–405 (2007)
42. Lee, L., Liang, C., Chang, H.: An Adaptive Task Scheduling System for Grid Computing. In: Proceedings of the Sixth IEEE international Conference on Computer and information Technology (CIT 2006), September 20-22, p. 57. IEEE Computer Society, Washington (2006)
43. Linderoth, L., Wright, S.J.: Decomposition algorithms for stochastic programming on a computational grid. Computational Optimization and Applications (Special issue on Stochastic Programming) 24, 207–250 (2003)
44. Maheswaran, M., Ali, S., Siegel, H.J., Hensgen, D., Freund, R.F.: Dynamic mapping of a class of independent tasks onto heterogeneous computing systems. Journal of Parallel and Distributed Computing 59(2), 107–131 (1999)
45. Di Gaspero, L., Schaerf, A.: EasyLocal++: an object-oriented framework for the flexible design of local search algorithms and metaheuristics. In: 4th Metaheuristics International Conference (MIC 2001), pp. 287–292 (2001)
46. Di Martino, V., Mililotti, M.: Sub optimal scheduling in a grid using genetic algorithms. Parallel Computing 30, 553–565 (2004)

47. Lee, Y.C., Zomaya, A.Y.: Practical Scheduling of Bag-of-Tasks Applications on Grids with Dynamic Resilience. IEEE Transactions on Computers 56(6), 815–825 (2007)
48. Michalewicz, Z., Fogel, D.B.: How to solve it: modern heuristics. Springer, Heidelberg (2000)
49. Moscato, P.: On evolution, search, optimization, genetic algorithms and martial arts: Towards memetic algorithms. Technical report No. 826, California Institute of Technology, USA (1989)
50. MacLaren, J., Sakellariou, R., Krishnakumar, K.T., Garibaldi, J., Ouelhadj, D.: Towards Service Level Agreement Based Scheduling on the Grid. In: Workshop on Planning and Scheduling for Web and Grid Services (held in conjunction with the 14th International Conference on Automated Planning and Scheduling (ICAPS 2004)), Canada (2004)
51. Newman, H.B., Ellisman, M.H., Orcutt, J.A.: Data-intensive e-Science frontier research. Communications of ACM 46(11), 68–77 (2003)
52. Othman, A., Dew, P., Djemame, K., Gourlay, K.: Adaptive Grid Resource Brokering. In: IEEE International Conference on Cluster Computing (CLUSTER 2003), p. 172 (2003)
53. Page, J., Naughton, J.: Framework for task scheduling in heterogeneous distributed computing using genetic algorithms. AI Review 24, 415–429 (2005)
54. Paniagua, C., Xhafa, F., Caballé, S., Daradoumis, T.: A parallel grid-based implementation for real time processing of event log data in collaborative applications. In: Parallel and Distributed Processing Techniques (PDPT 2005), Las Vegas, USA, pp. 1177–1183 (2005)
55. Perez, J., Kégl, B., Germain-Renaud, C.: Reinforcement learning for utility-based Grid scheduling. In: NIPS 2007 (Twenty-First Annual Conference on Neural Information Processing Systems) Workshops, Vancouver, Canada (2007)
56. Raman, R., Solomon, M., Livny, M., Roy, A.: The classads language. In: Nabrzyski, J., Schopf, J.M., Weglarz, J. (eds.) Grid Resource Management: State of the Art and Future Trends, pp. 255–270. Kluwer Academic Publishers, Norwell
57. Ritchie, G.: Static multi-processor scheduling with ant colony optimisation & local search. Master's thesis, School of Informatics, Univ. of Edinburgh (2003)
58. Ritchie, G., Levine, J.: A fast, effective local search for scheduling independent jobs in heterogeneous computing environments. Technical report, Centre for Intelligent Systems and their Applications, University of Edinburgh (2003)
59. Ritchie, G., Levine, J.: A hybrid ant algorithm for scheduling independent jobs in heterogeneous computing environments. In: 23rd Workshop of the UK Planning and Scheduling Special Interest Group (PLANSIG 2004) (2004)
60. Schwiegelshohn, U., Yahyapour, R.: Analysis of First-Come-First- Serve Parallel Job Scheduling. In: Proceedings of the 9th SIAM Symposium on Discrete Algorithms, January 1998, pp. 629–638 (1998)
61. Schopf, J.M.: Ten Actions when Grid Scheduling. In: Nabrzyski, Schopf, Weglarz (eds.) Grid Resource Management, ch. 2. Kluwer, Dordrecht (2004)
62. Steuer, R.E.: Multiple Criteria Optimization: Theory, Computation and Application. Series in Probability and Mathematical Statistics. Wiley, Chichester (1987)
63. Talbi, E.G.: A Taxonomy of Hybrid Meta-heuristics. J. Heuristics 8(5), 541–564 (2002)
64. Vengerov, D.: Adaptive Utility-Based Scheduling in Resource-Constrained Systems. In: Zhang, S., Jarvis, R. (eds.) AI 2005. LNCS (LNAI), vol. 3809, pp. 477–488. Springer, Heidelberg (2005)

65. Venugopal, S., Buyya, R., Winton, L.: A Grid service broker for scheduling e-Science applications on global data Grids. Concurrency and Computation: Practice and Experience 18(6), 685–699 (2006)
66. Wright, S.J.: Solving optimization problems on computational grids. Optima 65 (2001)
67. Wu, M.Y., Shu, W.: A high-performance mapping algorithm for heterogeneous computing systems. In: Proceedings of the 15th International Parallel & Distributed Processing Symposium, p. 74 (2001)
68. Xhafa, F.: A Hybrid Evolutionary Heuristic for Job Scheduling in Computational Grids, ch. 10. Studies in Computational Intelligence, vol. 75. Springer, Heidelberg (2007)
69. Xhafa, F.: A Hyper-heuristic for Adaptive Scheduling in Computational Grids. International Journal on Neural and Mass-Parallel Computing and Information Systems 17(6), 639–656 (2007)
70. Xhafa, F., Duran, B., Abraham, A., Dahal, K.P.: Tuning Struggle Strategy in Genetic Algorithms for Scheduling in Computational Grids. In: IEEE CelGrid Workshop, OOstrava, The Czech Republic, June 26-June 28 (to appear, 2008)
71. Xhafa, F., Alba, E., Dorronsoro, B., Duran, B.: Efficient Batch Job Scheduling in Grids using Cellular Memetic Algorithms. Journal of Mathematical Modelling and Algorithms (accepted, 2008) Published Online DOI: http://dx.doi.org/10.1007/s10852-008-9076-y
72. Xhafa, F., Barolli, L., Durresi, A.: An Experimental Study On Genetic Algorithms for Resource Allocation On Grid Systems. Journal of Interconnection Networks 8(4), 427–443 (2007)
73. Xhafa, F., Carretero, J., Abraham, A.: Genetic Algorithm Based Schedulers for Grid Computing Systems. International Journal of Innovative Computing, Information and Control 3(5), 1–19 (2007)
74. Xhafa, F., Carretero, J., Alba, E., Dorronsoro, B.: Design and Evaluation of a Tabu Search Method for Job Scheduling in Distributed Environments. In: The 11th International Workshop on Nature Inspired Distributed Computing (NIDISC 2008) held in conjunction with the 22th IEEE/ACM International Parallel and Distributed Processing Symposium (IPDPS 2008), Miami, Florida, USA, April 14-18 (2008)
75. YarKhan, A., Dongarra, J.: Experiments with scheduling using simulated annealing in a grid environment. In: Parashar, M. (ed.) GRID 2002. LNCS, vol. 2536, pp. 232–242. Springer, Heidelberg (2002)
76. Yu, J., Buyya, R.: A Taxonomy of Workflow Management Systems for Grid Computing. Journal of Grid Computing 3(3), 171–200 (2006)
77. Yu, K.-M., Zhou, J., Chou, C.-H., Luo, Z.-J., Chen, C.-K.: A Fuzzy Neural Network Based Scheduling Algorithm for Job Assignment on Computational Grids. In: Enokido, T., Barolli, L., Takizawa, M. (eds.) NBiS 2007. LNCS, vol. 4658, pp. 533–542. Springer, Heidelberg (2007)
78. Yu, J., Li, M., Li, Y., Hong, F.: An Economy-Based Accounting System for Grid Computing Environments. In: Web Information Systems – WISE 2004 Workshops, pp. 233–238. Springer, Heidelberg (2004)
79. Zhang, S., Zong, Y., Ding, Z., Liu, J.: Workflow-Oriented Grid Service Composition and Scheduling. In: Proceedings of the International Conference on information Technology: Coding and Computing (Itcc 2005), vol. II, pp. 214–219. IEEE Computer Society, Los Alamitos (2005)

80. Zhou, J., Yu, K.M., Chou, Ch.H., Yang, L.A., Luo, Zh.J.: A Dynamic Resource Broker and Fuzzy Logic Based Scheduling Algorithm in Grid Environment. ICANNGA 2007(1), 604–613 (2007)
81. Zomaya, A.Y., Teh, Y.H.: Observations on using genetic algorithms for dynamic load-balancing. IEEE Transactions on Parallel and Distributed Systems 12(9), 899–911 (2001)

2
Optimizing Routing and Backlogs for Job Flows in a Distributed Computing Environment

David Montana and John Zinky

BBN Technologies
10 Moulton Street, Cambridge, MA 02138
dmontana@bbn.com, jzinky@bbn.com

Summary. We address the problem of optimizing the flow of compute jobs through a distributed system of compute servers. The goal is to determine the best policy for how to route jobs to different compute clusters as well as to decide which jobs to backlog until a future time. We use an approach that is a hybrid of dynamic programming and a genetic algorithm. Dynamic programming determines the routing and backlog decisions about individual flows of homogeneous jobs, while a genetic algorithm optimizes the order in which the different flows are fed to the dynamic programming algorithm. We demonstrate the effectiveness of this approach on sample problems, some designed to yield a known correct answer and others designed to test the scaling.

Keywords: Distributed System, Job Flows, Cluster, Genetic Algorithms, Dynamic Programming, Routing and Backlog of Jobs.

2.1 Introduction

Distributed computing is the ability to share computational load among multiple computers across a network, and it is a powerful way to increase compute capacity. Effective usage of this joint compute power depends on making good decisions about how to assign the compute tasks to the compute resources. The question of how best to distribute the tasks to the resources can often be formulated as an optimization problem. However, this optimization problem can vary widely depending on the assumptions about the nature of the compute tasks, resources, connectivity, and criteria for what is a good set of assignments.

2.1.1 Overview of the Problem

We have defined an optimization problem based on the needs of a customer who controls a large distributed network of computing devices, also referred to as an *enterprise grid*. It is formulated in general enough terms that it is applicable to other enterprise grids, and not just that of the customer. Some distinguishing properties of this problem are:

- The jobs are aggregated into **job flows**, where the jobs in a flow are homogeneous in their properties and follow certain arrival statistics. The

optimization considers only flows and not individual jobs. The assumption is that there are enough jobs in a flow that they can be modeled accurately and more efficiently as an aggregate flow.
- The resources are aggregated into **clusters**, each with its own local scheduler that assigns individual jobs to individual resources. We assume that the local scheduling at a cluster is handled separately and focus on the high-level problem of routing the jobs flows between clusters. The resource configuration is an enterprise grid, and the problem is to design a meta-scheduler.
- Each job consists of a sequences of steps, also referred to as **tasks**. In general, the tasks for a job require different compute resources, and hence fully scheduling a job requires finding a sequence of different clusters for the job to visit, i.e. a route through the clusters. When considering a job flow rather than an individual job, this route is potentially multipath, as different jobs in the flow can follow different routes. Note that these routes are not routes in the networking sense, i.e. a set of intermediate points leading to a destination, but closer to routes in the vehicle routing sense, i.e. a sequence of destinations.
- The jobs can have varying **utilities** and **deadlines**, although the utilities and average deadlines need to be homogeneous among jobs within each job flow. This reflects the reality that some jobs, such as providing an interactive response to a human, require fast turnaround to be useful, while others, such as overnight batch jobs, do not. Similarly, some jobs are more important to complete than others based on the mission of the enterprise. Balancing the tradeoff between jobs with tighter deadlines and those with higher utility is an important functionality of the meta-scheduler.
- Jobs can be stored in **backlog** until a future time as a means of ceding resources to other jobs. This is generally used to prevent jobs with longer deadlines from blocking the execution of jobs with shorter deadlines when the latter jobs have lesser or equal utility than the former. One assumption is that the local schedulers simply schedule the higher-utility jobs first without considering the deadlines, leaving consideration of deadlines to the meta-scheduler (which makes sense because the meta-scheduler has the global view needed to make tradeoffs between utility and deadlines). A second assumption is that the meta-scheduler is provided predictions about what the future job flow loads and resource availabilities will be to provide a basis for these tradeoffs.

The details of the problem definition are given in Section 2.2.

2.1.2 Previous Work

There is a long history of work in distributed computing, and we do not attempt to summarize it all. Past research has addressed various aspects of distributed computing, including both how to write algorithms that execute on distributed infrastructure and how to create the infrastructure. Just on the infrastructure side, which is our focus, many issues have been studied, such as how hosts connect and communicate, how hosts coordinate to share tasks, and security. An example of an application that addresses a wide range of these issues is Condor [19]. We limit our attention to job scheduling, i.e. how best to share tasks among the assorted compute resources.

Many of the techniques for assigning compute tasks to resources are referred to as *load balancing*, since a key objective is to minimize the amount of time that resources are idle. *Adaptive load balancing* is when the assignment algorithm reacts online to the current situation. Many of the schemes for load balancing are application-specific and need to be revised for each usage, but others are more general [14]. *Scheduling* is potentially more general than load balancing, since scheduling can have more complex objectives than just the immediate correction of imbalances in processing loads.

One important issue to address is scaling. If there are a very large number of compute resources and tasks, it is challenging to efficiently use all the resources. A common approach to scaling is a hierarchical decomposition of the load balancing or scheduling responsibilities. The resources are divided into *clusters*. Each cluster has its own *local scheduler* that assigns tasks to resources, while a high-level scheduler dispatches tasks to clusters (essentially treating clusters as resources). Such a high-level scheduler is now often referred to as a *grid metascheduler*, with the distributed collection of compute resources under its control called a *grid* [12, 20]. There are different development environments for the creation of grid metaschedulers including Community Scheduler Framework, Gridway, and Condor-G [19]. Grimme [9] distinguishes three types of grids: a global grid is a loose confederation across a wide geography with different owners; a high-perfomance-computing grid is a tight clustering of resources; and an enterprise grid (which is the focus of our chapter) is a loose cluster like the global grid but with all machines under the ownership of a single organization [17]. There are different scheduling requirements for each type of grid.

We now mention some previous research that addresses aspects of the scheduling problem not commonly investigated but highly relevant to our work. Lo et al. [11], in their work on metascheduling, address the issue of the effect of time zones on scheduling, allowing the scheduler to anticipate lower loads on compute resources when it is night in their local time zones. In general, the ability to predict loads in the future can help inform scheduling decisions made in the present, particularly if some tasks can wait until the future to be executed [7,3]. Bose et al. [3] show that it is possible to use a genetic algorithm as a meta-scheduler and that it can execute fast enough to be used online under certain circumstances. This genetic algorithm uses a direct encoding with a chromosome that maps each task to its assigned resource. Andresen and McCune [1] define the concept of a *task chain*, a sequence of compute tasks that must be performed to complete a compute jobs, and scheduling a job means routing the job in sequence between resources. Stone [16] uses a network flow algorithm for determining routes of jobs through the resource; this differs from most techniques for scheduling of distributed computing, which consider each task in isolation rather than as part of a flow.

A flow-based view of scheduling leads to a fundamentally different scheduling problem, one that is less reactive and more predictive and one that focuses more on statistical trends rather than individuals jobs and tasks. While uncommon for scheduling of distributed processing, a flow-based approach is common for network routing, and some techniques for determining routes in networks are actually more similar to our approach than those from distributed computing. For example, Casetti et al. [4] has used hierarchical load balancing in the network routing context with the statically determined routing strategy based on the *offered loads*, which are the average flows of different types of traffic. Oueslati and Roberts [15] have demonstrated the benefits in networks of flow-aware routing, i.e. considering each packet as part of a larger flow, as opposed to flow-oblivious routing, i.e. treating each packet separately. Barolli et al. [2] use a genetic algorithm whose chromosomes directly encode a routing tree to determine optimal routes through a network. Okuhara et al. [13] also use a genetic algorithm whose chromosomes directly encode a route, or multiple routes, for optimizing flow-based routing. They include the concept of optimizing flow control, which is the prevention of certain flows from entering the network in order to prevent congestion and which is very similar to our use of backlog. Key and Massoullie [10] integrate the concept of utility associated with a flow into their optimization criterion used with their fluid model for network routing.

2.1.3 Overview of Our Approach

A distinguishing property of our solution to this problem is that it is a hybrid approach, i.e. a combination of different techniques for handling different parts of the problem. These different techniques are the following.

- A simple **greedy algorithm** selects the assigned resource cluster for a given task in a given job at a given time. For each cluster that can handle the task, the algorithm temporarily assigns the task to that cluster, propagates the consequences of this assignments, and determines which assignment minimizes the overall increase in the score.
- The question of when and for how long to place a given job in backlog is addressed using **dynamic programming**. For each task/step in a job, it explores different lengths of time for which to backlog the job at this step, creating a new branch in a search tree for each choice and pruning the tree to explore only the best possibilities. This cannot properly be done as a greedy search because the selection of the backlog times at earlier steps constrains the options at later steps, and the effects cannot be determined until handling the later steps.
- The order in which to consider the flows for routing and backlog decisions is determined by a **genetic algorithm**. The dynamic programming and greedy algorithms determine the routing and backlog decisions one flow at a time, and the order in which these flows are handled greatly affects these decisions. Genetic algorithms have proven good at rapidly searching spaces of permutations, and hence we use one to find the ordering of flows that produces the best overall score.

This general approach is a common one for genetic-algorithm-based scheduling and was first described by Whitley et al. [22] and Syswerda [18]. A fast schedule builder incrementally constructs a schedule one job/task at a time with different schedules resulting from different presentation orders of the jobs/tasks; a genetic algorithm optimizes the presentation order. The details of the optimization algorithm are presented in Section 2.3. Experiments that demonstrate the effectiveness and good scaling properties of the algorithms are described in Section 2.4.

2.2 Problem Definition

We now discuss the various components of the problem definition.

2.2.1 Jobs, Tasks and Flows

Compute jobs consist of a set of steps, or tasks, which are the atomic units of computational work to be performed. Each step must be executed in sequence, so a task cannot begin until its predecessor has completed. Fig. 2.1 illustrates a job with six tasks labeled A-F, which is the standard task breakdown for all jobs in the experiments described below. Each task is assigned to and executed by a single cluster, but the various tasks in a job can be, and generally will be, assigned to different clusters. Therefore, a job will in general visit a sequence of clusters, which means that it will follow a route through the distributed system of computational resources.

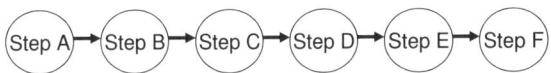

Fig. 2.1. A sample compute job consisting of six steps/tasks

Each job is part of a job flow. The jobs in a flow are homogeneous, i.e. they all possess the same properties. These include

- the sequence of tasks to execute
- the mean execution time of each task
- the mean lifespan of the job, i.e. the time between the arrival time and the deadline
- utility, i.e. how important it is to accomplish the job before the deadline
- the routing constraints, which specify for each task which clusters are allowed to be assigned to that task (The constraints can arise from a variety of causes including network connectivity and the inherent capabilities of the clusters.)

The jobs in a flow enter the system with a known mean time between arrivals.

2.2.2 Resources and Clusters

A cluster is an aggregation of individual compute resources together with a manager to distribute the tasks among the resources and a local scheduler that decides how to assign the tasks to resources. In the work described here, we are not performing the local scheduling but rather just the metascheduling, i.e. the routing of jobs between the clusters. To support the meta-scheduler, we do need a model of the behavior of a local scheduler.

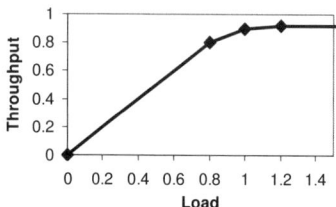

Fig. 2.2. Example load-to-throughput map for a cluster

Not all jobs/tasks that enter a cluster can be completed by the cluster, as each cluster has a finite capacity. We define the *capacity* of a cluster as the maximum number of task-seconds (where a task-second is the amount of computation accomplished on a single task by a "standard" compute engine) that can be completed every second. If the cluster consists of all "standard" resources, then the capacity equals the number of resources. The *load* on a cluster is the number of task-seconds entering the cluster per second, or alternatively, this quantity normalized by dividing by the capacity. The *throughput* of a cluster is the number of task-seconds of processing completed per second, or alternatively, this quantity normalized by dividing by the capacity. The average (normalized) throughput is constrained to be less than 100% and less than the average load.

We characterize the aggregate behavior of a cluster and its local scheduler using a piecewise-linear load-to-throughput mapping, which specifies the expected throughput for a given load. Fig. 2.2 shows an example of such a mapping, which is the one that we used for all the clusters of the experiments described below. To determine which tasks are the ones that are completed, we order the tasks according to the utility of their jobs. The highest-utility tasks are completed at a rate equal to the throughput-to-load ratio of just these tasks, the next-highest-utility tasks at a rate which is the ratio of the additional throughput to the additional load, and so on.

Each cluster has an *input queue* and an *output queue*. Jobs wait in the input queue until a resource becomes available. The output queue is for backlog, where jobs can be saved until a future time when they are released for the next step in their processing sequence. Any job whose deadline passes while waiting in a queue is removed from the queue and discarded.

2.2.3 Routing/Backlog Policies and Routing Constraints

A *routing/backlog policy* determines the decisions for where (i.e., at which cluster) and when the steps of each job are executed. A policy for a given job flow is represented as a set of probabilities. For each step of a job in the flow, the policy specifies for each cluster the probability that the step will be executed at that cluster, as well as the probability that the job will instead be held in backlog until a future time. Jobs held in backlog are released back into the system and re-evaluated when the routing policy is updated. Fig. 2.3 illustrates a sample routing policy for a job flow whose jobs have six steps.

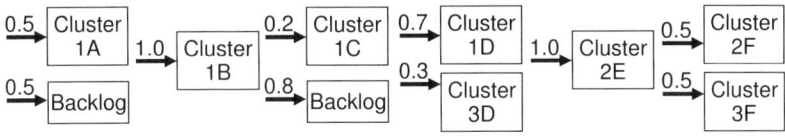

Fig. 2.3. Graphical depiction of a routing/backlog policy for a flow

The routing/backlog policy is the one aspect of the system under external control and is what we can vary to optimize the performance of the system. Section 2.3 discusses how to determine an optimal routing/backlog policy.

Routing constraints limit the possibilities of which clusters can have non-zero probabilities. For each step of each job flow, there is a list of legal clusters that are allowed to handle this step. These constraints can reflect underlying constraints of the system, such as limitations on network connectivity, or can serve to aid the optimization process (either automated or manual) by limiting the choices available and hence reducing the size of the search space.

2.2.4 Epochs and Time Dependence

An epoch is a time interval over which we can assume that all aspects of the system remain constant. This includes the job flows and all their properties, and the clusters and their capacities. We say that the routing policies will remain the same throughout an epoch, as there is no reason for them to change, although the policies will in general change at epoch boundaries. This assumption of piecewise constant behavior for the system allows us to apply a model based on mean-value analysis, as discussed in Section 2.2.5.

The changes in the properties of the job flows and clusters across epochs reflect predictions about how the loads and capacities will vary with time. For example, in Section 2.4 we will consider job flows whose arrival rates follow a 24-hour cyclical pattern, with arrival rates higher during the local daytime and lower during the local nighttime. Similarly, knowledge of future scheduled service times for a resource can be reflected by changing the capacity of its cluster in future epochs. Predictions of future conditions are important for determining not just future policy but also current policy, since backlogs and deadlines can extend across epoch boundaries.

2.2.5 Evaluation Function

To evaluate the effectiveness of a particular policy, we simulate the system with this policy in place. The simulation does not consider individual jobs but rather examines the aggregate flows using a mean-value analysis. The results of the simulation can be scored using an optimization criterion.

The simulator propagates each flow in each epoch over a multipath route through the network of clusters. It multiplies the system arrival rate of a flow by the probability of the first step being assigned to a cluster to obtain the rate of the flow entering this cluster at this step. Similarly, multiplying the system arrival rate by the probability of backlog for the first step yields the rate at which the flow is backlogged before its first step is executed. If more than one of the probabilities is non-zero, the flow will split along multiple paths, which is why we refer to the route as multi-path. The simulator uses the load-to-throughput map for a cluster to determine the output rate of the flow following this step. This process is continued for the subsequent steps, with the cumulative throughput for step N multiplied by the probabilities for step N+1 to give the input rates at the various clusters for step N+1.

While the simulator is relatively simple, there are a few details that complicate it. One is the need sometimes to retract throughput that has been allocated to a flow. If a cluster that handles a flow is then assigned a new flow of utility greater than or equal to that of the original flow, it may be that the original flow loses some of its throughput to the new flow. When this happens, there is in general a chain reaction, since downstream steps of the original job now have lower rates, which in turn can allow other flows to grab some of the forfeited throughput, and so on. The simulator propages these perturbations until they die out.

A second detail is what happens to jobs that are backlogged or waiting in queues across epoch boundaries. Such jobs are added to the flows at the appropriate step in the process for the new epoch, with flow rates that cumulatively across the epoch would integrate to the right number of jobs.

The goal is to minimize the number of dropped jobs, i.e. jobs not completed within their deadline, with an emphasis on not dropping higher-utility jobs. Therefore, the primary component of the optimization criterion is the sum of the utilities of all the dropped jobs. A secondary component of the optimization criterion is a penalty for delaying the execution of jobs into future epochs. The rationale is that the strategy of backlogging jobs for the future depends on the future occurring as predicted, which it often will not in a dynamic environment, so there is benefit to finishing a job earlier rather than later. (We originally discovered the need for such a penalty when the optimizer, faced with jobs identical except for utility, chose as often as not to defer the high-utility jobs to the future and execute the low-utility jobs first, which defies the logic that all else being equal it should schedule the high-utility jobs first.) Optionally, we can add other penalties, such as one for jobs traveling between clusters in order to minimize network traffic, but we do not consider these other types of penalties in this chapter.

We now provide a mathematical definition of the optimization criterion. From the viewpoint of individual jobs (as opposed to job flows), the criterion is

$$\sum_{j \in J_d} u(j) + \sum_{j \notin J_d} u(j)(1 - P^{t(j)}) \tag{2.1}$$

where J_d is the set of all dropped jobs (i.e. jobs that did not complete before their deadline), $u(j)$ is the utility of job j, $t(j)$ is the time in the future at which j is completed, and $P < 1$ is a constant whose role is to penalize the deferral of jobs to the future. This translates into the following formula at the flow level

$$\sum_{e \in E} P^{t(e)} \left(\sum_{f \in F(e)} u(f)[d(f) + b(f)(1 - P^{\tau(e)})] \right) \tag{2.2}$$

where E is the set of all epochs, $F(e)$ is the set of all job flows during epoch e, $t(e)$ is the start time of e, $u(f)$ is the utility of flow f, $d(f)$ is the rate at which jobs in flow f are dropped, $b(f)$ is the rate at which jobs in f are backlogged, and $\tau(e)$ is the duration of epoch e. Note that we choose the penalty for deferring completion of jobs to the future to have an exponential decay purely because this makes it easier mathematically to separate the effects of deferral across epochs.

2.3 Scheduling Algorithm

The policy optimization algorithm has three levels, with each of the lower two levels feeding results to the next higher level. We now present these.

2.3.1 Level 1: Single-Flow, Single-Epoch Optimization

This component determines a routing/backlog policy for the jobs from a single flow entering the system during a single epoch. If a flow is small enough, then a single set of decisions is used for all the jobs in the flow, i.e. for each step the policy has a single non-zero probability and hence all the jobs follow the same route. The test for whether the flow is small enough is whether the flow cannot load any cluster more than x% of its capacity, where we have used x=20%.

Alternatively, if the flow is large, i.e. can produce a load of more than x% on a cluster, it is instead split into N identical smaller subflows, where N is just large enough to reduce the maximum load on a cluster below the threshold. A single-path route is determined for each of these subflows independently and in succession. The results are then aggregated into a single policy, or equivalently a multipath route, using probabilities to specify what fraction of the flow follows each path. (If a flow is large enough to require a division into subflows and hence a multipath route, the discretization and stochastic effects of probabilistic routing will not usually produce large transient deviations from steady-state.) Splitting large flows allows the routing to distribute the load across multiple clusters, which may be necessary for efficiently handling the flow.

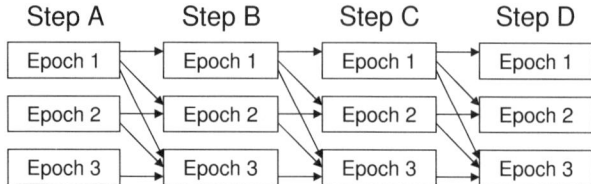

Fig. 2.4. Assignment of a single flow uses dynamic programming to select an epoch for each step. A step can be assigned to any epoch not earlier than that for the previous step.

We now discuss how to determine the single-path route for one of these indivisible subflows. For each step/task in the process, there are two decisions to make: (i) to which cluster to assign the tasks and (ii) in which epoch to execute the tasks (i.e., how long, if at all, to backlog the tasks). The former is done using a purely greedy approach; in the epoch of choice, select the cluster for which the overall penalty (i.e., the increase in the value of the optimization criterion) is minimized by the assignment. Note that the assignment of a flow to a cluster can result in another flow losing throughput at this cluster, and this effect is accounted for in the optimization criterion and hence the greedy selection.

The selection of the epoch in which to execute each step of a flow's processing chain (i.e., decisions about backlog policy) is done using dynamic programming. The rationale is that the choice to postpone the processing of one step can have large consequences for downstream steps that cannot be foreseen when deciding about the current step. So, instead of a greedy selection of the epoch for each step one at a time, we perform a more computationally intensive optimization over all combinations of legal selections of epochs for each step. Note that an epoch is legal for a step if it is not earlier than the epoch of the previous step and not later than the deadline of the flow.

The combinatorics of considering all possible combinations of epochs per step means that it is important to find an efficient optimization technique. Dynamic programming is such a technique because it (i) eliminates entire branches of the search tree early in the process and (ii) pursues the most promising branches first. The first branch point in the search tree is based on the selection of the epoch for the first step, with subsequent branch points under each of these branches based on the selection of the epoch for the second step, etc. The different branches correspond to the different paths through the graph shown in Fig. 2.4. For each epoch E and step S, the search procedure eliminates all but the single best path leading up to the selection of epoch E at step S, which quickly prunes many branches. Furthermore, since the penalty (i.e., change in the optimization criterion) is non-decreasing with each step, we can restrict the search to pursuing only the path with the lowest penalty so far, declaring the search finished when a path that has completed all the steps has a score less than or equal to the score of any partial path.

2.3.2 Level 2: Multi-flow, Single-Epoch Optimization

Using the single-flow route optimizer, we can define what we call the *rapid route builder*, which creates an entire set of routes, i.e. a full set of routing/backlog policies, for the jobs flows in an epoch. Given an ordering of the flows, the rapid route builder uses the single-flow optimizer to create the routes for each flow in succession in the order given. Due to interactions between the flows, the policies produced are potentially very different depending on the order in which the flows are routed. Therefore, finding the best ordering of jobs to feed the rapid route builder is an optimization problem we need to solve.

To perform this optimization, we use an order-based genetic algorithm. The development of order-based genetic algorithms [6, 8] was inspired by the recognition that for problems like the traveling salesman problem, the goal is to find the best ordering of N objects. Its chromosome is a direct representation of a permutation of N objects, labeled 1 through N, and its operators are designed to manipulate chromosomes of this type. Order-based genetic algorithms have been demonstrated to be very effective and efficient at searching the space of permutations, which is why we have chosen this technique.

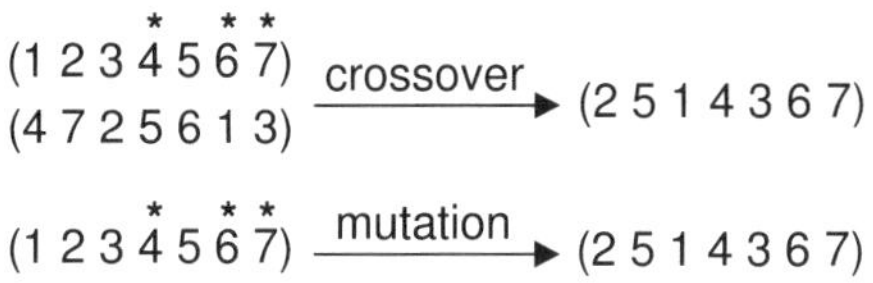

$$(1\ 2\ 3\ \overset{*}{4}\ 5\ \overset{*}{6}\ \overset{*}{7}) \\ (4\ 7\ 2\ 5\ 6\ 1\ 3)\ \xrightarrow{\text{crossover}}\ (2\ 5\ 1\ 4\ 3\ 6\ 7)$$

$$(1\ 2\ 3\ \overset{*}{4}\ 5\ \overset{*}{6}\ \overset{*}{7})\ \xrightarrow{\text{mutation}}\ (2\ 5\ 1\ 4\ 3\ 6\ 7)$$

Fig. 2.5. The crossover and mutation operators. The *'s indicate the randomly selected positions that remain fixed in the (first) parent.

The crossover operator used by the genetic algorithm is position-based crossover [18], and its operation is illustrated in Fig. 2.5. It works as follows. A set of positions is randomly selected (which in the example of Fig. 2.5 are positions 4, 6 and 7). The elements at these selected positions in the first parent (which in the example are the integers 4, 6 and 7) are maintained at these positions in the child. The remaining elements (which in the example are the integers 1, 2, 3 and 5) are used to fill in the remaining slots in the child, but will in general be at different positions in the child than in the first parent. The order of these elements in the child will be the same as their order in the second parent (which in the example means that 2 is placed in the first empty position, followed in order by 5, 1 and 3).

Also illustrated in Fig. 2.5 is the mutation operator. It works the same as the crossover operator except without a second parent to provide the ordering for the subset of elements that are reordered in the child. Instead, the new order of the shuffled elements is randomly selected.

Note that there are other possible definitions for the crossover and mutation operators with an order-based genetic algorithm. Some of these are discussed in

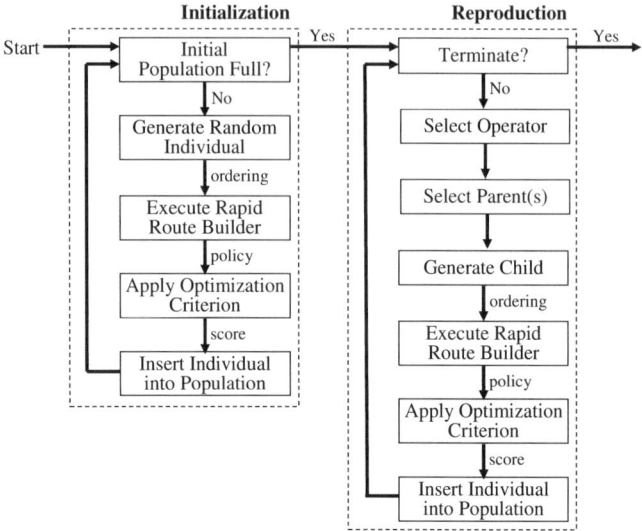

Fig. 2.6. The operation of the genetic algorithm

[21] and [5]. The differences in performance between the good ones are relatively minor, so we have not experimented with different operators.

Each member of the initial population is generated by selecting a random ordering. The flow of operations of the genetic algorithm is shown in Fig. 2.6.

The genetic algorithm is steady-state, which means that it generates and replaces one individual at a time rather than an entire population. The advantage of a steady-state replacement strategy is that the search generally proceeds faster, since the genetic algorithm can use good individuals as soon as they are created rather than waiting for generational boundaries. Since there are no generations, the amount of work done by the search algorithm is measured by the number of individuals evaluated.

Two key parameters that control performance are the population size and the number of evaluations. Increasing them increases the expected quality of the solution found, at the expense of increasing the search time. Hence, the selection of these parameters controls the inherent tradeoff between solution quality and search time. We have found empirically for this problem that it is generally good to have the number of evaluations five times the population size, since on average this provides enough time for the search to converge without spending too much time at the end of the run stuck without making progress. So, for each run, we specify the number of evaluations and automatically set the population size to be one-fifth of that quantity. (Ideally, we would vary both the population size and the number of evaluations, along with other parameters, to find the best combination. However, since we are running many experiments with different problem definitions, it is more practical to reduce the number of free parameters.)

In general, the number of evaluations (and population size) needs to be larger when there are more flows, since the search space is larger. However, we can choose a smaller number of evaluations and quicker search time in exchange for a worse expected solution. This ability to shorten the search is important, since the policy optimizer is potentially used adaptively to update the routing policy in real time (in response to an unexpected change in operating conditions such as a surge in load or disabled resources). Note that taking advantage of the inherent parallelism of genetic algorithms by using multiple processors can also improve the execution speed, but without sacrificing solution quality.

2.3.3 Level 3: Multi-epoch Optimization

This component of the optimization algorithm steps through the epochs one at a time and executes the Level 2 optimization for all the flows in the current epoch. It starts with the final epoch and works backwards in time, as shown in Fig. 2.7. The rationale for working backwards in time is that flows from a particular epoch can be postponed to the future, hence requiring knowledge of the future loads on the clusters to make good decisions about whether to backlog the flows or not. Furthermore, the earlier epochs are the more important ones to do correctly, since they will be the ones executed first without the opportunity for revision.

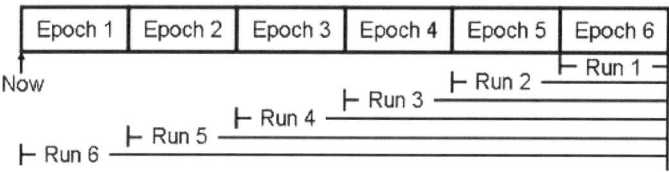

Fig. 2.7. The optimization algorithm starts by optimizing the routing policies for the last epoch and working backwards

The result of the entire process is a set of routing/backlog policies, one for each epoch. While these generally will be good policies, usually optimal or near optimal, there are three places in the process which can lead to suboptimality:

- The best policy for an epoch is not guaranteed to be generated by any job ordering fed to the rapid route builder.
- The genetic algorithm is not guaranteed to find an optimal ordering, since it is a heuristic search technique.
- Optimizing each epoch in succession rather than all in single large optimization is potentially suboptimal.

What our approach does provide is a good tradeoff between finding a good solution and keeping the search time relatively small, so that using this procedure is actually practical even for large distributed systems. In the next section, we demonstrate experimentally both the ability to find good policies and the relatively rapid execution times even as the problem size grows.

2.4 Experiments

We start with a set of experiments that show that the approach just described finds the right solution on a set of problems for which we can determine a good solution by analysis. The next experiments examine the scaling properties of the approach, i.e. how the performance, and in particular the execution speed, of the algorithm increases as the problem size increases.

2.4.1 Sample Scenario and Perturbations

This set of experiments involves a relatively small (though not trivially small) problem containing 24 job flows and 18 clusters and lasting for 6 epochs. Because of its symmetries, this particular problem lends itself to analysis by a human, so we can determine whether our approach finds a good solution. Perturbing the problem causes the optimal strategy to change. We introduce perturbations that include losses of resources in the present, anticipated losses of resources in the future, and surges in the loads, and we verify that the algorithm makes the proper adjustments to the policy.

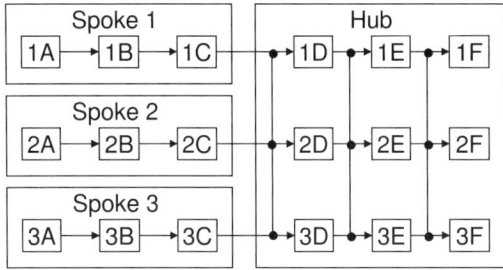

Fig. 2.8. The topology of the clusters is determined by the routing constraints on the flows. Steps A-C must all be performed in a prespecified one of the three spokes, while steps D-F can be performed in any of three clusters in the hub.

Baseline Problem - We now describe the initial problem on which we test our approach. There are 18 clusters in total. Each cluster is specialized to handle one of the six steps of the jobs, whose sequence of steps is shown in Fig. 2.1, with three clusters per step. Each cluster has a capacity of 13, with the underlying assumption that there are 13 identical compute resources aggregated at each cluster. The load-to-throughput map for each cluster is that shown in Fig. 2.2.

The routing constraints of the job flows induce an inherent connectivity on the clusters, which is the hub-and-spokes configuration shown in Fig. 2.8. The first three steps, i.e. steps A-C of a job flow, are constrained to be executed in one of the three spokes. For example, some of the job flows are constrained to spoke 1, and hence must be assigned to clusters 1A, 1B and 1C for their first

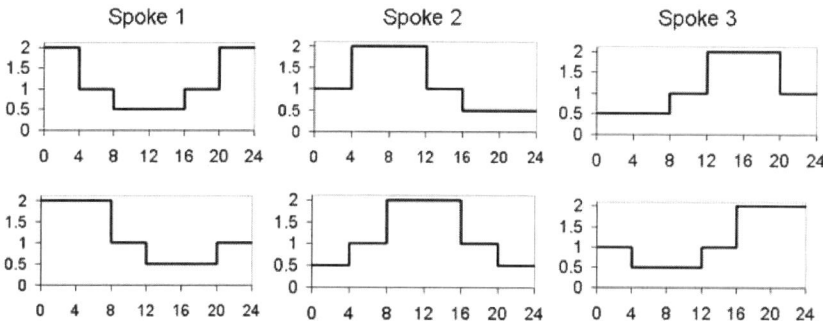

Fig. 2.9. The six different arrival rate patterns and their associated spikes

three steps. The final three steps, steps D-F, are handled in the hub, and the job flow is free to be assigned to any of the three clusters specializing in that step.

There are 24 different job flows. The job flows have all different combinations of the following three properties.

- There are two different utilities, high (numerical value = 2) and low (numerical value = 1).
- There are two different deadlines, short (numerical value = 1 hour) and long (numerical value = 16 hours).
- There are six different arrival rate patterns, i.e. arrival rates as a function of time. These are pictured in Fig. 2.9. The six patterns are all cyclical over 24 hours and all essentially the same pattern with different offsets, so that the peaks and valleys of each are at different times. (This captures in an idealized form the daily cycles in usage requests, with more requests during the local daytime.) Each pattern is associated with a particular spoke, with two patterns adjacent in their offsets assigned to each spoke.

The six steps in every job flow each require one minute to complete.

There are six epochs each of duration four hours. The entire problem covers a 24-hour period. The result of the optimization will be six routing/backlog policies, one for each epoch.

An analysis of this scenario yields the following. During epoch 1, spoke 1 is overloaded. There are eight flows associated with spoke 1, and each of the flows has arrival rate of 2 jobs/minutes. Therefore, there is an aggregate arrival rate of 16 jobs/minute that are constrained to use the clusters in spoke 1. These clusters have a capacity of 13 jobs/minute, and a maximum throughput of even less. So, not all these jobs can be processed during the first epoch. If all these jobs are allowed to enter the clusters, as opposed to being backlogged until future epochs, the high-utility flows will receive most of the throughput, with the low-utility jobs waiting in the input queues. Most of the low-utility, short-deadline jobs will time out and hence be dropped.

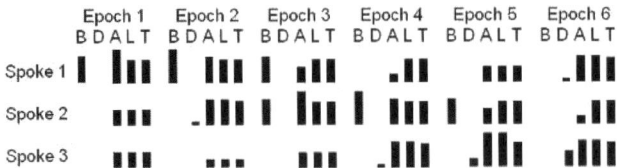

Fig. 2.10. A graphical depiction of the results for the baseline problem set. For each spoke in each epoch, the figure shows the relative size of the backlog (B), dropped jobs (D), aggregate arrival rate (A), cluster loads (L), and cluster throughputs (T). The units for A, L and T are jobs/minute, while those for B and D are jobs.

So, a better strategy is to backlog enough long-deadline flows from spoke 1 at the entry to the system to allow the short-deadline flows to all complete in the first epoch. These long-deadline jobs are released from backlog into the clusters of spoke 1 during epochs 3 and 4, when the there is spare capacity compared to the load due purely to arrivals.

Spoke 2 is similarly overloaded in epochs 2-4, with a peak in epoch 3. Hence, the best strategy is to backlog the flows from spoke 2 during epoch 3 and release them during epochs 5 and 6, when the arrival load is lightest.

The arrivals for spoke 3 peak during epoch 5. Because there are no epochs included beyond epoch 6, there is no advantage to backlogging the flows here, and hence the best strategy is to let all the jobs into the system and allow the local schedulers to give first priority to the high-utility jobs. [A lesson here is that it is important to include enough epochs beyond the last epoch whose optimized policy might actually be used so that all policies of interest are not influenced by this type of "boundary effect".]

As partially illustrated in Fig. 2.10, the results from the optimization were as expected from the analysis, so the algorithm found an approximately optimal set of policies.

Perturbation 1: Current Loss of Hub Cluster - This scenario is the same as the baseline problem except with the capacity of cluster 1E set to zero for epochs 1 and 2. In the hub, unlike in the spokes, there is a choice of multiple clusters for each step of the job flows, and tasks that would have been assigned to the missing cluster can instead be sent to the two alternative clusters, 2E and 3E. Because the two clusters cannot quite handle the full load, some of the long-deadline jobs are backlogged until the anticipated return of the disabled cluster. Optimizing the policies produces the expected behavior; this demonstrates how our approach can be used to modify the routing policy to adapt to changes in the distributed system.

Perturbation 2: Future Loss of Spoke Resources - This scenario is the same as the baseline except with the capacity of cluster 1B set to 6 instead of 13 in epochs 3-6. This anticipated future loss of resources changes the current (i.e., epoch 1) optimal backlog policy for the job flows associated with spoke 1. Since there no longer will be excess capacity available in the future, the best current policy is to complete the high-utility jobs and allow some of the low-utility jobs to be dropped. Our algorithm finds this new optimal policy, demonstrating the ability to adapt current policy to anticipated future changes in the distributed system.

Perturbation 3: Surge in Load in a Spoke - This scenario is the same as the baseline except one of the high-utility, long-deadline job flows in spoke 1 has an arrival rate that is increased from 2 jobs/minute to 5 jobs/minute during epochs 1 and 2. This surge means that the optimal backlog policy for spoke 1 now has to focus on completing all the high-utility jobs, letting the low-utility jobs be dropped during the first four epochs. Some of the additional high-utility jobs that are part of the surge are immediately sent to the clusters for processing, while others are backlogged until there is excess capacity in fuure epochs. Note that low-utility jobs continue to be dropped even after the surge has ceased in order to handle the backlog of high-utility jobs accumulated during the surge. Our approach finds this new policy, demonstrating the ability to adjust policy to adapt to changes in the load.

The solutions are generated within roughly 12 seconds on a single 2.8GHz CPU, showing that the approach not only finds a good solution but does so in a reasonably short time.

2.4.2 Scaling Properties

It is important to understand how our approach performs not just on relatively small problems but also on larger problems. A second set of experiments investigate the scalability of the algorithm, i.e. how increasing the size of the problem affects the algorithm's performance.

The problem size can vary along multiple different dimensions. We have identified what the different dimensions of problem scale are, and we have developed the capability for varying the problem size along one dimension at a time. This allows us to investigate the effects of changing only one, or some subset, of these dimensions, as well as all of them simultaneously. We now enumerate these different dimensions along with our theoretical analysis of how they effect search time:

- **(average) number of legal clusters per task -** For each task, the greedy selection process needs to evaluate the effect on the optimization criterion of assigning that task to each legal cluster. Since each such evaluation is independent of the others, the total time required is proportional to the number of legal clusters to evaluate. Therfore, the overall search time should scale linearly along this dimension.
- **total number of epochs -** For each epoch, the algorithm needs to perform a separate genetic algorithm run. These runs are independent, so the execution time should scale linearly in this dimension.
- **(average) number of epochs before a job's deadline -** For each step of the dynamic programming process, i.e. each task in the job, the algorithm

maintains potentially one branch for each epoch before the deadline. For the next step of each branch, it can explore a number of branches that is on average half as many as the number of epochs before the deadline. So, the algorithm potentially scales as a square of this dimension. However, in practice, dynamic programming should eliminate most of these potential branches, and the scaling could be closer to linear.
- **(average) number of steps per job** - The dynamic programming process needs to take one more step in its chain for each step in the job. Since these steps are largely independent, we would predict linear scaling in this dimension.
- **number of job flows** - There are two ways that the number of job flows effects search time. For each individual in the genetic algorithm, the rapid route builder needs to route this many flows. Each flow is mostly independent (although not entirely independent because of competition for throughput at the clusters), so the time should increases linearly. Secondly, increasing the number of job flows increases the number of possible orderings of these flows, and hence the size of the search space for the genetic algorithm. This will increase the number of individuals the genetic algorithm must evaluate to find a near optimal one. Based on past experience with order-based genetic algorithms, we predict that the increase in the required number of evaluations is between linear and quadratic, but this is problem-dependent and can only be determined empirically.

Additionally, there can be assorted costs or savings due to secondary interactions. For example, a decrease in the capacity of each cluster can cause the flows to be split into more subflows during the rapid schedule building process, hence leading to longer execution times.

We have devised methods to increase the scale in one of these dimensions at a time maintaining approximately the same optimization problem.

- To increase the number of legal clusters per task by a factor of N, replace each cluster in the original problem with N clusters, each with 1/N times as much capacity as the original. For every task for which the original cluster was legal, make all N new clusters be legal.
- To increase the number of epochs, convert each epoch in the original problem into N epochs identical to the original except with duration 1/N as long. Note that this changes both the total number of epochs and the number of epochs before the deadline by a factor of N.
- To increase the number of steps per job, convert each step of each job flow of the original problem into N steps identical to the original, in particular allowing the same legal clusters, except each requiring 1/N the time to complete.
- To increase the number of job flows, convert each job flow into N job flows each identical to the original except with arrival rate 1/N of the original rate.

Table 2.1. Results of the scaling experiments

Dataset Description	100-Eval Time (secs)	Evals for Optimum	Time for Optimum
baseline	12	100	12
10x clusters	166	100	166
10x flows	352	4000	14080
10x epochs	1091	100	1091
10x steps	358	100	358
10x clusters 10x flows	720	4000	28800
2x clusters 2x flows 2x epochs 2x steps	406	250	837

We have applied these transformations to the baseline data set described in Section 2.4.1. For each transformed data set, we measure three quantities. One is the amount of time required for the full optimization to execute with the number of evaluations for the genetic algorithm specified to be 100. This measures the change in exeucution time of the rapid route builder. The second quantity is the number of evaluations required of the genetic algorithm to reach a near optimal solution. We determine this value by executing with different numbers of evaluations and finding at what point the result stops improving significantly (no more than 1%). The third quantity is the time required to reach this near optimal solution, which should approximately equal the product of the first two quantities divided by 100. These three quantities are shown in the following table for each of the datasets. Note that all runs are performed on the same single machine with a single 2.8GHz processor.

The results are largely as predicted with a few exceptions, which we now discuss. Perhaps the biggest deviation from predicted behavior is when the number of flows is increased by a factor of ten (10x flows). This leads to an 100-evaluation execution time that is 30 times larger than that for the baseline problem. This execution time was predicted to be linear in the number of flows, and hence we would have instead expected a factor of 10. A possible explanation is that increasing the flows without increasing the number of clusters resulted in ten times as many flows at each cluster, leading to overhead in accounting, most importantly the propagation of retracted throughput, for all these flows. Note that when the number of flows and the number of clusters are both increased by a factor of ten, the 100-evaluation execution time is only increased by a factor of 60, which is even less than the factor of 100 predicted.

If the problem is such that the time it takes to optimize is longer than the desired time, there are some techniques to reduce the search time. The simplest is just to reduce the number of evaluations of the genetic algorithm, accepting the lesser quality of the solution. A similar method that may sacrifice less of the solution quality is based on the recognition that epochs further in the future are less important to optimize well than epochs closer to the present. Therefore, using less evaluations for the genetic algorithm on these future epochs leads to faster execution with an acceptable decrease in solution quality. An alternative method to decreasing optimization time is to decrease the number of epochs, number of flows, etc. by merging them, blurring some of the finer distinctions of the model (and hence the quality of the solution when applied to the real system), but decreasing the problem size. Again, this technique can be applied more heavily to the epochs further in the future to reduce the effects on the policies that need to be in place soon.

2.5 Conclusions and Future Work

We have defined a problem involving optimizing the routing and backlog policy of a large distributed computing system. Our approach to this scheduling problem involves a combination of dynamic programming and a genetic algorithm. This approach allows the optimization to proceed rapidly over a large search space while still finding good solutions. One set of experiments has proven the ability of the approach to find an optimal policy, while another set of experiments has demonstrated its scalability.

We have integrated this policy optimization algorithm into a prototype design tool and demonstrated its effectiveness on sample problems; the next steps involve moving this tool into an operational setting. Initially, it would be used off-line with data collected from a functioning enterprise grid used to define the optimization problem. The ultimate goal is to integrate this policy optimization algorithm into an online adaptive controller that adjusts routing/backlog policies in real time based on automated data feeds.

References

1. Andresen, D., McCune, T.: Towards a Hierarchical Scheduling System for Distributed WWW Server Clusters. In: Proceedings of the The Seventh IEEE International Symposium on High Performance Distributed Computing (1998)
2. Barolli, L., Koyama, A., Matsumoto, K., Suganuma, T., Shiratori, N.: A Genetic Algorithm Based Routing Method Using Two QoS Parameters. In: Proceedings of the 13th International Workshop on Database and Expert Systems Applications (2002)
3. Bose, A., Wickman, B., Wood, C.: MARS: A Metascheduler for Distributed Resources in Campus Grids. In: Proceedings of the Fifth IEEE/ACM International Workshop on Grid Computing (2004)
4. Casetti, C., Cigno, R., Mellia, M.: QoS-Aware Routing Schemes Based on Hierarchical LoadBalancing for Integrated Services Packet Networks. In: Proceedings of the IEEE International Communication Conference (1999)
5. Chen, S., Smith, S.: Improving Genetic Algorithms by Search Space Reduction (with Applications to Flow Shop Scheduling). In: Proceedings of the Genetic and Evolutionary Computation Conference, pp. 135–140 (1999)

6. Goldberg, D., Lingle, J.R.: Alleles, Loci, and the Traveling Salesman Problem. In: Proceedings of the First International Conference on Genetic Algorithms, pp. 154–159 (1985)
7. Goswami, K., Devarakonda, M., Iyer, R.: Prediction-Based Dynamic Load-Sharing Heuristics. IEEE Transactions on Parallel and Distributed Systems (1993)
8. Grefenstette, J., Gopal, R., Rosmaita, B., van Gucht, D.: Genetic Algorithms for the Traveling Salesman Problem. In: Proceedings of the First International Conference on Genetic Algorithms, pp. 160–165 (1985)
9. Grimme, C.: Grid Metaschedulers: An Overview and Up-to-date Solutions. PowerPoint presentation (2007)
10. Key, P., Massoullie, L.: Fluid Models of Integrated Traffic and Multipath Routing. Queueing Systems: Theory and Applications 53(1-2), 85–98 (2006)
11. Lo, V., Zhou, D., Zappala, D., Liu, Y., Zhao, S.: Cluster Computing on the Fly: P2P Scheduling of Idle Cycles in the Internet. In: International Workshop on Peer-to-Peer Systems (2004)
12. Mausolf, J.: Grid in Action: Managing the Resource Managers. IBM developerWorks (2005)
13. Okuhara, K., Tanaka, T., Ishii, H.: Routing and Flow Control by Genetic Algorithm for a Flow Model. Systems and Computers in Japan 34(1), 11–20 (2003)
14. Othman, O., Schmidt, D.: Issues in the Design of Adaptive Middleware Load Balancing. In: Proceedings of the ACM SIGPLAN Workshop on Optimization of Middleware and Distributed Systems, pp. 205–213 (2001)
15. Oueslati, S., Roberts, J.: Comparing Flow-Aware and Flow-Oblivious Adaptive Routing. In: 40th Annual Conference on Information Sciences and Systems, pp. 655–660 (2006)
16. Stone, H.: Multiprocessor Scheduling with the Aid of Network Flow Algorithms. IEEE Transactions on Software Engineering SE-3(1), 85–93 (1977)
17. Strong, P.: Enterprise Grid Computing. ACM Queue 3(6) (2005)
18. Syswerda, G.: Schedule Optimization Using Genetic Algorithms. In: Davis, L. (ed.) Handbook of Genetic Algorithms, pp. 332–349. Van Nostrand, Reinhold (1991)
19. Thain, D., Tannenbaum, T., Livny, M.: Distributed Computing in Practice: The Condor Experience. Concurrency and Computation: Practice and Experience 17(2-4), 323–356 (2005)
20. Vadhiyar, S., Dongarra, J.: A Metascheduler for the Grid. In: Proceedings of the 11th IEEE International Symposium on High Performance Distributed Computing (2002)
21. Vazquez, M., Whitley, D.: A Comparison of Genetic Algorithms for the Static Job Shop Scheduling Problem. In: Deb, K., Rudolph, G., Lutton, E., Merelo, J.J., Schoenauer, M., Schwefel, H.-P., Yao, X. (eds.) PPSN 2000. LNCS, vol. 1917, pp. 303–312. Springer, Heidelberg (2000)
22. Whitley, D., Starkweather, T., Fuquay, D.: Scheduling Problems and Traveling Salesmen: The Genetic Edge Recombination Operator. In: Proceedings of the Third International Conference on Genetic Algorithms, pp. 133–140 (1989)

3

Robust Allocation and Scheduling Heuristics for Dynamic, Distributed Real-Time Systems

Dazhang Gu and Lonnie Welch

Center for Intelligent, Distributed, and Dependable Systems
School of Electrical Engineering and Computer Science
Ohio University, Athens, Ohio 45701 U.S.A.
gud@ohio.edu, welch@ohio.edu

Summary. A challenge facing real-time computing is the need to deploy real-time systems in dynamic operational environments. The systems have explicit deadline requirements, but their execution times are often affected by unpredictable environmental inputs that cannot be known a priori and have no worst-case estimates. As a result, traditional real-time task allocation and scheduling techniques do not apply.

This research proposes a new task allocation and scheduling approach for these dynamic, distributed real-time systems. The approach offers these systems explicit real-time guarantees as well as maximized tolerance (robustness) of unpredictable changes in environmental inputs. This work consists of (1) a real-time computing model that incorporates environmental factors, (2) metrics that characterize robustness, and (3) algorithms that find robust allocations with feasible schedules for local schedulers. Analytical bounds were derived to guarantee the performance of the algorithms. The work produces a dependable foundation for task allocation and scheduling so that real-time systems may be designed and deployed for many time-critical but unpredictable real world environments.

Keywords: Real Time Computing, Real Time Allocation, Scheduling, Dynamic Environment, Heuristics.

3.1 Introduction

Distributed real-time systems provide guarantees on timing requirements while boosting performance through concurrency in computing resources. They are used to build large and complex real-time applications. In order to both achieve timing requirements and maximize resource utilization, resource management for these systems involves both the allocation of tasks to processors and the local scheduling of tasks on each processor. Usually, the allocation and scheduling algorithms make use of a set of predetermined task and processor parameters, and when exact values cannot be obtained, worst-case estimates are often used.

However, a new problem arises when real-time systems need to be engineered for those real world environments where such parameters have no worst-case estimates, and the values only become known after each event. The execution times of algorithms in real-time tasks generally depend on input sizes, but in these environments (labeled "dynamic environments"), some of the inputs originate in the environment external to the system. Values of these inputs are a result of man or nature that no one can predict. An example in case is the number of aircrafts tracked by a radar system. Real-time systems deployed in these environments may experience bursts of inputs from time to time which drive up the execution times of certain tasks. If traditional allocation and scheduling approaches were applied, the systems would suffer deadline misses due to an ill-conceived allocation. Intuitively, it is not desirable to allocate those tasks heavily influenced by one environmental input onto the same processor, even if they initially seem to fit feasibly. This research follows up on that intuition.

The notion of tasks that have dependencies on environmental factors originated from the study of a generic air defense system [1]. The *detect* task identifies threats to a defended entity. The task runs periodically and performs the functions of filtering and evaluating radar tracks. When a threat is detected, the *detect* task triggers the *engage* task, which fires a missile at the threat. After a missile is in flight, the *guide missile* task keeps the missile on the correct course. The *guide missile* task executes periodically; uses sensor data to track the threat; recalculates the flight path for the missile; and issues guidance commands to the missile. During operation, there may be multiple replicas of the three tasks running concurrently. When the number of radar tracks grows too large for a single replica of the *detect* task to process within the required time, one or more replicas are created and the radar tracks are partitioned among them. In a similar manner, the *guide missile* task is replicated as necessary to meet deadlines, and replication is also used for the *engage* task when heavy workloads are anticipated. All three of the tasks have resource needs that are environment dependent. The execution time of the *detect* task is primarily workload-dependent. Since the task evaluates each radar track to determine if it is a potential threat, its execution time is a function of the number of radar tracks in the environment. The workload of the *engage* task is also variable since it is activated by the number of tracks deemed as threats. Similarly, the work performed by the *guide missile* depends on the number of missiles in flight. Thus, an important problem to solve for this system is how to allocate resources to the tasks and schedule them in a manner that allows real-time constraints to be met and that minimizes the need for reallocations (which create overhead in the system). Also, it is desirable to know the maximum numbers of missiles and radar tracks that can be sustained by a given configuration.

Execution times of tasks in these *dynamic* distributed real-time systems must be regarded as functions of unpredictable environmental factors because the running times of their algorithms depend on sizes of these environmental inputs [2, 3]. Researchers engineering these systems have realized that employing the systems in unpredictable environments may affect these input sizes and result in varying execution times of tasks that cannot be known in advance [4]. Meaningful worst-case execution times (*WCET*'s) cannot be obtained. Therefore, many have pointed out that traditional periodic task scheduling and allocation based on worst-case estimation are not applicable [5, 6, 7, 8].

Existing approaches to address the problem in these systems include adaptive resource allocation, proactive robust allocation, and probabilistic deadline guarantee. However, these approaches have not been satisfactory.

Adaptive resource allocation reacts to changes in environment and a system's resource needs by passively reallocating the system [4, 9, 10]. Thus it is vulnerable to frequent environment changes that trigger costly reallocations and result in thrashing, and no guarantees can be made. Further, it is often infeasible to reallocate *stateful* applications in real-time.

A current area of active research is proactive resource allocation with an objective of robustness. It seeks to maximize an allocation's tolerance of unpredictable environment changes without jeopardizing feasibility. Such robust allocation reduces the necessity of reallocations, which are time-consuming both to compute and to enact. Additionally, reallocations are not appropriate for stateful real-time applications whose complex states are costly to recover. The problem is being studied and solutions are still primitive. An approximation algorithm was developed to maximize the allowable workload in an allocation [2], but the optimization was limited to one environmental input. The algorithm and analysis were developed to produce robust allocation in the case of multiple environmental inputs [11]; however, rate monotonic scheduling was required and the analysis was pessimistic about execution time functions. [12] reported a set of heuristic algorithms useful in finding robust allocations for independent, periodic real-time tasks. A mixed-integer programming approach was proposed by [13] to maximize the allowable increase in load for a static allocation. However, heuristics were applied before MIP and processors were assumed to be fair-shared. Several heuristic algorithms were used in [14] to find robust allocations for periodic task strings. A special scheduler based on tightness was assumed, but no guarantee was developed. An l-2 norm based robustness metric was introduced in [15] for multiple inputs. The metric partially characterizes feasible regions by using an intangent sphere, and no algorithm was developed to optimize it.

The probabilistic model characterizes unpredictable task execution times as random variables, and the objective is to derive statistical confidence in deadline misses [5, 6]. The task allocation problem was studied for systems with dependencies and multiple processors by [8]. However, the allocation search and evaluation were expensive; certain knowledge of input (such as distribution) has to be assumed.

The solution from this research addresses these shortcomings. First, a new model that explicitly incorporates environmental input factors is introduced based on a widely accepted model of real-time systems [16]. Task execution times are characterized as functions of multiple environmental inputs. Second, metrics characterizing the robustness of allocations are defined, and the problem of robust task allocation and scheduling is defined based on the metrics. Third, allocation and scheduling algorithms are designed that use the metrics to find feasible and robust allocations. The heuristics build upon results from classic real-time scheduling algorithms (*RMS* and *EDF* [17]), as well as real-time task allocation algorithms [18, 19]. The algorithms have fast running time

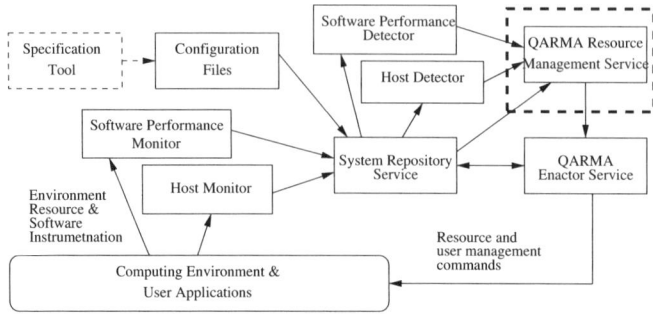

Fig. 3.1. The QARMA architecture

and scalability, which are crucial for modern distributed systems that may contain hundreds of processors and thousands of tasks. In addition to experimental evaluation of the performance, theoretical bounds for the solution quality are also derived, warranting the robustness that may be achieved by the systems in operation while meeting all deadline requirements.

Besides the impact on engineering and deployment of real-time applications, allocation and scheduling algorithms from this research will also benefit resource management software. For example, *QARMA*, short for the Quality-based Adaptive Resource Management Architecture in Fig. 3.1, serves as a basis for integration of existing *CORBA* services and management mechanisms into a single, coherent framework for resource management, and it is easily extendable to allow the use of new resource management mechanisms [20]. It consists of three major components: the System Repository Service, the Resource Management Service, and the Enactor Service. The *System Repository* stores both static and dynamic information that describe the software systems and resources in the computing environment. Information is provided from specification files and gathered by various monitors at run-time. The *Resource Management Service* (dashed box) is responsible for using information in a system repository to decide what actions should be performed to ensure that performance requirements are satisfied and system performance is optimized. The *Enactor Service* receives instructions from the resource management service about actions to perform and enacts them. The actions may include the adjustment of quality settings, the assignment or migration of tasks, and the replication of tasks. The algorithms incorporated into the *Resource Management Service* would offer both feasibility and robustness in these adjustments.

Rest of the chapter is organized as follows. Section 3.2 re-examines a traditional real-time computing model and makes the necessary extensions to incorporate environment factors. Based on the extended model, section 3.3 formally poses the research problem and presents the solution of robust allocation and scheduling heuristics. Section 3.4 then provides experimental validations for the theoretical analysis. Finally, section 3.5 brings this work into perspective by reviewing related research.

3.2 System Model

As a first step, a new real-time computing paradigm is introduced. The model captures features relevant to the problems. It is based on periodic real-time tasks characterized by environment-dependent execution time functions, as opposed to the traditional model with hard periodic real-time tasks characterized by worst case execution times. In this paradigm, occasional deadline misses can be tolerated when unpredictable environmental factors drive a demand beyond the limit of available resources. This section introduces the new system model by beginning with a traditional model, and then making necessary extensions to incorporate environmental factors. An example will follow to illustrate the model.

Traditional Model

The system model used in this work is derived from the standard periodic task model given in [16]. In this model, the software system consists of a set of n periodic tasks $S = \{T_1, T_2, \ldots, T_n\}$. Each T_i is released periodically with period p_i and has a deadline equal to its period. The execution time of each task $T_i \in S$ is a constant e_i that represents the worst case execution time of T_i. There is a set of m processors $H = \{P_1, P_2, \ldots, P_m\}$.

Model Extensions

Traditional models do not capture dynamic environmental factors, since the execution times are modeled as constants. The models are inadequate for some real-time systems operating in dynamic environments. For example, a general distributed control system is depicted in Fig. 3.2. It has filter, analysis, action planning, and actuation tasks. One or more of the tasks may contain algorithms and execution times that are affected by unpredictable environmental factors. Systems with these properties include building surveillance, air defense, and intelligent vehicles. Accurate modelling of such systems will supply the necessary information to allocate resources to them so that they are resilient to many unpredictable scenarios.

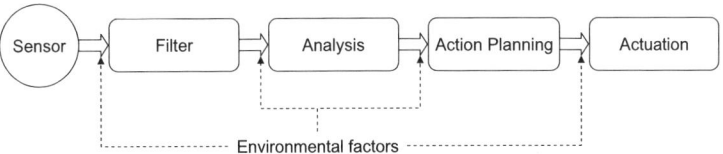

Fig. 3.2. Example of a general control system

The changing environmental factors that affect a system are modeled as l environmental variables, $\mathbf{w} = (w_1, w_2, ..., w_l)$, where w_i ($1 \leq i \leq l$) is a non-negative integer. Each task $T_i \in S$ has an execution time $e_i(\mathbf{w})$ that is a function of these environmental variables. System utilization is the resource demand from all tasks and thus a function of their execution times. It can be written as $U(\mathbf{w}) = \sum_{i=1}^{n} \frac{e_i(\mathbf{w})}{p_i}$. Like execution times, it becomes a function of the environmental variables. $U(0)$ corresponds to the portion of system utilization that is independent of the environment. Each task T_i can be allocated to any $P_j \in H$. All processors P_j are assumed to be identical, and rate monotonic scheduling (or earlier deadline first scheduling) is assumed to be used on every processor. An allocation M of the system is a many-to-one mapping of the task set to the processor set, $S \rightarrow H$.

Task execution times are modeled as functions because they may be dependent on environmental conditions. For example, a building surveillance system may contain a task implementing an object-identification algorithm, whose running time depends on the number of objects in a room. According to the model, $l = 1$, and w_1 is the number of objects. As another example, an air defense system contains filtering, situation assessment, and missile guidance tasks. These tasks can depend on such factors as the number of radar tracks and the number of missiles. Thus, $l = 2$, w_1 is the number of radar tracks and w_2 is the number of missiles. The model can make the following characterization: T_1, the filtering task, has an execution time function $e_1(w_1, w_2) = w_1 + w_2$; for T_2, the situation assessment task, $e_2(w_1, w_2) = w_1$; T_3, the missile guidance task, $e_3(w_1, w_2) = w_2$. Tasks T_1, T_2, T_2 have periods $p_1 = p_2 = p_3 = 2$ seconds. There are 3 processors: P_1, P_2, P_3. In an allocation M, task T_1 is on P_1, task T_2 is on P_2, and task T_3 is on P_3.

3.3 Robust Task Allocation for Dynamic Distributed Real-Time Systems

In this section, a task allocation strategy is developed that not only guarantees the feasibility of allocations but also explicitly maximizes the robustness of feasible allocations in unpredictable environments. Intuitively, a robust allocation allows the system to absorb a large amount of environmental variation, while continuing to deliver feasible real-time services. Significance of the robustness lies in the fact that no new allocation has to be recomputed, and no reallocation needs to be enacted. Both actions can be very time consuming, and deadline violations may result from a poorly allocated system, when changes in environmental variables frequently trigger reallocations. The problem definition is introduced next.

Problem definition: Define a robustness metric $R = R(W_1, W_2..., W_l)$ ($R \in \mathbb{R}^+$) that characterizes the ranges, $[0, W_k]$ ($1 \leq k \leq l$), of environmental variable values allowed simultaneously by an allocation. For any instantaneous value of $(w_1, w_2, ..., w_l)$ falling within these ranges, the allocation should be feasible (no deadline misses). Next find the allocation $S \to H$ under which the robustness metric R is maximized, while the allocation satisfies the following feasibility constraints,

- when *EDF* scheduling is used, $\forall j : 1 \leq j \leq m, \sum_{i: T_i \to P_j} \frac{e_i(\mathbf{w})}{p_i} \leq 1$,
- when *RMS* scheduling is used, $\forall j : 1 \leq j \leq m, \sum_{i: T_i \to P_j} \frac{e_i(\mathbf{w})}{p_i} \leq n_j(2^{\frac{1}{n_j}} - 1)$,

where n_j is the number of tasks allocated to processor P_j. Return the set of maximum allowable ranges $[0, W_k^{max}]$ $(1 \leq k \leq l)$. (The acronym *EDF* refers to Earliest Deadline First scheduling, and *RMS* refers to Rate Monotonic Scheduling. They will be used in the remainder of the chapter for convenience.)

The problem calls for the maximization of a robustness measure against multiple environmental variables while the allocation stays feasible on every processor. This involves two issues: the choice of a meaningful robustness metric to maximize, and the development of an algorithm that finds the robust allocation based on the metric. It is also necessary to guarantee the quality of robustness it finds. These issues will be addressed in this section.

The problem is not a regular constrained optimization problem, because the constraints are not given explicitly but are functions of allocations and they need to be discovered. Exhaustively enumerating all allocations is not desirable because of the expensive running time. Therefore, to still be able to obtain assurance on quality, allocation strategies are used that may help provide explicit bounds on these constraints. One such strategy is the well-known greedy first-fit heuristic. The analysis will be able to leverage the existing results such as the worst case utilization bound by [18] for rate monotonically scheduled systems and a similar bound developed by [19] for *EDF* systems.

The section is organized as follows. Section 3.3.1 discusses finding robust allocations in the case of one environmental variable. An approximation ratio is developed for the robustness found by a greedy first-fit based allocation algorithm. The result will build the foundation for subsequent extensions to multiple environmental variables. Section 3.3.2 develops a robustness metric appropriate for the case of multidimensional environmental variables, and then properties of the metric are examined. Section 3.3.3 proposes an allocation algorithm that works for the multi-dimensional case, and an approximation ratio is then derived.

3.3.1 Robust Allocation for the One-Dimensional Problem

When a dynamic distributed real-time system (*DDRTS*) is affected by a *single* environmental variable w, the robust allocation means a feasible allocation (all deadlines are met) that maintains feasible for any possible value of w in the range $[0, W^{max}]$. Thus W^{max} is regarded as the robustness metric of an allocation and maximized. An allocation algorithm has been developed based on greedy first-fit that maximizes robustness using this metric, named the maximal allowable workload [12]. The algorithm is next referred to as *RAFF-1*, or the Robust Allocation algorithm based on First-Fit under 1 environmental variable.

A two-component approximation ratio of the algorithm was developed in [21] for rate-monotonic (RM) systems by assuming two well-behaved conditions. In this section, the approximation ratio for the robustness of allocations by $RAFF$-1 will be derived for more general cases and with better results. Processors may be scheduled either by RMS or EDF, and both cases will be discussed. The section begins with an examination of a $DDRTS$'s dependency on the environment, and then the order of such dependency is characterized. The characterization will later be helpful to show that systems of higher orders can attain much better approximation ratios.

Order of dynamic real-time systems

In the one-dimensional case, the environment affects a $DDRTS$ through the variable w that alters execution times of tasks according to profile functions, $e_i(w)$. The function characterizes the time complexity of algorithms implemented in each task and provides a new type of task information useful for the optimization problem. Before using these profile functions in subsequent discussions, we make some mathematical preparations and qualifications on them.

When there is no workload ($w = 0$), a task should only have a constant workload independent execution time $e_i(0) = c$. To fulfill this requirement when a profile function involves a logarithmic function, it may be shifted by 1 as $log(w+1)$ to satisfy this requirement. The domains of profile functions $e_i(w)$ are expanded from $\mathbb{Z}^+$ to $\mathbb{R}^+$, since the profile function forms (e.g. polynomial, logarithmic, exponential...) generally have mathematical definition on $\mathbb{R}^+$. For instance, $e_i(w) = cwlog(w+1)$ that profiles a sorting task is indeed well defined for $w \in \mathbb{R}$ and $w \in [0, \infty)$. Usually their domains are restricted to integers due to practical considerations. However, working with the full domain of a function enables us to conveniently exploit intrinsic function properties via derivatives, which are useful to speculate global function behaviors. The profile functions are assumed at least twice differentiable. This is not a problem for normal functions. If a profile function is gathered in some non-continuous manner, it is always possible to bound or interpolate it using polynomial functions [22]. Two definitions are introduced next to differentiate $DDRTSs$ based on derivative properties of their system utilization functions. This helps to characterize the level of influence that the environment has on the $DDRTSs$.

Definition 3.1. *A function $f(w)$ is well-behaved to the k-th order, $k \in \{1, 2, ..\}$, if there are, $f^{(i)}(w) \geq 0$ for $1 \leq i \leq k+1$, and $f^{(k)}(w) \neq 0$.*

Definition 3.2. *A DDRTS is order-k dependent on the environment via variable w (or "order-k dynamic" for short), if the execution time profile function of every task, $e_i(w)$, is non-decreasing in w, and its total utilization, $U(w) = \sum_{i=1}^{n} \frac{e_i(w)}{p_i}$, is well-behaved to the k-th order by definition 3.1.*

These k constants will later be shown to contribute to approximation ratios about the robustness of $DDRTSs$ achievable by $RAFF$-1 allocations. To illustrate how a *DDRTS* may easily be classified as order-k dynamic, a few simple examples of various kinds are listed.

- If $\frac{e_i(w)}{p_i} = c_i w^i$ for $1 \leq i \leq 10$, then this system of 10 tasks is order-10 dynamic.
- If $\frac{e_i(w)}{p_i} = c_i w$ for $1 \leq i \leq 9$, and $\frac{e_{10}(w)}{p_{10}} = c_{10} w^2$, then this system of 10 tasks is order-2 dynamic.
- If $\frac{e_i(w)}{p_i} = c_i w$ for $1 \leq i \leq 9$, and $\frac{e_{10}(w)}{p_{10}} = c_{10} e^w$, then this system of 10 tasks is order-∞ dynamic.
- If $\frac{e_i(w)}{p_i} = c_i w$ for $1 \leq i \leq 9$, and $\frac{e_{10}(w)}{p_{10}} = c_{10} w \log_2(w+1)$, then this system of 10 tasks is order-1 dynamic.
  If $\frac{e_i(w)}{p_i} = c_i w$ for $1 \leq i \leq 9$, and $\frac{e_{10}(w)}{p_{10}} = c_{10} w^2 \log_2(w+1)$, then this system of 10 tasks is order-2 dynamic. The two examples above show that functions like $w^n \log_2 w$ are allowed.
- If $\frac{e_i(w)}{p_i} = w \log_2(w+1)$ for $1 \leq i \leq 5$, and $\frac{e_j(w)}{p_j} = \log_2(w+1)$ for $6 \leq j \leq 10$, then this system of 10 tasks is order-1 dynamic. This example shows that tasks with logarithmic execution times may also be allowed if other tasks in the system with super-linear execution time can outweigh them. If a system does have tasks with $\log w$ execution time yet cannot be outweighed, the next treatment is to bound it with a linear function to render it well-behaved. Consequently, the extra savings of execution time would be ignored.

Robustness analysis

After characterizing the environment dependency order of a *DDRTS*, we proceed to analyze its robustness quality when allocated by *RAFF-1*. As shown in Algorithm 3.1, *RAFF-1* combines a binary search with a first-fit allocation (see Algorithm 3.3) to find the allocation with maximum robustness. It uses the robustness metric defined as the upper limit of the allowable workload range by an allocation. The algorithm performs binary search along the workload value and uses the greedy first-fit allocation algorithm by [18] to assign tasks to processors using execution time values at that workload; the search terminates when the workload value cannot be increased any further. The largest feasible workload is the *MAW* (acronym for Maximal Allowable Workload) and the allocation by first-fit is the most robust allocation.

If this algorithm is used and the *DDRTS* being allocated is order-k dynamic, the system is found to have an *absolute approximation ratio* and an *asymptotic performance ratio*. For this maximization problem, the former is the tightest bound on $R_A(I) = \frac{OPT(I)}{A(I)}$ for all instances I, and the latter is the bound under large instances [23]. Let $OPT(I)$ be the optimal W^{max} that exists in the allocation problem I, and $FF(I)$ be the W^{max} produced by the *RAFF-1* algorithm; let r_{FF}^1 denote the absolute approximation ratio for 1 dimension (single environmental variable) and r_{FF}^{1*} be the asymptotic ratio, then there is:

Lemma 3.1. *If a DDRTS system has utilization function such that $U'(w) \geq 0$ and $U''(w) \geq 0$, and* RAFF-1 *is used to allocate the system,*

Algorithm 3.1. *RAFF-1*

1: **Input:** $\langle S, H \rangle$.
2: **Output:** The max workload w for a feasible allocation $S \to H$.
3: set $w = 1$;
4: **while** FirstFitAllocation($S(w)$, H) = "Feasible" **do**
5:    set $w = w \cdot 2$;
6: **end while**
7: set $low = \frac{w}{2}$, $high = w$;
8: **while** $high \neq low$ **do**
9:    set $w = \frac{low+high}{2}$;
10:   **if** FirstFitAllocation($S(w)$, H) = "Feasible" **then**
11:      set $low = w$;
12:   **else**
13:      set $high = w$;
14:   **end if**
15: **end while**
16: **return** w

(a) *if processors are scheduled by* RMS*, then* $r^1_{FF(RM)} < \frac{2-2\delta}{\sqrt{2}-1-\delta}$*, where* $\delta = \frac{U(0)}{m}$ *and m is the number of processors; asymptotically when* $OPT(I) \to \infty$, $r^{1*}_{FF(RM)} \leq \frac{1-\delta}{\sqrt{2}-1-\delta}$.

(b) *if processors are scheduled by* EDF*, then* $r^1_{FF(EDF)} < \frac{4-4\delta}{1+1/m-2\delta}$*; asymptotically when* $OPT(I) \to \infty$, $r^{1*}_{FF(EDF)} \leq \frac{2-2\delta}{1+1/m-2\delta}$.

The technical proof is provided in Appendix A. Lemma 3.1 results in the following theorem.

Theorem 3.1. *If a DDRTS is order-k dynamic ($k \geq 1$), and* RAFF-1 *is used to allocate the system,*

(a) *if processors are scheduled by* RMS*, then* $r^1_{FF(RM)} < \frac{2-2\delta}{\sqrt{2}-1-\delta}$*, where* $\delta = \frac{U(0)}{m}$ *and m is the number of processors; asymptotically when* $OPT(I) \to \infty$, $r^{1*}_{FF(RM)} \leq \frac{1-\delta}{\sqrt{2}-1-\delta}$.

(b) *if processors are scheduled by* EDF*, then* $r^1_{FF(EDF)} < \frac{4-4\delta}{1+1/m-2\delta}$*; asymptotically when* $OPT(I) \to \infty$, $r^{1*}_{FF(EDF)} \leq \frac{2-2\delta}{1+1/m-2\delta}$.

Proof. By definition of order-k dynamic (Definition 3.2), the DDRTS's utilization function $U(w)$ is well-behaved to the k-th order. According to Definition 3.1, it has $U^{(i)}(w) \geq 0$ for $1 \leq i \leq k+1$ ($k \geq 1$). Thus $U'(w)$ and $U''(w) \geq 0$ for $i = 1$ and 2. Therefore, conditions of Lemma 3.1 are sufficiently satisfied and its results follow.

Next we proceed to show that for systems that are higher order dynamic ($k > 1$), it turns out an even better asymptotic approximation ratio can be derived.

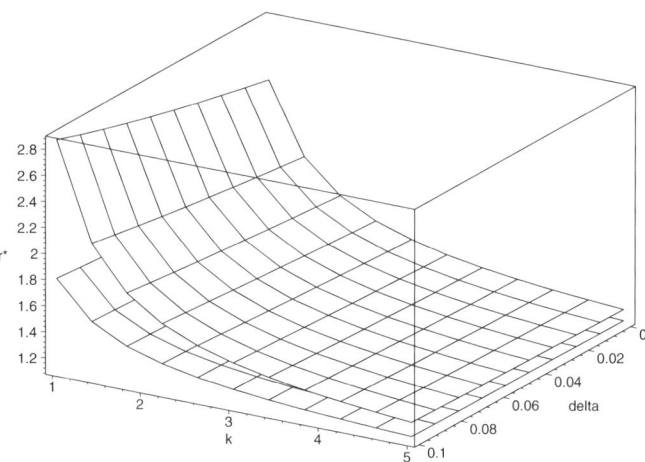

Fig. 3.3. Asymptotic approximation ratio as a function of order k and independent utilization δ

Theorem 3.2. *If a DDRTS is order-k dynamic ($k > 1$), then asymptotically when $OPT(I) \to \infty$, $r^{1*}_{FF(RM)} \leq \left(\frac{1-\delta}{\sqrt{2}-1-\delta}\right)^{\frac{1}{k}}$, and $r^{1*}_{FF(EDF)} \leq \left(\frac{2-2\delta}{1+1/m-2\delta}\right)^{\frac{1}{k}}$.*

Proof for the theorem is listed in Appendix B. The result can be visualized in Fig. 3.3. The asymptotic approximation ratio is plotted as a function of two variables: the order of the system, k, and the workload independent system utilization as a fraction of overall resource, δ. The upper surface corresponds to the ratio for systems scheduled by *RMS*, and the lower surface corresponds to the ratio for systems scheduled by *EDF* (at $m = 5$).

3.3.2 A Multi-dimensional Robustness Metric

When a system is affected by multiple environmental variables, the range of a single variable is obviously inadequate to capture the overall robustness of the system, and a more comprehensive multi-dimensional robustness metric needs to be defined. Although the set of environmental variables may be regarded as a vector, norm of the vector is not a good choice for robustness metric. For instance, l_2 norm is defined as: $l_2 \equiv \|x\|_2 = (\sum_{i=1}^{n} x_i^2)^{1/2}$, where the x_i corresponds to environmental variables. This norm, however, implies that a given robustness value may potentially be achieved by any point of environmental variables on a sphere of the radius $\|x\|_2$, even if the point has all components equal to 0 but just one, $x_k = \|x\|_2$. This is not desirable since environmental variables may not be tradable with each other in value. For example, (a) the ability to handle 15 missile tracks alone cannot substitute (b) the ability to handle 5 missile tracks and 5 torpedo tracks, even though scenario (a) may appear to offer more robustness than (b) with the norm of $15 > \sqrt{5^2 + 5^2} = 7.07$. It can misguide

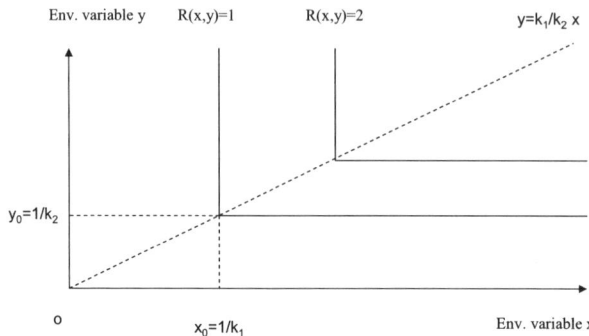

Fig. 3.4. Contour lines of robustness metric in two dimensions

the optimization algorithm to prefer the case (a) and cause undesirable results. The *non-tradability* between environmental variables has to be considered in the metric definition.

Another consideration in the choice of metric is the relative importance among the components. Naturally, certain environmental variable dimension may be regarded more important than others. For instance, military statistics may indicate that for a certain combat type A, incoming missiles are twice as likely as torpedoes. Then handling more missile tracks becomes more important than torpedo tracks. Therefore, the metric definition needs to take weights into consideration.

The robustness metric is defined similar to l_∞, but rather than the maximal component, the minimal component is chosen. To address the relative importance, each workload dimension is weighted. Let $k_i > 0$, and $W_i \geq 0$, the robustness metric R is defined as,

$$R(\mathbf{W}) \equiv \min_{1 \leq i \leq l}(k_i|W_i|) = \min_{1 \leq i \leq l}(k_i W_i). \tag{3.1}$$

Similar choice was made by [13, 24, 25]. The minimal range of value among weighted environmental variables is maximized. A system optimized by this metric will be able to absorb maximum changes in any dimension of environmental variable. When this metric is used in scenarios (a) and (b), now (a) would have a robustness of 0 while (b) would have a robustness of 5 (equal weights), thus (b) would be preferred.

To illustrate properties of this metric, the equi-value contour lines for $R(\mathbf{W}) = 1, 2$ are plotted in Fig. 3.4 in two dimensions. As shown, the metric value increases as the lines spread out, and the lines are shaped as right angles parallel to the coordinate frame. In addition, it is easy to see that 'origins' of these contour lines reside on a diagonal line, $y = \frac{k_1}{k_2}x$, which can be generalized in the multi-dimensional space to,

$$k_i W_i = t, \tag{3.2}$$

where t is the parameter in the parametric format of the line.

In the graph, the point $\mathbf{W}_0 = (x_0, y_0) = (1/k_1, 1/k_2)$ was shown on this line. It has $k_1 x_0 = k_2 y_0 = 1$ and thus a metric $R(\mathbf{W}) = min(1, 1) = 1$. Generally, if $\mathbf{W}$ is on the line, $R(\mathbf{W}) = min_{1 \leq i \leq l}(k_i W_i) = t$, and thus parameter t turns out to be the value of robustness metric itself. So the line may also be given as,

$$W_i = \frac{R(\mathbf{W})}{k_i} \qquad (1 \leq i \leq l). \tag{3.3}$$

As shorthand, the diagonal line is referred to as the *ray of origins*. Geometrically, any point on it helps determine the whole $R(\mathbf{W}) = c$ contour line through the point.

Search space reduction

Next the robustness metric will be employed in the optimization problem. This subsection will show that it suffices to concentrate the search for the maximum metric value on a subset of the total search space. This allows to design an efficient algorithm to maximize robustness in the next section.

Generally, an algorithm's running time does not decrease as its input size grows larger, thus it was assumed that execution time functions are non-decreasing in one dimension [12]. When an execution time function depends on multiple environmental variables, it is reasonable to assume the function is also non-decreasing along every dimension, or mathematically,

$$\frac{\partial e_i(\mathbf{w})}{\partial w_j} \geq 0 \qquad (1 \leq j \leq l, 1 \leq i \leq n). \tag{3.4}$$

For instance, a bin-packing algorithm may run in $O(mn)$, where m and n is the numbers of bins and blocks. Clearly, its execution time function is non-decreasing in either dimension. This yields an important property.

Lemma 3.2. *If tasks of a DDRTS have execution time functions that are non-decreasing in all dimensions, then for every allocation of the DDRTS, there can exist only one tangent point between the boundary line (or surface) of its feasible area, $f(\mathbf{W}) = c$, and the robustness metric's contour line (or surface), $R(\mathbf{W}) = c'$. The tangent point occurs at the metric contour's origin.*

Proof. Without loss of generality, the proof is given for the two-dimensional case. Let the feasibility boundary of an allocation be described by $f(W_1, W_2) = c$, or equivalently $W_2 = g(W_1)$.

We first show that for every allocation, everywhere on the feasibility boundary line there is $\frac{dW_2}{dW_1} \leq 0$. Assume that somewhere $\frac{dW_2}{dW_1} > 0$. Let there be $W_2 = g(W_1)$ and $W_2' = g(W_1')$. Then it is possible that when $W_1' = W_1 + \Delta$ (for any $\Delta > 0$), $W_2' > W_2$. It follows that since the allocation is feasible at (W_1', W_2'), it should also be feasible at (W_1, W_2'), because $W_1 < W_1'$ decreases execution times of all tasks and tasks allocated on every processor stays schedulable. However, since (W_1, W_2) is on the feasibility boundary, (W_1, W_2') should not be feasible

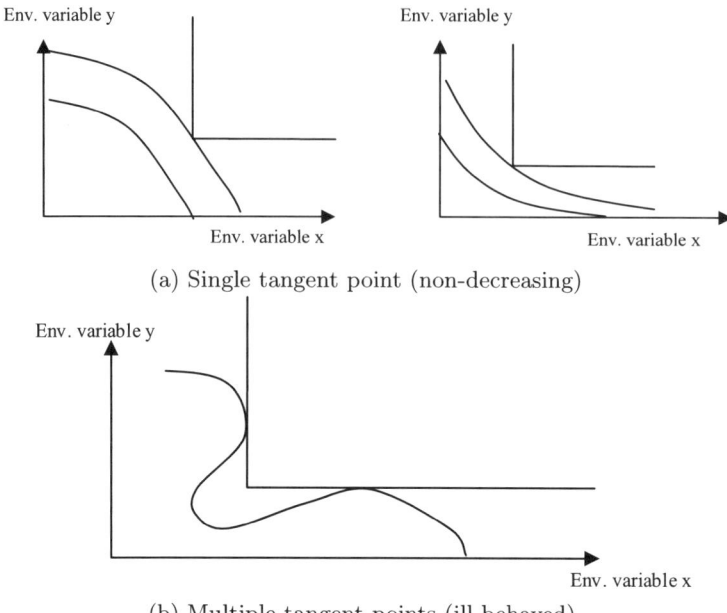

Fig. 3.5. Different tangents between feasibility boundary and robustness metric contour

because $W_2' > W_2$. Thus we have reached a contradiction and the assumption was false. There is $\frac{dW_2}{dW_1} \leq 0$ everywhere on the feasibility boundary line.

This means that at any point on the feasibility boundary line of each allocation, the angle is greater than $\frac{\pi}{2}$, and a tangent contact with the metric's right angle contour at its origin will prevent any further crossing of the two lines.

Here are some ideas to prove the generalized n-dimensional case. The orientation of the surface of $f(\mathbf{W})$ can be abstracted by the vectors normal to it, $\nabla f(\mathbf{W})$. Due to the non-decreasing assumption, these vectors $\nabla f(\mathbf{W}) = \sum_i \mathbf{e}_i \frac{\partial f}{\partial w_i}$ have positive projections on all coordinate frame base vectors $\mathbf{e}_i$. If there is a tangent point other than the origin, the convexity there will lead to vector $\nabla f(\mathbf{W})$ with negative projections, which results in a contradiction to the assumption.

An illustration is given in Fig. 3.5(a). An illustration of an ill-behaved feasibility boundary without the non-decreasing condition is given in a counterexample in Fig. 3.5(b). As can be seen in (a), the tangent point corresponds to the origin of the metric contour and is easily found; while in the ill-behaved case, the tangent points have to be searched in general or solved for in special functions. Every allocation of a non-decreasing $DDRTS$ contains one such special point, which resides on the same "ray of origins" line. Next, one of the points will be shown to correspond to the maximum robustness metric value, and the corresponding allocation will be the robust allocation.

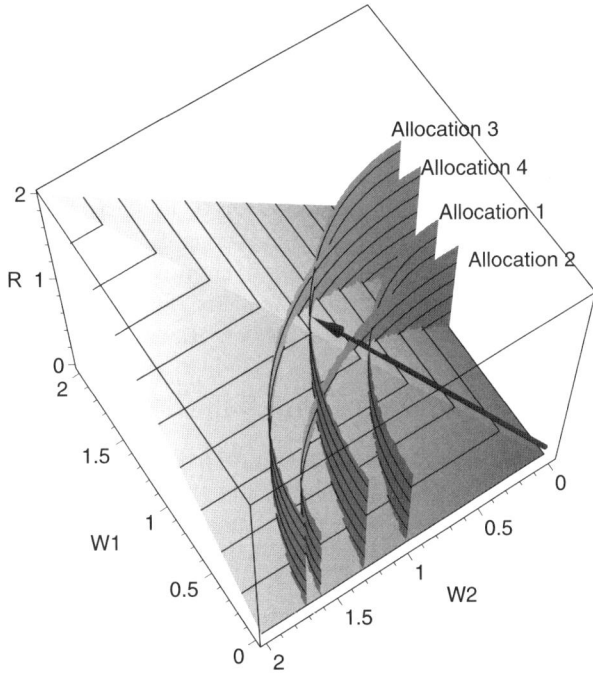

Fig. 3.6. Illustration of the search for an allocation with the maximum robustness metric

Theorem 3.3. *If a DDRTS has tasks with non-decreasing execution time functions, then its maximum achievable robustness metric value, R^{max}, can be found by incrementally searching along the "ray of origins" line of the metric, $W_i = \frac{1}{k_i}t$ $(1 \leq i \leq l)$, until the metric value fails to have any feasible allocation. The resulting maximum allowable values for environmental variables are $W_i^{max} = \frac{R^{max}}{k_i}$ $(1 \leq i \leq l)$.*

Proof. For each allocation, the tangent point(s) between the feasibility boundary and the metric contour mathematically means that the metric value is the maximum at the point(s). Because such point is unique now and resides on the "ray of origins" line (by the lemma above), to find the overall maximum of this value among all allocations, the line should be searched incrementally starting from $t = R(\mathbf{W}) = 0$. The search proceeds while at each value a feasible allocation is found (with task execution times fixed by the value). If no feasible allocation can be found for $R(\mathbf{W})+1$, then $R^{max} = R(\mathbf{W})$. This is because if $R(\mathbf{W})+2$ or more were to produce a feasible allocation (say A) again, then $R(\mathbf{W})+1$ must have a feasible allocation (at least the same A) due to the non-decreasing condition (a contradiction). Therefore it must be $R^{max} = R(\mathbf{W})$, and the resulting maximum allowable values for environmental variables are $W_i^{max} = \frac{R^{max}}{k_i}(1 \leq i \leq l)$ since the point resides on the "ray of origins."

The search literally "pushes the boundary" of allocations at their maximally valuable points until they cannot be expanded any further. An illustration is given in Fig. 3.6. The pyramid shape is the robustness metric function surface in two dimensions. The boundaries of four allocations were plotted and the maximum robustness of each allocation is its interception point with metric surface where the R value is highest (the tangent point mentioned earlier). The arrow ("ray of origins") points at the direction of the search which pushes the feasibility boundary and discovers the allocations $1, 2, 3, 4$ in increasing order of their maximum robustness value, until it cannot be increased any further under available resource. The observation helps to develop the robust allocation algorithm for multiple dimensions in the next section.

3.3.3 Robust Allocation for the Multi-dimensional Problem

The previous section showed that, when the robustness metric is used, it is unnecessary to exhaustively explore the whole l dimensional space of environmental variables to find the maximum metric value. In fact, it is sufficient to just search a diagonal line across the space. This valuable result can significantly reduce the running time of the search.

In this section, an approximation algorithm is developed to find robust allocations under multiple dimensions of environmental variables, and its performance is analytically evaluated. It was demonstrated in the one-dimensional case that a binary search coupled with first-fit allocation efficiently produces good maximal allowable workload [12, 26]. A linear search along the *ray of origins* will now serve as a driver to determine a set of constant environmental variable values used to evaluate task's execution times. With these constant execution-time values, a greedy first-fit allocation algorithm will test for the existence of a feasible allocation. If the allocation algorithm finds a feasible allocation, the search advances to the next set of environmental variable values; if the allocation algorithm does not find any feasible allocation, the search terminates, and the largest feasible values of environmental variables are returned along with the corresponding allocation.

Algorithm 3.2. *RAFF-n*

1: **Input:** $\langle S, H \rangle$
2: **Output:** A set of environmental variable (EV) values found to have maximal robustness metric and their associated allocation
3: $metric = 0$
4: **while** FirstFitAllocation$(S(\{\frac{metric+1}{k_i}\}), H)$="Feasible" **do**
5:    $metric = metric + 1$
6:    save FeasibleAllocation
7: **end while**
8: $\{w_i^{max}\} = \{\frac{metric}{k_i}\}$
9: **return** $\{w_i^{max}\}$, FeasibleAllocation

Algorithm 3.3. FirstFitAllocation
1: **Input:** $\langle S(\{w_i\}), H \rangle$
2: **Output:** "Feasible" or "Not Feasible," and $FeasibleAllocation : S \longrightarrow H$
3: **for** each task i **do**
4:    set $j = 1; n = |H|$;
5:    **while** job i has not been allocated and $j \leq n$ **do**
6:      set $n_j = |\{T_k | alloc(T_k) = P_j\}| + 1$;
7:      **if** processors scheduled by RMS **then**
8:        set $unibound = n_j(2^{\frac{1}{n_j}} - 1)$;
9:      **else if** processors scheduled by EDF **then**
10:       set $unibound = 1$;
11:      **end if**
12:      **if** $\left(\sum_{alloc(T_k)=P_j} \frac{T_k \cdot e(\{w_i\})}{T_k \cdot p} \right) + \frac{e_i(\{w_i\})}{p_i} \leq unibound$ **then**
13:        set $alloc(T_i) = P_j$;
14:      **else**
15:        set $j = j + 1$;
16:      **end if**
17:    **end while**
18:    **if** $j > n$ **then**
19:      **return** "Not Feasible";
20:    **end if**
21: **end for**
22: FeasibleAllocation = alloc
23: **return** "Feasible"

The first part of the algorithm is listed in Algorithm 3.2. A linear search is coupled with the greedy first-fit algorithm, shown in Algorithm 3.3, which works for variable execution times and processors scheduled by either RMS or EDF. The two algorithms together are referred to as $RAFF$-n. The quality of the algorithm will be given next as approximation ratios.

Earlier in section 3.3.1, approximation ratios have been derived for the one dimensional case of environmental variable. The results will be extended next to the multi-dimensional case. As before, the extension begins with a definition of order-k dynamic systems, under a higher dimension of environmental variables.

Definition 3.3. *A DDRTS is order-k dynamic in the l-dimensional environmental variable space if the execution time function of every task, $e_i(\mathbf{w})$, is non-decreasing in every dimension, w_i $(1 \leq w_i \leq l)$; define derivatives on the system utilization to be,*

$$U^{(i)}(\mathbf{w}) \equiv \frac{\partial^i U(\mathbf{w})}{\partial w_{m_1} \partial w_{m_2} \ldots \partial w_{m_i}},$$

and there are:

- $\forall m_1, m_2..m_i \in [1,l]$, $U^{(i)}(\mathbf{w}) \geq 0$ for $1 \leq i \leq k+1$, and
- $\exists m_1, m_2, ..m_i \in [1,l]$ such that $U^{(k)}(w) \neq 0$.

Many functions satisfy these conditions. For example, $O(n^2 + mn)$ is order-2 dynamic in the dimension of m and n.

Theorem 3.4. *If a DDRTS is order-k dynamic in l dimensions and allocated by the algorithm of RAFF-n, its robustness, as measured by the metric, has an absolute approximation ratio of $r^l_{FF(RM)} < \frac{2-2\delta}{\sqrt{2}-1-\delta}$ and $r^l_{FF(EDF)} < \frac{4-4\delta}{1+1/m-2\delta}$. Asymptotically, it has $r^{l*}_{FF(RM)} \leq \left(\frac{1-\delta}{\sqrt{2}-1-\delta}\right)^{\frac{1}{k}}$, and $r^{l*}_{FF(EDF)} \leq \left(\frac{2-2\delta}{1+1/m-2\delta}\right)^{\frac{1}{k}}$.*

Proof. Intuitively, since the algorithm searches along the "ray of origins," which is actually one dimensional, previous approximation ratios seem helpful for obtaining the approximation ratio in the l-dimensional case. Notice that execution time functions of tasks in order-k dynamic systems are non-decreasing, therefore the previous linear search approach is applicable.

We first express the ray of origins in vector form, i.e.,

$$\mathbf{l} = \sum_{i=1}^{l} \frac{t}{k_i} \mathbf{e}_i, \tag{3.5}$$

where $t \in R$ and $t \geq 0$. The system utilization is a function of the environmental variable vector, thus its change with respect to t along the ray of origins is,

$$dU(\mathbf{w}) = \nabla U(\mathbf{w}) \cdot d\mathbf{l} = \sum_{i=1}^{l}(\mathbf{e}_i \frac{\partial U(\mathbf{w})}{\partial w_i}) \cdot \sum_{j=1}^{l}(\mathbf{e}_j \frac{dt}{k_j})$$

$$= \sum_{i=1}^{l} \frac{1}{k_i} \frac{\partial U(\mathbf{w})}{\partial w_i} dt \quad \Rightarrow \quad \frac{dU(\mathbf{w})}{dt} = \sum_{i=1}^{l} \frac{1}{k_i} \frac{\partial U(\mathbf{w})}{\partial w_i}, \tag{3.6}$$

In addition, its double derivative is

$$\frac{d^2 U(\mathbf{w})}{dt^2} = \sum_{i=1}^{l} \frac{1}{k_i} \frac{d}{dt}(\frac{\partial U(\mathbf{w})}{\partial w_i})$$

$$= \sum_{i=1}^{l} \frac{1}{k_i} \sum_{j=1}^{l} \frac{\partial^2 U(\mathbf{w})}{\partial w_j \partial w_i} \frac{dw_j}{dt} = \sum_{i=1}^{l} \sum_{j=1}^{l} \frac{1}{k_i} \frac{\partial^2 U(\mathbf{w})}{\partial w_i \partial w_j} \frac{1}{k_j}, \tag{3.7}$$

similarly, the ith derivative is,

$$\frac{d^i U(\mathbf{w})}{dt^i} = \sum_{m_1=1}^{l} \cdots \sum_{m_i=1}^{l} \frac{1}{k_{m_1}} \cdots \frac{1}{k_{m_i}} \frac{\partial^i U(\mathbf{w})}{\partial w_{m_1} \ldots \partial w_{m_i}}. \tag{3.8}$$

When a *DDRTS* is order-k dynamic in l-dimension, it follows from the above results and definition 3.3 that, $\frac{d^i U(\mathbf{w})}{dt^i} \geq 0$ for $1 \leq i \leq k+1$ and $\frac{d^k U(\mathbf{w})}{dt^k} \neq 0$. Therefore, the system utilization function $U(\mathbf{w}(t)) = U(t)$ is well-behaved to the k-th order in the dimension of t, which is effectively identical to the workload variable earlier. Further, we note that it was previously shown in equation (3.3) that parameter t is equivalent to the robustness metric $R(\mathbf{W})$. Therefore, by Theorem 3.1 and Theorem 3.2, there are: $r^l_{FF(RM)} < \frac{2-2\delta}{\sqrt{2}-1-\delta}$ and $r^l_{FF(EDF)} < \frac{4-4\delta}{1+1/m-2\delta}$; asymptotically, there are: $r^{l*}_{FF(RM)} \leq \left(\frac{1-\delta}{\sqrt{2}-1-\delta}\right)^{\frac{1}{k}}$, and $r^{l*}_{FF(EDF)} \leq \left(\frac{2-2\delta}{1+1/m-2\delta}\right)^{\frac{1}{k}}$.

3.4 Experiments

Three experiments are designed to test the performance and scalability of *RAFF-n* algorithm. **Experiment 1** compares solutions of the algorithm against optimal solutions and validates the theoretical approximation ratios using the solutions. **Experiment 2** compares performance of the *RAFF-n* with several baseline algorithms for large problem instances. The purpose is to show that the algorithm has both good and scalable performance. **Experiment 3** illustrates a use of the algorithm to solve a toy problem of missile-defense system.

The experiments are performed using simulations on a Pentium 4 PC running Fedora Linux. Problem instances are generated in which various system parameters are set. These include task period, deadline, number of processors, processor speed, number of environmental variables, number of environmental variable-dependent tasks, number of environmental variable-independent tasks, and the execution time function of each task. Given an order-k *DDRTS*, the execution-time function of each task is constructed by means of linear combinations of basis functions chosen from the set, $\{w_i^j \log w_i, w_i^j | 1 \leq i \leq l, 1 \leq j \leq k\}$. The elements used are chosen randomly according to a user-defined distribution, and their coefficients are randomly generated. By doing so, it is intended to represent general problem instances by the randomly produced samples. Two environmental variables are assumed. Results of the first experiment are presented next.

3.4.1 Experiment 1

This experiment compares solutions of the *RAFF-n* algorithm with optimal solutions, and it also demonstrates the theoretical approximation ratios based on the solutions. To obtain optimal solutions, an exhaustive search algorithm, named *RABB-n*, has been developed. Problem instances are generated with 5 processors, and the number of tasks varies from 5 to 20. Systems of order 1 to 3 are generated. Both cases of RM and EDF scheduling are tested for each problem instance.

The robustness values of allocations produced by the *RAFF-n* and *RABB-n* for each instance are recorded and plotted. In addition, in order to visualize the approximation ratios on the same graph, each ratio is computed and then converted to the unit of robustness by multiplying with the optimal robustness value by *RABB-n*. Thus, the approximation ratios are represented as robustness lower bounds to *RAFF-n*, and obviously, the lower bounds corresponding to asymptotic ratios are tighter than those of absolute ratios.

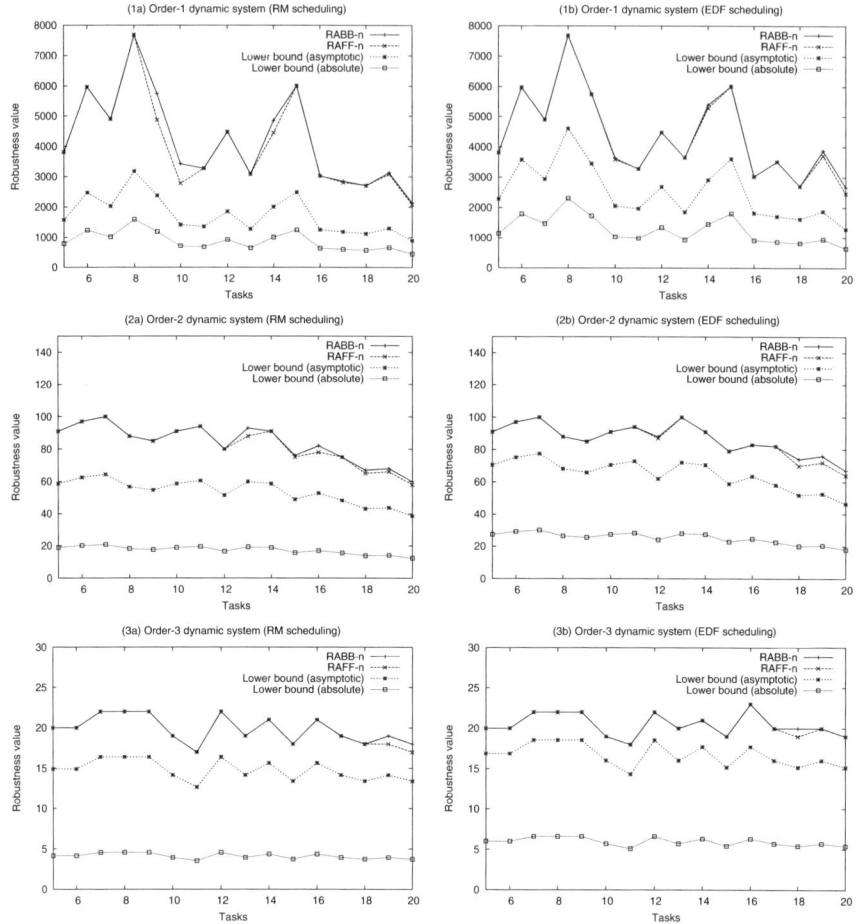

Fig. 3.7. Comparisons of robustness values found by *RAFF-n*, *RABB-n*, and bounds from absolute and asymptotic approximation ratios under *RMS* or *EDF* scheduling for three orders of *DDRTS*

Robustness results are shown in Fig. 3.7. In the figure, the row number corresponds to the order of the system, and the column tells whether a system is scheduled by *RMS* (left) or *EDF* (right). As shown, the robustness value of *RAFF-n* is 97% of the optimal value (by *RABB-n*) on average, overlapping on many occasions. This shows *RAFF-n* is near-optimal under these small instances (in the next experiment, large instances will also be examined). When EDF-scheduling is used, *RAFF-n* achieves higher robustness compared to RM-scheduling, and this is due to a higher schedulable utilization on every processor. When the order of a system increases, its robustness value generally drops. This is because utilization of a higher-order system rises much faster under the same

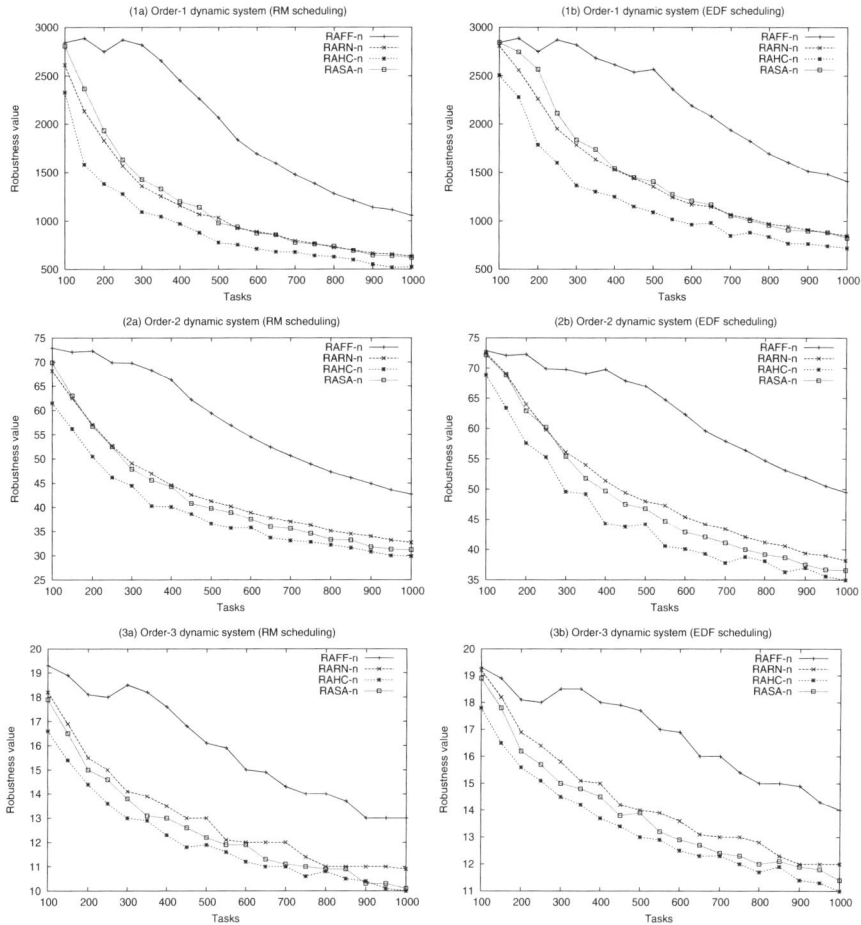

Fig. 3.8. Comparisons of robustness values found by *RAFF-n*, *RARN-n*, *RAHC-n*, and *RASA-n* for large problem instances under *RMS* and *EDF* scheduling for three orders of *DDRTS*

workload change, compared to a lower-order system, and thus it takes much less workloads to reach resource limits. The approximations ratios have correctly bounded the performance of *RAFF-n*. On average, lower bounds by the absolute ratio are 25% of optimal, while lower bounds by the asymptotic ratio are 66% of optimal. The lower bounds can be seen to further improve as the system order becomes higher. This is because a k-th root can be taken in the ratio expression (k is the system order).

The experiment has demonstrated the good performance of the *RAFF-n* algorithm and verified its approximation ratios in small problem instances. The

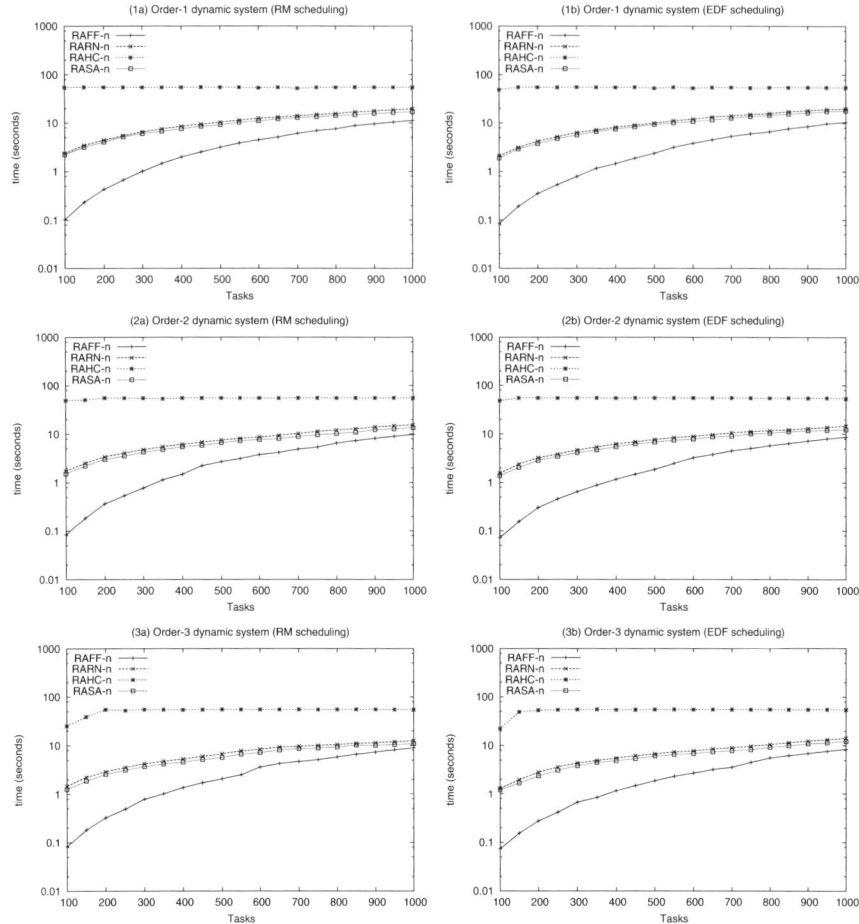

Fig. 3.9. Comparisons of running times of *RAFF-n*, *RARN-n*, *RAHC-n*, and *RASA-n* for large problem instances under *RMS* or *EDF* scheduling for three orders of *DDRTS*

small problem instances are necessary for the use of the *RABB-n* algorithm to find optimal solutions. In order to demonstrate that its good performance is also scalable, the algorithm will be tested next under much larger problem instances.

3.4.2 Experiment 2

This experiment aims at demonstrating *RAFF-n* has both good and scalable performance. Problems are generated that have 100 processors, and the number of tasks ranges from 100 to 1000. As before, the order of systems varies from 1 to 3, and both cases of RM and EDF scheduling are tested. Since the

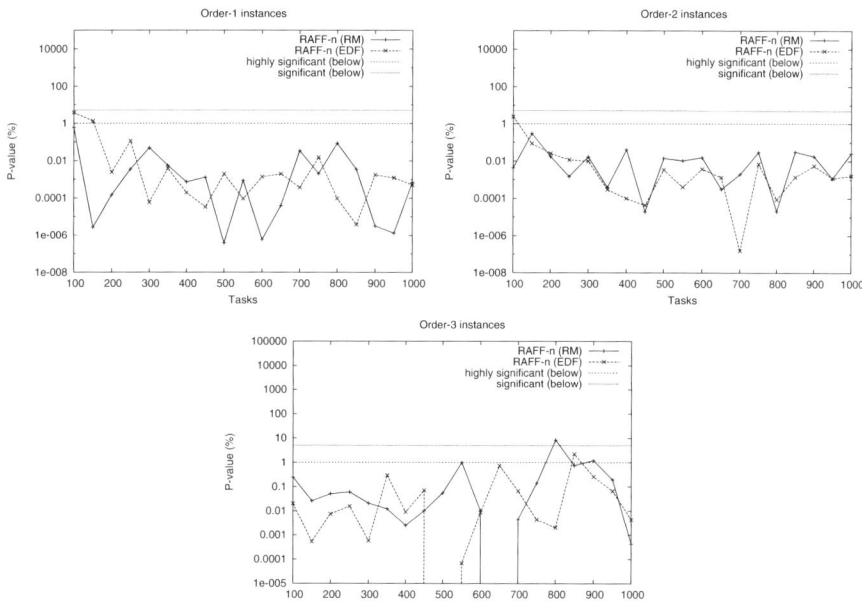

Fig. 3.10. Significance test for $RAFF$-n with RMS and EDF scheduling systems under three orders of $DDRTS$

problem size renders $RABB$-n inapplicable, several standard optimization algorithms have been developed and used as baselines for comparison. These are random-search (RARN-n), hill-climbing (RAHC-n), and simulated-annealing (RASA-n). For each problem size, 10 random instances are generated. Robustness and running time data from every algorithm are gathered and averaged among 10 instances. The comparison of maximum robustness values produced by these algorithms is shown in Fig. 3.8. The comparison of their running times is shown in Fig. 3.9.

Plots in Fig. 3.8 show that the robustness quality from $RAFF$-n is consistently higher than the other algorithms. As can be seen, its advantage is significant and more than 50% better on average. Besides the advantage in robustness quality, Fig. 3.9 shows that $RAFF$-n also has the most efficient running-time, costing 0.1 to 10 seconds among all instances. It is followed by random-search and simulated-annealing. Hill-climbing has the worst running time, because it has to explore a large number of neighbors every step. A cutoff time of one minute has been implemented to speed up the experiment, so its running times in the figure all stop at 60 seconds.

Significance tests are performed for the data above. The samples from the $RAFF$-n are compared with the samples from the second best $RARN$-n. The purpose is to ensure the mean of the entire population of $RAFF$-n is larger than that of $RARN$-n, that is, its performance advantage is statistically significant.

The null hypothesis H_0 is that the mean of RAFF-n is equal to the mean of RARN-n; the alternative hypothesis H_a is that the mean of RAFF-n is larger than the mean of RARN-n (one-tailed). It is assumed the variances of the two are unknown but the same, and therefore, t-test for two samples is used.

For each problem size, the P-value is calculated based on the *test statistic* value. The results are shown in Fig. 3.10. P-value is shown on the y-axis as a percentage in log scale. The acceptable significance value of 1% and 5% are plotted as straight lines. Mathematically, when the P-value is 1%-5%, it is considered significant; when the P-value is below 1%, it is considered highly significant. As shown, all instances are significant except just one at 800 tasks. A vast majority of the instances are highly significant. As a result, the alternative hypothesis is accepted. This means although 10 samples are tested for each problem size, there is the statistical confidence to say that RAFF-n performs better than RARN-n for all population.

The experiments have shown that the algorithm of RAFF-n is both a good and scalable robust allocation solution in the presence of multiple environmental variables.

3.4.3 Experiment 3

In this experiment, the algorithms are employed for the problem in the *Dynbench* missile defense testbed [1]. There are three types of tasks in the system: Detect, Engage and Guidance, and their execution times depend on two environmental variables: r and m, which stands for the number of radar tracks and real threats. Their execution time functions are determined experimentally in [27]:

Detect: $e_1(r, m) = 0.0869r^2 + 15.4374r + 615$ μs,
Engage: $e_2(r, m) = 12897m + 45610$ μs,
Guidance: $e_3(r, m) = 0.0869r^2 + 15.4374r + 12903.909m + 46476$ μs.

The objective is to produce an allocation that maximizes the minimum of r and m values. Three scenarios have been tested. As shown in Table 3.1, the scenarios

Table 3.1. Max robustness value found with RARN-n, RAFF-n, RABB-n algorithms under three scenarios

#	Detect	Engage	Guide	Procs	RARN-n	RAFF-n	RABB-n(OPT)
1	20	5	10	8	170	178	178
2	6	5	3	5	98	105	105
3	8	8	8	5	53	54	54

differ in the number of replicas of each task and the number of processors. It is assumed that for Detect tasks, the deadlines and periods are 0.1 sec; for Engage tasks, the deadlines and periods are 0.01 sec; for Guidance tasks, the deadlines and periods are 0.05 sec. Results show that similar to the previous experiment, RAFF-n has produced very good robustness, in this case same as the optimal achieved by *RABB-n*, while random search does not perform as well. This experiment has corroborated previous findings.

3.5 Literature Review

This section compares and contrasts this work with related research in the area. A problem of maximum allowable increase in load was introduced in [13]. Its goal was to find a feasible allocation that maximizes a system runtime parameter. A mixed-integer-programming (MIP) algorithm was designed to find the allocation. The work used end-to-end latency requirements and considered both computation and communication latencies. However, for these end-to-end real-time tasks, no real-time scheduling issues were considered to guarantee feasibility, such as sub-release time and deadline assignments in [28, 29]. It was assumed that all execution times can vary, but the variances were treated as linear functions of just one system parameter. The proposed algorithm solved the problem using MIP but a simplifying heuristic had to be applied first. No provable optimality bound was provided. Its performance was tested through simulations. A norm-based robustness metric was proposed in [15]. Multiple runtime parameters were considered. The metric was defined as a radius for maximum allowable perturbation along any direction in the parameter space without violating system performance bounds. The upper and lower bounds defined for system performance form a hyperplane in the space, and the radius is the perpendicular distance from the operating point to the plane. However, the metric directly used l_2 norm of the parameters, which cannot reflect importance differences among parameters. More seriously, only the generic robustness metric was proposed, and no optimization algorithm was actually developed.

[12] studied a robust task allocation problem for dynamic systems with one workload variable. The problem was to find the maximum allowable workload (MAW) for the system, when real-time tasks can be feasibly scheduled on every processor using rate-monotonic scheduler. The task's execution times were modeled to have a workload-independent portion and a dependent portion, which is non-decreasing with respect to workload. An algorithm was developed that searches the workload using binary search and allocates tasks under each of the workload using greedy first-fit. The schedulable utilization of rate-monotonic scheduling was used to test feasibility on each processor. Several heuristic algorithms were introduced to experimentally compare with the performance of this approach, and they were simulated annealing, hill-climbing and random search. An approximation ratio for the maximal allowable workload was proved in [21] for identical processors and tasks with well-behaved execution times defined by two conditions, which are satisfied by many algorithmic running-time functions. [26] expanded the set of allocation algorithms to include tabu search, genetic algorithm, dynamic programming, and optimal branch-and-bound search. The experimental results indicated that the greedy first-fit algorithm performs quite well in finding the MAW compared to these heuristic or exhaustive algorithms, and it offers great cost-efficiency in time and space complexity. However, the

work was limited to only one workload variable, although having multiple variables is very likely in many systems. [30] further considered the problem of robust allocation when tasks are replicable, and several heuristics were developed.

A different approach regarded task's unpredictable execution times as random variables and statistical guarantees for deadline misses were derived. The semi-periodic task model by [5] assumed execution times are profiled with probability histograms, and two methods were presented to compute each task's deadline miss probability. The first method assumed fix-priority scheduler in order to develop a probabilistic time-demand analysis. Traditional time-demand analysis was extended to consider randomness in task's execution time and probabilistic bounds were developed. The second method separated the semi-periodic task into a fixed execution time task and a sporadic task, which arrives with a probability depending on the partitioning. Then the former was scheduled with RM or EDF normally, while the latter was run in a sporadic server or slack stealer. However, the analysis was restricted to a single processor, and task allocation was treated by simply evening the worst-case total utilization on every processor. [8] similarly viewed execution time as a probability-density distribution. In addition, tasks had dependencies and there are multiple processors. Given probability thresholds for deadline misses in tasks, the optimization goal was to minimize their sum of deviation beyond the thresholds. Tabu-search was used to explore the space of processor assignment and task priority. The approach was shown to achieve a much better goal than allocating simply based on average execution times. However, the search and deviation evaluation were expensive.

Event stream model by [7] attributed varying execution times to a finite set of event types that incur different workloads. It was argued when there is certain regularity among events, it can be exploited to give better workload estimation than using worst-case estimation. The workload estimation was based on type-rate curves characterizing event regularity, which can be analytically generated using finite state machine. The capture of the regularity in MPEG-2 I-P-B frame sequence was demonstrated. In practice however, the type-rate curves have to be profiled when no pre-knowledge is given, and they may be as poor as worst-case estimation.

Varying task execution times were treated by [31] as distributed generalized multi-frame tasks (DGMF). Each task has multiple frames with different execution times, deadlines, and periods. The release of a latter frame must be later than the release of a former frame plus period of the frame. Each frame of a task can be executed on a different host. A deadline was given separately for each frame. An optimal but computationally intensive feasibility test was given for the schedulability of a set of the DGMF tasks, using concept similar to critical instance. A less expensive schedulability criterion was also given. It derived solutions for priority assignment in several special cases: the deadline-monotonic assignment is optimal for tasks with same frame deadlines and for tasks with equal frame execution times; the rate-monotonic assignment is optimal for tasks with frame deadline equaling period. However, the sequence of frames with varied execution times had to be known in advance.

3.6 Conclusions

Real-time systems are expected to operate predictably even when they are deployed in highly dynamic and unpredictable environments. The average throughput of computation is often not as important as the guarantee that work is always completed on time. The allocation and scheduling of the systems are key to offering such predictability. Research introduced in this chapter addresses the problem by making the following contributions: (1) an accurate model of environment-dependent systems, (2) metrics characterizing an allocation's robustness against environment variations, (3) allocation and scheduling algorithms solving the robustness optimization problem, and the theoretical performance guarantees.

Traditionally, real-time computing models do not incorporate environmental factors and are not suitable for systems sensitive to environment changes. This research has introduced a new model that captures the environmental factors so that this valuable information is an active factor of consideration in the resource optimization problem. For the systems, sources of environmental influences are identified as environmental variables that are unpredictable, and tasks' resource needs are modeled as functions of the variables. Resource allocation problems are posed that not only find allocations that are feasible but also maximize their robustness against these unpredictable variables.

Robustness metrics have been defined to quantitatively characterize an allocation's robustness. The metric guides optimization algorithms to the most robust allocations. A metric is defined based on max-of-min, which is found to significantly reduce the necessary search space and boost search speed.

Algorithms to find robust allocations have been developed, and they are found to have good and scalable performance. The fast running times meet the needs of modern distributed real-time systems that may contain hundreds of processors and thousands of tasks. Meanwhile, the quality of their allocations is warranted by approximation ratios and lower bounds that have been derived. Experiments demonstrated that they deliver the theoretical guarantee and have good performance and fast running time. The advantage over baseline algorithms is found statistically significant.

This is an interesting but difficult research problem to work on. Assumptions have been made to make the problem manageable, such as execution time functions are non-decreasing, identical processors are used, and only cpu resource is considered. What comes out is a simple yet power algorithm, and we believe it can serve as a solid foundation to be expanded on. It is certainly our goal that as more research is done on this problem, these restrictions will be gradually removed in the future work.

References

1. Welch, L.R., Shirazi, B.A.: A dynamic real-time benchmark for assessment of qos and resource management technology. In: Proceedings of the 5th IEEE Real-Time Technology and Applications Symposium, pp. 36–45 (1999)

2. Juedes, D., Drews, F., Gu, D., et al.: Approximation algorithm for periodic real-time tasks with workload-dependant running-time functions. Journal of Real-Time Systems 34(3), 173–194 (2006)
3. Gu, D., Drews, F., Welch, L.: A characterization of task allocation problems for dynamic distributed real-time systems. In: The 16th IASTED International Conference on Parallel and Distributed Computing and Systems, Cambridge, MA (2004)
4. Ravindran, B., Welch, L.R., Shirazi, B.A.: Resource management middleware for dynamic, dependable real-time systems. The Journal of Real-time Systems 20(2), 183–196 (2000)
5. Tia, T.-S., Deng, Z., Shankar, M., Storch, M., Sun, J., Wu, L.-C., Liu, J.W.-S.: Probablistic performance guarantee for real-time tasks with varying computation times. In: The IEEE Real-Time Technology and Applications Symposium (1995)
6. Hu, X.S., Zhou, T., Sha, E.H.-M.: Estimating probabilistic timing performance for real-time embedded systems. IEEE Transactions on Very Large Scale Integration Systems 9, 833–844 (2001)
7. Wandeler, E., Maxiaguine, A., Thiele, L.: Quantitative characterization of event streams in analysis of hard real-time applications. In: The 10th IEEE Real-Time and Embedded Technology and Applications Symposium (2004)
8. Manolache, S., Eles, P., Peng, Z.: Optimization of soft real-time systems with deadline miss ratio constraints. In: The 10th IEEE Real-Time and Embedded Technology and Applications Symposium (2004)
9. Welch, L.R., Ravindran, B., Shirazi, B.A., Bruggeman, C.: Specification and modeling of dynamic, distributed real-time systems. In: Proceedings of the 19th IEEE Real-Time Systems Symposium, December 1998, pp. 72–81 (1998)
10. Welch, L.R., Werme, P.V., Fontenot, L.A., Masters, M.W., Shirazi, B.A., Ravindran, B., Mills, D.W.: Adaptive qos and resource management using a posteriori workload characterizations. In: Proceedings of the 5th IEEE Real-Time Technology and Applications Symposium, June 1999, pp. 266–275 (1999)
11. Gu, D., Drews, F., Welch, L.: Robust task allocation for dynamic distributed real-time systems subject to multiple environmental parameters. In: The 25th International Conference On Distributed Computing Systems (ICDCS), Columbus, Ohio (June 2005)
12. Juedes, D., Drews, F., Welch, L., Fleeman, D.: Heuristic resource allocation algorithms for maximizing allowable workload in dynamic, distributed, real-time systems. In: The 12th Workshop on Parallel and Distributed Real-Time Systems (2004)
13. Gertphol, S., Yu, Y., Gundala, S.B., Prasanna, V.: A metric and mixed-integer-programming-based approach for resource allocation in dynamic real-time systems. In: The 16th International Parallel and Distributed Processing Symposium (2002)
14. Shestak, V., Chong, E.K.P., Maciejewski, A.A., Siegel, H.J., Benmohamed, L., Wang, I.-J., Daley, R.: Resource allocation for periodic applications in a shipboard environment. In: The Heterogeneous Computing Workshop in the 19th IEEE International Parallel & Distributed Processing Symposium, Denver, Co (April 2005)
15. Ali, S., Maciejewski, A.A., Siegel, H.J., Kim, J.-K.: Definition of a robustness metric for resource allocation. In: The 17th International Parallel and Distributed Processing Symposium (2003)
16. Liu, J.W.S.: A Reference Model of Real-Time Systems. In: Real-time Systems, pp. 34–59. Prentice-Hall, Inc., Upper Saddle River (2000)

17. Liu, C.L., Layland, J.W.: Scheduling algorithms for multiprogramming in a hard real-time environment. Journal of the Association for Computing Machinery 20(1), 46–61 (1973)
18. Oh, D.-I., Baker, T.P.: Utilization bounds for N-processor rate monotonic scheduling with stable processor assignment. Real Time Systems Journal 15(2), 183–193 (1998)
19. Lopez, J.M., Diaz, J.L., Garcia, D.F.: Utilization bounds for edf scheduling on real-time multiprocessor systems. Journal of Real-Time Systems 28, 39–68 (2004)
20. Fleeman, D., Gillen, M., Lenharth, A., Delany, M., Welch, L., Juedes, D., Liu, C.: Quality-based adaptive resource management architecture (qarma): A corba resource management service. In: The 12th Workshop on Parallel and Distributed Real-Time Systems in the 18th International Parallel and Distributed Processing Symposium (2004)
21. Juedes, D., Welch, L., Drews, F., Fleeman, D.: Resource allocation algorithms for maximizing allowable workload in dynamic, distributed real-time systems. Technical report, Center for Intelligent, Distributed, and Dependable Systems, Ohio University (2003)
22. Kincaid, D.R., Cheney, E.W.: Numerical Analysis: Mathematics of Scientific Computing, 3rd edn. Brooks Cole, Pacific Grove (2001)
23. Blazewicz, J., Ecker, K.H., Pesch, E., Schmidt, G., Weglarz, J.: Scheduling Computer and Manufacturing Processes. Springer, Heidelberg (2001)
24. Di Natale, M., Stankovic, J.A.: Dynamic end-to-end guarantees in distributed real time systems. In: Proceedings of Real-Time Systems Symposium, pp. 216–227 (1994)
25. Ali, S., Kim, J.-K., Yu, Y., Gundala, S.B., Gertphol, S., Siegel, H.J., Maciejewski, A.A., Prasanna, V.: Utilization-based heuristics for statically mapping real-time applications onto the HiPer-D heterogeneous computing system. In: The 11th IEEE Heterogeneous Computing Workshop (HCW 2002) in the 16th International Parallel and Distributed Processing Symposium (2002)
26. Aber, E., Drews, F., Gu, D., Juedes, D., Lenharth, A., Parrott, D., Welch, L., Zhao, H., Fleeman, D.: Experimental comparison of heuristic and optimal resource allocation algorithms for maximizing allowable workload in dynamic, distributed real-time systems. In: The 6th Brazilian Workshop on Real-Time Systems (2004)
27. Tan, Z.: Producing application CPU profiles in DynBench via curve fitting (manuscript, 2003)
28. Bettati, R., Liu, J.W.S.: End-to-end scheduling to meet deadlines in distributed systems. In: Proceedings of the 12th International Conference on Distributed Computing Systems, pp. 452–459 (1992)
29. Liu, J.W.S.: Multiprocessor Scheduling, Resource Access Control, and Syncronization. In: Real-time Systems, Prentice-Hall, Inc., Upper Saddle River (2000)
30. Zhao, H., Gu, D., Welch, L., Drews, F.: Stable allocations in distributed real-time systems with multiple environmental parameters and replicable applications. In: The Workshop on Parallel and Distributed Real-Time Systems in the 19th IEEE International Parallel & Distributed Processing Symposium, Denver, Co (April 2005)
31. Chen, D., Mok, A., Baruah, S.: Scheduling distributed real-time tasks in the dgmf model. In: Proceedings of the Real-time Technology and Applications Symposium, pp. 14–22 (2000)

A Proof for Lemma 3.1

Intuition of the proof was to use the knowledge of derivatives in system utilization function $U(w)$ to speculate its global behavior without knowing its exact form. A good mathematical tool to exploit this is the Taylor expansion. $U(w)$ can be accurately expanded as a Taylor series when the *integral remainder* is used [22]. The expansion has this general form:

$$f(x) = \sum_{k=0}^{n} \frac{1}{k!} f^{(k)}(x-c)^k + R_n(x), \text{ where} \qquad (3.9)$$
$$R_n(x) = \frac{1}{n!} \int_c^x f^{(n+1)}(t)(x-t)^n dt.$$

It will be used in the proof next.

Let w_0 be the real value for which $U(w_0) = \sum_{i=1}^{n} \frac{e_i(w_0)}{p_i} = (\sqrt{2}-1)m$, which is the worst case utilization bound for first-fit allocation on rate-monotonically scheduled processors [18]. We may assume $w_0 \geq 1$ since otherwise only zero workload ($\lfloor w_0 \rfloor = 0$) can satisfy this utilization for feasible allocation, which becomes the problem of fixed execution time task allocation, and there is no value to consider the feasible range for allowed workload. Now for all $w \leq w_0$, there is $U(w) \leq U(w_0)$ since the system utilization function has $U'(w) > 0$. Therefore, by [18], all workload value $w \leq \lfloor w_0 \rfloor$ has a feasible allocation by first fit. Thus $FF(I) \geq \lfloor w_0 \rfloor$.

If we use the Taylor expansion to expand the system utilization $U(w)$ to first order about $w = 0$ and evaluate at w_0, we have:

$$U(w_0) = U(0) + U'(0)w_0 + \int_0^{w_0} U''(t)(w_0 - t)dt. \qquad (3.10)$$

Similarly, if we evaluate at point cw_0 where $c \geq 1$ and $c \in \mathbb{R}$:

$$U(cw_0) = U(0) + U'(0)cw_0 + \int_0^{cw_0} U''(t)(cw_0 - t)dt. \qquad (3.11)$$

Then,

$$\frac{U(cw_0) - U(0)}{cw_0} - \frac{U(w_0) - U(0)}{w_0}$$
$$= \frac{1}{cw_0} \int_0^{cw_0} U''(t)(cw_0 - t)dt - \frac{1}{w_0} \int_0^{w_0} U''(t)(w_0 - t)dt \qquad (3.12)$$
$$= \ldots = \frac{c-1}{cw_0} \int_0^{w_0} U''(t) t\, dt + \frac{1}{cw_0} \int_{w_0}^{cw_0} U''(t)(cw_0 - t)dt.$$

The first term $\frac{c-1}{cw_0} \int_0^{w_0} U''(t) t\, dt \geq 0$ because $c \geq 1$, $U''(t) \geq 0$, and $t \geq 0$ for $t \in [0, w_0]$; the second term $\frac{1}{cw_0} \int_{w_0}^{cw_0} U''(t)(cw_0 - t)dt \geq 0$ because $U''(t) \geq 0$ and $cw_0 \geq t$ for $t \in [w_0, cw_0]$. As a result, $\frac{U(cw_0) - U(0)}{cw_0} - \frac{U(w_0) - U(0)}{w_0} \geq 0$, thus:

$$U(cw_0) \geq U(0) + c[U(w_0) - U(0)]. \qquad (3.13)$$

If we choose $c = \frac{m-U(0)}{U(w_0)-U(0)}$, then $U(cw_0) \geq m$. Because it is impossible to utilize processors more than full, and $U(w)$ is non-decreasing, we have:

$$OPT(I) \leq cw_0 = \frac{m-U(0)}{U(w_0)-U(0)} w_0. \tag{3.14}$$

Since $FF(I) \geq \lfloor w_0 \rfloor$, there is:

$$\frac{OPT(I)}{FF(I)} \leq \frac{OPT(I)}{\lfloor w_0 \rfloor} \leq \frac{m-U(0)}{U(w_0)-U(0)} \cdot \frac{w_0}{\lfloor w_0 \rfloor} < \frac{m-U(0)}{U(w_0)-U(0)} \cdot 2 = \frac{2-2\delta}{\frac{U(w_0)}{m}-\delta}. \tag{3.15}$$

We have expressed the workload independent system utilization $U(0)$ as δm, and used the fact that $\frac{w_0}{\lfloor w_0 \rfloor} < 2$ since $w_0 \geq 1$. For the $RAFF$-1 on rate-monotonically scheduled processors, we recall $\frac{U(w_0)}{m} = \sqrt{2}-1$. Thus the absolute approximation ratio $r^1_{FF(RM)} < \frac{2-2\delta}{\sqrt{2}-1-\delta}$.

Asymptotically when $OPT(I) \to \infty$, the bound becomes tighter. From equation (3.14), there is $w_0 \geq OPT(I)/c$. When $OPT(I) \to \infty$, $w_0 \to \infty$. This leads to $\frac{w_0}{\lfloor w_0 \rfloor} = 1$. Plugging it into equation (3.15), we have:

$$\frac{OPT(I)}{FF(I)} \leq \frac{1-\delta}{\frac{U(w_0)}{m}-\delta}, \tag{3.16}$$

and thus the asymptotic approximation ratio $r^{1*}_{FF(RM)} \leq \frac{1-\delta}{\sqrt{2}-1-\delta}$.

In the case when first-fit allocation is used on processors scheduled according to EDF, let w'_0 be the real value for which,

$$U(w'_0) = \sum_{i=1}^{n} \frac{e_i(w'_0)}{p_i} = \frac{m+1}{2}.$$

Then there is,

$$U(\lfloor w'_0 \rfloor) \leq U(w'_0) = \frac{m+1}{2} \leq \frac{\beta m+1}{\beta+1} \quad (\beta \geq 1).$$

According to the worst case utilization bound by [19], the utilization $U(\lfloor w'_0 \rfloor) \leq \frac{\beta m+1}{\beta+1}$ sufficiently guarantees feasible EDF schedules using first-fit allocation (m is the number of processors and β is the max number of the task with largest utilization schedulable on one processor). As a result,

$$FF(I) \geq \lfloor w'_0 \rfloor.$$

Based on the same argument leading to Equation (3.15) and substituting $U(w_0)$ by $U(w'_0)$, there is,

$$\frac{OPT(I)}{FF(I)} < \frac{2-2\delta}{\frac{1}{m} \cdot \frac{m+1}{2} - \delta} = \frac{4-4\delta}{1+1/m-2\delta}.$$

Similarly in the asymptotic case, Equation (3.16) leads to,

$$\frac{OPT(I)}{FF(I)} \leq \frac{2-2\delta}{1+1/m-2\delta}.$$

Thus, the *EDF*-based robustness bound for first-fit is: $r^1_{FF(EDF)} < \frac{4-4\delta}{1+1/m-2\delta}$, and asymptotically, $r^{1*}_{FF(EDF)} \leq \frac{2-2\delta}{1+1/m-2\delta}$.

Notice that there also exists a special case where $FF(I) = OPT(I)$ (a ratio of 1). That is when there is a task T_i for which $e_i(w_p)/p_i = 1$ and $U(w_p) \leq (\sqrt{2}-1)m$ (or $\frac{\beta m+1}{\beta+1}$). Then $w_p = OPT(I)$. Since first-fit can always find an allocation when system utilization is less than Oh and Baker (or Lopez) bound, it will have feasible allocation for w_p: $FF(I) = w_p$. Thus $FF(I) = OPT(I)$, and the approximation ratios equal 1.

B Proof for Theorem 3.2

We notice that by definition order-k dynamic systems automatically have $U'(w) \geq 0$ and $U''(w) \geq 0$. Therefore, for $k > 1$ all results from the proof of lemma 3.1 still hold. We will make use of these during the proof.

If we use the Taylor expansion to expand the system utilization $U(w)$ to kth order about $w = 0$ and evaluate at w_0, we have:

$$U(w_0) = U(0) + \sum_{i=1}^{k} \frac{1}{i!} U^{(i)}(0) w_0^i + \frac{1}{k!} \int_0^{w_0} U^{(k+1)}(t)(w_0-t)^k dt, \quad (3.17)$$

and equivalently,

$$U(w_0) = U(0) + w_0^k \left(\sum_{i=1}^{k-1} \frac{U^{(i)}(0)}{i!} \frac{1}{w_0^{k-i}} + \frac{U^{(k)}(0)}{k!} \right) + \frac{1}{k!} \int_0^{w_0} U^{(k+1)}(t)(w_0-t)^k dt. \quad (3.18)$$

From equation 3.14, when asymptotically $OPT(I) \to \infty$, $w_0 \to \infty$. Therefore, $\frac{1}{w_0^{k-i}} \to 0$ ($k-i \geq 1$) against the constant $\frac{U^{(k)}(0)}{k!}$ where $U^{(k)}(0) \neq 0$ (by definition). So asymptotically,

$$U(w_0) = U(0) + \frac{1}{k!} U^{(k)}(0) w_0^k + \frac{1}{k!} \int_0^{w_0} U^{(k+1)}(t)(w_0-t)^k dt. \quad (3.19)$$

Similarly, if we evaluate at point cw_0 where $c \geq 1$ and $c \in R$, asymptotically:

$$U(cw_0) = U(0) + \frac{1}{k!} U^{(k)}(0)(cw_0)^k + \frac{1}{k!} \int_0^{cw_0} U^{(k+1)}(t)(cw_0-t)^k dt. \quad (3.20)$$

Then,

$$\begin{aligned}
&\frac{U(cw_0) - U(0)}{(cw_0)^k} - \frac{U(w_0) - U(0)}{w_0^k} \\
&= \frac{1}{w_0^k k!}\left[\frac{1}{c^k}\int_0^{cw_0} U^{(k+1)}(t)(cw_0 - t)^k dt - \int_0^{w_0} U^{(k+1)}(t)(w_0 - t)^k dt\right] \\
&= \frac{1}{w_0^k k!}\left[\frac{1}{c^k}\int_0^{w_0} U^{(k+1)}(t)[(cw_0 - t)^k - (cw_0 - ct)^k]dt \right. \\
&\quad \left. + \frac{1}{c^k}\int_{w_0}^{cw_0} U^{(k+1)}(t)(cw_0 - t)^k dt\right] \geq 0.
\end{aligned}$$
(3.21)

Since $(cw_0 - t)^k > (cw_0 - ct)^k$ for $c > 1$, and $U^{(k+1)}(t) \geq 0$, the first integration is positive; it is also easy to see the second integration is also positive. Then,

$$U(cw_0) \geq U(0) + c^k[U(w_0) - U(0)]. \tag{3.22}$$

When $c = \left(\frac{m - U(0)}{U(w_0) - U(0)}\right)^{\frac{1}{k}}$, $U(cw_0) \geq m$. Therefore,

$$OPT(I) \leq cw_0 = \left(\frac{m - U(0)}{U(w_0) - U(0)}\right)^{\frac{1}{k}} w_0. \tag{3.23}$$

The asymptotic approximation ratio becomes,

$$\frac{OPT(I)}{FF(I)} \leq \frac{OPT(I)}{\lfloor w_0 \rfloor} \leq \left(\frac{m - U(0)}{U(w_0) - U(0)}\right)^{\frac{1}{k}} \frac{w_0}{\lfloor w_0 \rfloor} = \left(\frac{1 - \delta}{\frac{U(w_0)}{m} - \delta}\right)^{\frac{1}{k}},$$

where $\delta = \frac{U(0)}{m}$. Since w_0 tends to ∞, the ratio is 1. We used $FF(I) \geq \lfloor w_0 \rfloor$ as before because for systems running RMS schedulers, we would choose w_0 such that $\frac{U(w_0)}{m} = \sqrt{2} - 1$ based on [18]; for systems running EDF schedulers, we would choose w_0 such that there is $\frac{U(w_0)}{m} = \frac{m+1}{2m}$, using the previous discussion based on [19]. Therefore, $r^{1*}_{FF(RM)} \leq \left(\frac{1-\delta}{\sqrt{2}-1-\delta}\right)^{\frac{1}{k}}$, and $r^{1*}_{FF(EDF)} \leq \left(\frac{2-2\delta}{1+1/m-2\delta}\right)^{\frac{1}{k}}$.

4

Supercomputer Scheduling with Combined Evolutionary Techniques

A. LaTorre[1], J.M. Peña[1], V. Robles[1], and P. de Miguel[1]

CeSViMa, Department of Computer Architecture and Technology
Universidad Politécnica de Madrid, Campus de Montegancedo S/N
Madrid, 28660, Spain
atorre@fi.upm.es, jmpena@fi.upm.es, vrobles@fi.upm.es,
pmiguel@fi.upm.es

Summary. Scheduling is a very important problem in many real-world scenarios. In the case of supercomputers it is even more important because available resources are limited and expensive. The optimal use of supercomputer facilities is a critical question. We have found that the definitions of traditional scheduling problems do not provide an appropriate description for Supercomputer Scheduling (SCS). Thus, a new definition for this kind of problems is proposed. The research already done in the field of other scheduling problems can be modified to be applied in this new scenario. Nevertheless, new techniques can also be developed. Thus, we have proposed a theoretical framework to combine multi evolutionary algorithms called Multiple Offspring Sampling (MOS). We have used this approach to combine multiple codings and genetic operators in this scheduling problem. To summarise: first, we introduce a formal definition of supercomputer scheduling; second, we propose Multiple Offspring Sampling formalism; and third, we have carried out an experimental test to compare the performance of this formalism to solve SCS problems against traditional (non-combinatorial) techniques and single genetic algorithms.

Keywords: Scheduling, Supercomputing, Evolutionary Algorithms, Multiple Offspring Sampling, Genetic Algorithms.

4.1 Introduction

Scheduling problems are part of many real-world scenarios, such as logistics, manufacturing and engineering. Although on each of these scenarios the scheduling problem appears with different characteristics and restrictions, this kind of problems has been divided into a classical taxonomy: flow-shop scheduling (FSS), job-shop scheduling (JSS), multiprocessor scheduling (MPS), and so on.

Job scheduling for supercomputers presents particular characteristics from the ones named before and, therefore, it should be defined in a different way. Many current supercomputers are large cluster systems, as ranked on the *TOP500 supercomputers sites*[1]. They are composed by hundreds or thousands of processors interconnected by a high-speed network. These facilities are designed to run

---
[1] http://www.top500.org

parallel programs that are partitioned into a set of concurrent tasks. In general, only one of these tasks should be running on each processor (there should be no competing tasks on the same processor). Task scheduling for these systems consists in the partition of the resources (processors, in most of the cases) for the sequence of jobs to be run on the system with the objective of minimising the total execution time.

Supercomputers are designed to run jobs divided into hundreds of parallel tasks. Tasks of the same job are concurrent and interact among them exchanging messages while running. A clear example are MPI parallel programs, common in many scientific (e.g., physics simulations, protein docking) and engineering (e.g., fluid dynamics, finite elements calculus) fields. This means a significant difference regarding the classical scheduling (e.g., multiprocessor scheduling) where the set of interdependent sequenced tasks were described originally by means of a direct acyclic graph (DAG).

Traditional solutions to schedule jobs in a supercomputer have been taken from batch process scheduling algorithms. These algorithms are deterministic criteria to order waiting jobs and to submit them into execution. They are usually implemented on scheduling services, sometimes called resource managers. The most advanced systems provide the possibility to be configured to use many different algorithms, or even ad-hoc-defined variants that are more appropriate to the administrative policies of a given site.

State-of-the-art cluster-based supercomputers are equipped with interconnected multiprocessor nodes. Each node has two, four or, in the near future, more processors (or multiple cores) sharing the same main memory (RAM). Under this configuration, scheduling policies should satisfy an additional restriction: the total amount of memory required by the processes running on a specific node must not exceed its available shared memory. Although swapping virtual memory is common in modern operating systems, if there is only one process running on a system, swapping in and out memory pages significantly influences on its performance.

Considering more than two restrictions (requested processes and free shared memory on the available nodes) increases also the complexity of the scheduling problem. Evolutionary algorithms have been used to solve complex optimisation problems, including scheduling problems. The evolutionary paradigm is inspired by natural selection, and genetic algorithms are the most representative approach.

This chapter presents the formal definition of supercomputer scheduling (SCS) for parallel programs. To solve scheduling within this environment, traditional methods have been applied, as well as an hybrid evolutionary approach (based on genetic algorithms). For scheduling problems with moderate complexity, simple evolutionary methods provide better results in both resource usage and timespan, compared to classical approaches. But in the case of complex problems, with many waiting jobs and supercomputers with thousands of nodes, simple evolutionary techniques might not find the best scheduling as it is not easy to select the appropriate coding and genetic operators among all the available ones.

In these cases, combined heuristic methods are a quite interesting alternative. Results show how these combined methods help finding the best representation and genetic operators. Thus, the combined algorithm outperforms both simple traditional and evolutionary methods.

The experimental results shown in this chapter has been taken from the regular operations of the Magerit supercomputer hosted at the CeSViMa (*Centro de Supercomputación y Visualización de Madrid*)[2]. This system manages a queue of a hundred waiting jobs (in average), and the experimental datasets have been taken from the log files of previous executions.

Section 4.2 presents the state-of-the-art on different scheduling problems, while Section 4.3 proposes the definition of the new supercomputer scheduling problem. Section 4.4 presents traditional (non-combinatorial) methods. In Section 4.5, the Multiple Offspring Sampling formalism is introduced. Section 4.6 shows the experimentation performed on Magerit supercomputer system. Finally, Section 4.7 concludes this study.

4.2 Related Work on Scheduling Problems

There are several kinds of scheduling problems defined in the literature. Although none of them fits perfectly in our specific scheduling problem, it is important to review them, as we borrow from scheduling literature the most relevant codings and operators.

Scheduling problems deal with the allocation of resources over time to perform a set of tasks and they are characterised by three main components:

- A number of machines and a number of jobs that must be submitted.
- A set of constraints that must be satisfied.
- A target function that must be optimised.

In this section we analyse the flow shop, job shop and multiprocessor scheduling problems, which are relevant for the supercomputer scheduling introduced in this work. There are other variants and subtypes which are not included in this review.

4.2.1 Flow-Shop Scheduling Problem

The general flow-shop problem, defined by [1], is denoted as $n/m/C_{max}$ in the literature. It involves n jobs, each requiring operations on m machines, in the same machine sequence. The processing time for each operation is p_{ij}, where $i \in \{1, 2, \ldots, n\}$ denotes a job and $j \in \{1, 2, \ldots, m\}$ a machine. The problem is to determine the sequence of these n jobs that produces the smallest makespan assuming no preemption of operations. In the simplest situation, all jobs are available and ready to start at time zero. In more realistic situations jobs are released at different times.

---

[2] http://www.cesvima.upm.es

Scheduling literature has a lot of solution procedures for the general flow-shop scheduling problem. An excellent review about heuristic approaches can be found in [1]. Furthermore, several meta-heuristic methods like tabu search, genetic algorithms, simulated annealing and ant-colony have been used to solve this problem.

The best results have been obtained with tabu search [14, 31] and genetic algorithms [4, 36], which are the most popular methods. Besides, simulated annealing [32] and ant-colony [50] have also been applied.

4.2.2 Job-Shop Scheduling Problem

The $n \times m$ minimum-makespan general job-shop scheduling problem can be described by a set of n jobs $\{J_i\}, i \in \{1, 2, \ldots, n\}$ which has to be processed on a set of m machines $\{M_j\}, j \in \{1, 2, \ldots, m\}$. Each job must be processed in a sequence of machines. The processing of the job J_i on machine M_j is called the operation O_{ij}. Operation O_{ij} requires the exclusive use of M_j for an uninterrupted duration p_{ij}, its processing time. A *schedule* is a set of completion times for each operation that satisfies those constraints.

In many different works have been proposed to solve the job shop scheduling problem. The best results seem to be reached with tabu search [3, 31, 43]. An explanation of this behaviour can be found in [47].

GAs have also been applied to the job-shop scheduling problem in a number of ways. The first attempt for solving the problem using evolutionary methods is carried out by [11]. One of the most successful GAs for scheduling is the GA3 algorithm by [27]. Other relevant works are [5, 26, 49] and [6]. In all cases it is shown that conventional GAs are limited for this problem. Several improvements over different elements are proposed to produce results comparable to the most competitive methods. Typically these articles present methods that: (i) include hill-climbers, (ii) take into account the application of problem specific knowledge, or (iii) use more advanced evolutionary models.

4.2.3 Multiprocessor Scheduling Problem

The problem of scheduling a set of dependent or independent tasks to be processed in a parallel fashion is a field where some authors, such as [48], have done interesting works. A program can be decomposed into a set of smaller tasks. These tasks can have dependencies or precedence requirements, defined by a direct acyclic graph (DAG). The goal of the scheduler is to assign tasks to available processors such that dependencies are satisfied and the makespan is minimised.

In [21], it is possible to find a good review and classification of deterministic or static scheduling algorithms.

Genetic algorithms have been widely applied to the multiprocessor task scheduling problem [2, 24, 42, 44, 48]. Two main approaches are: (i) methods that use a GA in combination with other scheduling techniques and (ii) methods that use a GA to evolve the actual assignment and order of tasks over the processors.

4.2.4 Other Packing and Knapsack Problems

Together with the scheduling problems reviewed above, other combinatorial problems share a similar structure.

Packing problems try to minimise the size of a container which is able to contain a certain number of items. Some packing problems are actually puzzles in which finding the minimal size of a 2D shape to contain a given number of items with other shapes. Beside this type of problems, there are other interesting variants of packing problems, such as: (i) bin packing (N objects of different sizes must be packed into a finite number of bins of capacity V in a way that the number of used bins is minimised), (ii) multi-dimensional bin packing (objects have 2 or more dimensions and containers have different sizes depending on the dimension), (iii) set packing (several subsets of the same set of elements are provided and the objective is to maximise the number of selected subsets such as all pair-wise intersections between each two selected subsets are empty).

Knapsack problems try to maximise the value of the objects carried in a knapsack. Each of the objects has a certain weight and the knapsack has a weight limit. There are different variants of knapsack problems: (i) bounded knapsack (one object may be chosen several times), (ii) multiple-choice knapsack (items are subdivided into k classes and exactly one item must be taken from each class), (iii) subset sum knapsack (for each item the value and weights are identical), (iv) multiple knapsack (there are m knapsacks with capacities Wi), to name a few. A review of several knapsack problems can be found in [34].

4.3 Supercomputer Scheduling Problem

Supercomputer scheduling is introduced as a new scheduling problem. Real world problems belonging to supercomputing have shown that none of the traditional scheduling problems match with the requirements of this scenario. Jobs in a supercomputer are usually a set of tasks that must be executed in parallel. This means that tasks have no sequential dependencies but must be run concurrently. Execution restrictions are based on memory and processor availability.

Each job running on the system is defined as: $J_i = (T_i, M_i, t_i)$, where T_i is the number of parallel tasks, M_i is the amount of memory per task and t_i is the total execution time of all and each of the tasks.

A supercomputer in this domain is defined as a set of n nodes: $S = (N_1, N_2, ..., N_n)$. Each node is defined as: $N_j = (P_j, A_j)$, where P_j are the available processors on the same node and A_j is the available shared main memory.

When a job is running on the supercomputer, several tasks are assigned to a subset of the supercomputer nodes: $k = assign(J_i, N_j)$, that means that k tasks from job J_i are running on node N_j. The number of tasks running on each node should not exceed the number of processors in that node. The number of assigned tasks should be equal to the number of total tasks belonging to this job, T_i. All the tasks must be run at the same time (in parallel).

$$T_i = \sum_{j=1}^{n} assign(J_i, N_j) \quad / \quad \sum_{i=1}^{n} assign(J_i, N_j) \leq P_j \qquad (4.1)$$

In this context, job scheduling is an ordered sequence of jobs to be run on the supercomputer. Jobs are dispatched to the supercomputer if it has available resources to process the first job on this sequence (job queue), depending on two restrictions:

① There are enough free processors on the system

$$\forall j \in [1, n] : \sum_{i} assign(J_i, N_j) \leq P_j \qquad (4.2)$$

② There is enough free memory in each of the nodes

$$\forall j \in [1, n] : \sum_{i} M_i \times assign(J_i, N_j) \leq M_j \qquad (4.3)$$

SCS could be considered as a particular case of multi-dimensional packing problems. Time, number of CPUs and memory are the three different dimensions to take into account. Although the general structure is similar to these types of packing problems, memory usage restrictions are defined not for all the system but for a partition of the system (the nodes). There exists the possibility of defining SCS as a multi-dimensional packing problem with additional constraints, but SCS is considered as a different problem subtype based on the real-world application that inspired it.

4.4 Related Work on Cluster and Supercomputer Scheduling

Scheduling is a key policy in the performance of expensive HPC facilities. The goal of this policy is to reduce the execution time required for a parallel job (when only one of these jobs is considered) or to maximise the resource processor usage (when multiple jobs are taken into account). Additional criteria could also be considered as, for example, the minimisation of the turnaround time (time during different submission and termination). Scheduling policy should be aware of the requirements of the jobs. These requirements are usually expressed as execution constraints (termination deadline) or, mainly, resource requirements.

Valid scheduling policies must ensure both resource provision and constraint boundaries; but once these basic restrictions are provided, the policies have significant degrees of freedom to arrange the jobs. Commonly-used process scheduling and workload managers use non-combinatorial deterministic strategies implemented by commercial or free software solutions. An extended overview on non-combinational scheduling can be referred in [13].

4.4.1 Non-combinatorial Policies

Non-combinatorial policies are implemented for a broad range of scheduling systems in HPC clusters. Combinatorial techniques are those which use permutation-based functions (such as insertion, deletion or swapping), either deterministically (brute force) or stochastically. Non-combinatorial techniques use greedy approaches or other direct methods.

FCFS (First-Come-First-Serve) policy [40] is one of the most popular approaches. In this case, parallel jobs are scheduled in the same order they arrive. The order in the queue of waiting jobs is used to dispatch them. If there are enough resources available in the system, these resources are allocated and the first job in the queue starts its execution. This selection is repeated while the requirements of the next job can be fulfilled. If not, the next job in the queue waits until one of the already running jobs finishes and its resources are released.

Backfilling [25] improves FCFS policy allowing small jobs to be scheduled before their actual order if there are only few resources available. This policy prevents the waste of idle resources, in short term, but they would lead large jobs to starvation. EASY, the Extensible Argonne Scheduling sYstem, developed by IBM for SP1 clusters, reduces this unbalanced effect by means of a reservation mechanism, for the jobs waiting in the queue. Reservations are computed using the expected time when the resources, required by the first waiting job, will be available. This deadline is used to avoid the execution of smaller jobs that will finish after this deadline. The number of reservations may be parameterised. For example, in Conservative Backfilling [30], reservations are made for all of the waiting jobs in the queue. Reservations are computed using the expected time when the required resources by the first waiting job will be available.

Another important alternative in job scheduling is SJF (Shortest-job-first) [8]. SJF changes the order in the queue according to the expected execution time. This model can be generalised by the amount of any of the resources required, instead of execution time. If there are more than one resource considered (processors or memory), then different ordering criteria in the queue would be possible.

Many of these alternatives consider the runtime estimation as one important input to the scheduler. This assumption is quite usual, and scheduling system implementations use different alternatives to deal with it. In many cases, jobs exceeding their expected execution time are directly killed. This policy encourages users to be as exact as possible in their estimations.

4.4.2 Scheduling Tools

The implementation of the theoretical aspects of the different scheduling policies is also an important decision for cluster and supercomputer facilities. Based on the general policies mentioned above, many different toolkits and systems have been developed for job scheduling and workload management.

PBS family of resource managers (PBS, OpenPBS, and Torque)[3] implements a default FCFS scheduler, but also provide mechanisms to implement other simple schedulers. PBS includes a resource manager that acts as an interface for users to access the cluster. It can accept jobs and let users view the status of the queue. The scheduler reads the state of this queue, makes a scheduling decision, and informs the resource manager of its decision.

The Maui Cluster Scheduler offers compatibility with the Torque resource manager. It comes with a wide variety of scheduling policies to try to accommodate different scheduling needs. The Moab Cluster Suite[4] is the successor to Maui. Among other improvements, it provides a large set of graphical tools which help an administrator to monitoring the state of the cluster and the queue. These tools are strategy-dependent, allowing administrators to monitor and change information contained in strategy-dependent parameters of jobs.

Simple Linux Utility for Resource Management (SLURM)[5] [16] is an open source, fault-tolerant, and highly scalable cluster management and job scheduling system for Linux clusters of thousands of nodes. Components include machine status, partition management, job management, scheduling, and stream copy modules.

A well-known commercial system is IBM's LoadLeveler [19]. LoadLeveler provides basic queue management, based on priorities. This system allows administrators to configure dynamic priority updates using different policies. LoadLeveler is sometimes used as a lower-level interface to more advanced resource managers, like Maui. Any higher-level system can interact with LoadLeveler using an application programming interface (API).

4.5 Multiple Offspring Sampling and the Supercomputer Scheduling Problem

From the point of view of Evolutionary Algorithms, the Supercomputer Scheduling problem can be considered as a combinatorial problem where the goal is to find the most suitable ordered sequence of jobs according to a given criterion. In this case, the objective is to maximise the CPU usage. There exist a huge amount of works in the literature where Evolutionary Algorithms have been successfully applied to solve this type of problems [7, 33, 41].

Many different codings have been proposed for combinatory problems. In [22], the authors propose the following, not exhaustive, taxonomy:

- Binary representation.
- Path representation.
- Adjacency representation.
- Ordinal representation.
- Matrix representation.

---
[3] http://www.openpbs.org/
[4] http://www.clusterresources.com/
[5] http://www.llnl.gov/linux/slurm/download.html

All these different ways for representing an individual in the context of an Evolutionary Algorithm could be used to code the SCS problem. Selecting the appropriate coding for individuals among all the available representations without any previous knowledge about their performance is a critical decision. If we consider the different genetic operators that have been proposed for each of these encodings, it is even harder to make this decision.

As an alternative to traditional Evolutionary Algorithms, we propose Multiple Offspring Sampling as an effective technique to solving combinatorial problems. This new approach proposes the simultaneous use of different *techniques* (a proper definition of technique in the context of *MOS* will be given in subsection 4.5.3) to create new individuals (candidate solutions).

In this work we have focused on Path representation and Ordinal representations. These encodings should be different enough so that they could contribute with different properties to the overall optimisation process. A more detailed explanation of these two codings and their specific related genetic operators can be found in subsection 4.6.2.

To show how *MOS* modifies the behaviour of classic Evolutionary Algorithms (EA), we should first present a general schema of EA functioning, which will be given in the next subsection, and then give a functional formalisation in subsection 4.5.2. Afterwards, Multiple Offspring Sampling will be briefly presented in subsection 4.5.3.

4.5.1 Evolutionary Algorithms

Evolutionary algorithms and, in particular, Genetic Algorithms, are bio-inspired algorithms based on Darwin's Theory of Evolution. [15] was one of the first to use this approach to solve search and optimisation problems. Since then, they have been applied to many different domains with a remarkable success.

Evolutionary algorithms are population-based meta-heuristic optimisation algorithms. They evolve an initial population of candidate solutions to the problem being solved by means of certain recombination operators under the principle of the survival of the fittest.

Generally, the operation of these algorithms can be divided into different phases:

① Creation of the initial population P_0.
② Evaluation of the initial population P_0.
③ Checking of the algorithm termination (convergence or generation limit), if so then finish, otherwise continue.
④ Generation, using some individuals from P_i, of new individuals for the next generation, called offspring population O_i.
⑤ Evaluation of the new individuals in O_i.
⑥ Combination of offspring and previous populations to define the next population P_{i+1}.
⑦ Go back to ③.

Based on this schema, different evolutionary algorithms and approaches have been developed. For example, in step ⑥ classical GAs take the offspring as the next population ($P_{i+1} = O_i$). Other approaches, such as steady state algorithms generate only one offspring individual that replaces the worst individual in P_i, and intermediate approaches, based on elitism, take the best individuals from both O_i and P_i to generate P_{i+1}.

For step ④, the literature also offers a wide array of approaches, such as selecting different genetic operators. Other evolutionary algorithms, such as estimation of distribution algorithms [23], use statistical approaches for modelling the population and later sampling the offspring.

4.5.2 Functional Formalisation of an Evolutionary Algorithm

In order to introduce the contributions of this chapter, some preliminary formalisations should be defined.

In the context of evolutionary computation, for the description of one problem two different sets of elements should be considered:

- **S** is the set of all possible phenotypes (candidate solutions to the problem).
- **C** is the set of all possible combinations of the coding format (genotypes). This set denotes the search space of the evolutionary algorithm.

It should be taken into account that, in the general schema mentioned above, operations are performed on different sets of elements. For example, the evaluation of the solutions is a phenotype operation: the individual is the one that behaves well or bad in the environment. On the other hand, recombination of individuals to generate offspring is based on the genotype codification.

For an evolutionary algorithm there must also be a coding function *code* that transforms elements of the genotype set into elements in the phenotype set (coding into solutions):

$$\mathbf{C} \xrightarrow{code} \mathbf{S} \tag{4.4}$$
$$c \longrightarrow code(c) = s$$

This function can be extended to operate on a set of elements. The function ***code*** generates a set of solutions ($S \subset \mathbf{S}$) from a set of genotypes ($C \subset \mathbf{C}$):

$$\mathcal{P}(\mathbf{C}) \xrightarrow{code} \mathcal{P}(\mathbf{S}) \tag{4.5}$$
$$C \longrightarrow \boldsymbol{code}(C) = S$$
$$\boldsymbol{code}(C) = \{s \in \mathbf{S} / \exists c \in C : s = code(c)\} \tag{4.6}$$

The phenotype and genotype pair, $(s,c) \in \mathbf{S} \times \mathbf{C} \wedge s = code(c)$, identifies both the individual and the coding used for this solution.

To drive the search mechanism, all the evolutionary algorithms require the existence of an evaluation function that determines the individual's chances of survival in the environment, a fitness function *fit*:

$$\mathbf{S} \xrightarrow{\mathit{fit}} \mathbb{R}$$
$$s \longrightarrow \mathit{fit}(s) \tag{4.7}$$

This approach is quite restrictive, as some methods, specially co-evolutionary algorithms, define order relations to compare the quality of the phenotypes.

To complete these definitions, the construction of iterative generations should be formalised.

Let **off** be the Offspring Sampling Function. This function defines how new individuals are generated by recombination of the individuals in previous generations. This is a genotype-level function. The Offspring Sampling Function in GAs is defined as the combinations of gene operators (crossover, mutation and selection). In other approaches, such as EDAs, these functions are statistical modelling and model sampling.

$$\mathcal{P}(\mathbf{C}) \xrightarrow{\mathit{off}} \mathcal{P}(\mathbf{C})$$
$$C_i \longrightarrow \mathit{off}(C_i) = C_{i+1} \tag{4.8}$$
$$\text{Offspring size restriction:} \quad \forall i: \quad |\mathit{off}(C_i)| = \sigma \tag{4.9}$$

Let σ be the size of the new offspring population. Usually this size does not vary during different generations.

Finally, a method to combine the previous generation and the new individuals generated from this offspring should be included, resulting in the population combination function **comb**:

$$\mathcal{P}(\mathbf{S}) \times \mathcal{P}(\mathbf{S}) \xrightarrow{\mathit{comb}} \mathcal{P}(\mathbf{S})$$
$$(S_i, O_i) \longrightarrow \mathit{comb}(S_i, O_i) = S_{i+1} \tag{4.10}$$
$$\text{Previous population:} \quad S_i \subset \mathbf{S}/S_i = \mathit{code}(C_i)$$
$$\text{Offspring population:} \quad O_i \subset \mathbf{S}/O_i = \mathit{code}(\mathit{off}(C_i)) \tag{4.11}$$

There are many different population combination functions, for example classical elitism function is $\mathit{comb}(S_i, O_i) = S_{i+1}$:

$$S_{i+1} = \{s \in S_i \cup O_i / \nexists t \in S_i \cup O_i : t \notin S_{i+1} \wedge \mathit{fit}(t) \succ \mathit{fit}(s)\} \tag{4.12}$$

In this formula, $\succ$ represents better-fitness-than, which is "greater than" or "less than" depending on the sense of optimisation, maximisation or minimisation, respectively.

4.5.3 Multiple Offspring Sampling Formalism

Multiple Offspring Sampling Basics

We introduce the Multiple Offspring Sampling (MOS) approach as a combined alternative to the way steps ④ and ⑥ of an Evolutionary Algorithm are performed. MOS proposes the definition of multiple techniques to generate new individuals, and makes them compete during the evolutionary process. Each technique creates its own offspring $O_i^{(j)}$ (i is the generation and j is the technique).

The fitness of the individuals generated by each technique is used to evaluate the quality of this particular recombination technique. The most obvious measure that could be used for this purpose is the average fitness of the population, but more sophisticated measures can be proposed to take into account not only the current performance of the technique but its potentiality.

Finally, in phase ⑥, previous population P_i and all the offsprings $O_i^{(j)}$ are merged to produce the next population P_{i+1}. This process is usually done by using an elitist population merge function.

These MOS techniques, or techniques, as they are referred at the beginning of Section 4.5, could be defined as a technique to create new individuals, i.e., (a) a particular evolutionary algorithm model, (b) with an appropriate coding, (c) using specific operators (if required), and (d) configured with its necessary parameters.

According to the definition above we can consider different parameters and thus divide MOS into several categories. A rough taxonomy of how MOS can be divided could be:

- Algorithm-based MOS: different algorithms (GAs, EDAs) are used to create new individuals.

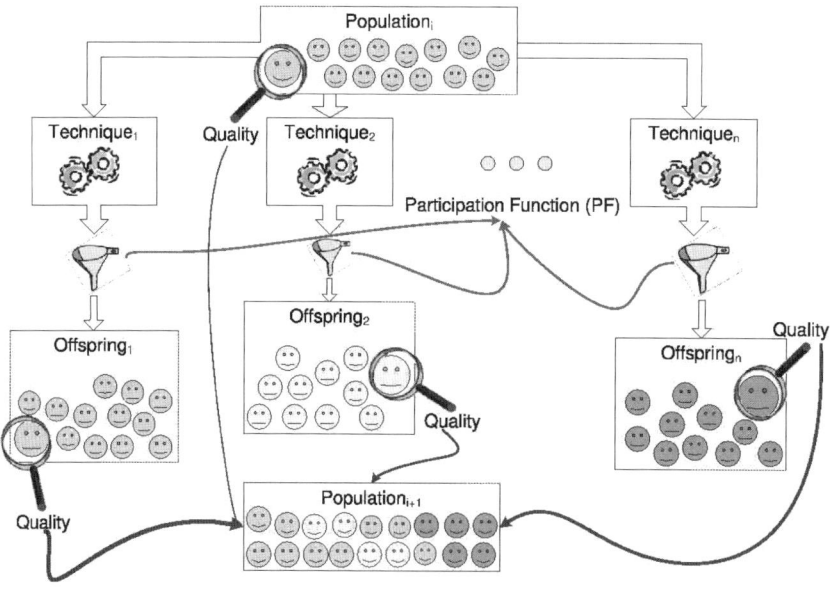

Fig. 4.1. MOS overview

- Coding-based MOS: different codings (genotypes) can be used to represent one candidate solution (phenotype) of the problem.
- Operator-based MOS: for a single coding of candidate solutions there could exist different genetic operators (if using GAs) that could be used simultaneously.
- Parameter-based MOS: different values for evolutionary parameters (crossover and mutation ratios, selection techniques, etc.) are used within each technique.
- Hybrid MOS: a combination of any of the previous.

A general view of MOS functioning is depicted in Fig. 4.1.

Genotype Codes as a Different Offspring Production Method

The selection of a genetic coding (genotype format) influences on how individuals are recombined to generate the next offspring. Recombination is a genotype operation previous to the evaluation of the actual individuals (phenotypes). There are many problems in which solutions could be coded in several different ways, considering for them different genotype formats. In all the cases, all of these formats are able to represent individuals, points on the solution space.

There are different approaches to profit from this solution space translation. Variable neighbourhood search (VNS), introduced by [29], has been used as a meta-heuristic for solving combinatorial and global optimisation problems whose basic idea is the systematic change of neighbourhood within a local search. VNS has been used in conjunction with other heuristic methods, like tabu search or GRASP.

In our approach, a genetic algorithm will combine the individuals in the population (in generation i) using all the different encodings available with their respective recombination operators. In order to generate the next offspring, each of the encoding methods is allowed to produce a fixed amount of individuals. To give more chances to the mechanisms that are producing the best individuals, the amount of offspring individuals is adjusted on each generation. This approach reduces the participation of worse methods and increases the number of individuals produced by better methods.

The effect of coding on the heuristic search techniques is very important due to the relationship between the coding criteria and the fitness landscape. The fitness landscape is a representation of the whole solution space assigning fitness values to each point on this space.

Studies, carried out by [17], on the difficulty of different optimisation problems have measured one of the aspects of problem complexity by the correlation of the difference between fitness function values and the Euclidean distance on the solution space.

Although considering that representation issues is one of most important aspects in the performance of the algorithm, only few references can be found in the literature. [39] introduced the use of multiple representations in evolutionary algorithms providing both Cartesian and pseudo-polar representations to solve real-valued minimisation functions. Each element of the offspring takes the same representation (in 95% of the cases) that the first parent, and only in 5% of it selects the opposite. This approach adapts the most suitable representation based on the selection pressure.

In this sense, for example, a pure binary coding landscape differs from a Gray code one for the same fitness function. This could make the problem easier or more difficult to be solved as mentioned by [35]. Similar studies on the effect of different codings for a common problem have been published by [9, 18, 38, 46] and [20].

Coding-based MOS: Case Study

In this section we will analyse one of the most interesting alternatives within MOS framework which is the use of several genotype encodings.

Using MOS techniques, one genotype encoding can be transformed into another, making the translation of a point from one solution space (and fitness landscape) into another solution space.

To generalise, it could be considered that the different offspring techniques use also different genotype encodings. So, let $\mathbf{C}^{(j)}$ be the encoding space produced by the mechanism j.

As different genotype formats are allowed, there must also be different coding functions (for both, a genotype code $code^{(j)}$ and a set of codes $\boldsymbol{code}^{(j)}$):

$$\mathbf{C}^{(j)} \xrightarrow{code^{(j)}} \mathbf{S}$$
$$c \longrightarrow code^{(j)}(c) = s \qquad (4.13)$$

$$\mathcal{P}(\mathbf{C}^{(j)}) \xrightarrow{\boldsymbol{code}^{(j)}} \mathcal{P}(\mathbf{S})$$
$$C \longrightarrow \boldsymbol{code}^{(j)}(C) = S \qquad (4.14)$$

$$\boldsymbol{code}^{(j)}(C) = \{s \in \mathbf{S} / \exists c \in C : s = code^{(j)}(c)\} \qquad (4.15)$$

In MOS, a solution (the phenotype) would have the possibility to participate in multiple genotype recombination mechanisms. If the different mechanisms also use different genotype formats, then, once an individual is created (and evaluated), it could be coded back to take part on different possible genotype formats (and their operators). To manage these transformations, a group of functions is required to transform genotypes between the two different encodings ($trans_{i,j}$).

4 Supercomputer Scheduling with Combined Evolutionary Techniques

$$\begin{array}{ccc} \mathbf{C}^{(i)} & \xrightarrow{trans_{i,j}} & \mathbf{C}^{(j)} \\ \downarrow code^{(i)} & & \downarrow code^{(j)} \\ \mathbf{S} & = & \mathbf{S} \end{array} \qquad (4.16)$$

Unique phenotype encoding : $\quad \forall i,j \quad code^{(i)}(c^{(i)}) = code^{(j)}(trans_{i,j}(c^{(i)}))$ (4.17)

Individual definition, in the case of MOS algorithms, should include information from the phenotype, but also from all the genotype encodings. The individual identification in MOS is $(s, c^{(1)}, c^{(2)}, cdots, c^{(n)}) \in \mathbf{S} \times \mathbf{C}^{(1)} \times \mathbf{C}^{(2)} \times \cdots \times \mathbf{C}^{(n)}$. This tuple should also validate $\forall j : s = code^{(j)}(c^{(j)})$ as well as the Unique Phenotype Encoding restriction (equation 4.17, mentioned above).

Additionally, the previous formalism should be extended to include different offspring sampling functions:

$$\begin{array}{ccc} \mathcal{P}(\mathbf{C}^{(j)}) & \xrightarrow{\mathit{off}^{(j)}} & \mathcal{P}(\mathbf{C}^{(j)}) \\ C_i & \longrightarrow & \mathit{off}^{(j)}(C_i) = C_{i+1} \end{array} \qquad (4.18)$$

In the case of single offspring functions there is a restriction on the size of the offspring production $|\mathit{off}(C_i)| = \sigma$. In the case of multiple offspring functions, this restriction could change dynamically to balance the offspring sampling according to the strategy defined by the algorithm. By this feature, MOS can select among different offspring generation alternatives in a generation-by-generation way. Section 4.5.3 presents how this sampling sizes should be defined.

In MOS, the population merge function $\mathit{comb}^\star$ should also be defined in order to combine multiple offspring populations with the population from the previous generation:

$$\begin{array}{ccc} (\mathcal{P}(\mathbf{S}))^{n+1} & \xrightarrow{\mathit{comb}^\star} & \mathcal{P}(\mathbf{S}) \\ (S_i, O_i^{(1)}, O_i^{(2)}, \cdots, O_i^{(n)}) & \longrightarrow & \mathit{comb}^\star(S_i, O_i^{(1)}, O_i^{(2)}, \cdots, O_i^{(n)}) = S_{i+1} \end{array} \qquad (4.19)$$

$$\begin{array}{rl} \text{Previous population:} & S_i \subset \mathbf{S}/S_i = \mathit{code}(C_i) \\ \text{Offspring population:} & O_i^{(j)} \subset \mathbf{S}/O_i^{(j)} = \mathit{code}(\mathit{off}^{(j)}(C_i)) \end{array} \qquad (4.20)$$

Definition of Offspring Sampling Sizes

The calculation of the amount of new individuals created on each generation, for the n different offspring sampling methods, is obtained using a Participation Function (***PF***). A Participation Function evaluates the quality of the offspring populations generated on each generation and defines the sampling size for the next one.

$$(\mathcal{P}(\mathbf{S}))^n \xrightarrow{\boldsymbol{PF}} \mathbb{N}^n$$
$$(O_i^{(1)}, O_i^{(2)}, \cdots, O_i^{(n)}) \longrightarrow \boldsymbol{PF}(O_i^{(1)}, O_i^{(2)}, \cdots, O_i^{(n)}) = (\sigma_1, \sigma_2, \cdots, \sigma_n) \tag{4.21}$$

$$\text{Offspring sampling size limit}: \quad |\boldsymbol{off}^{(j)}(C_i)| = \sigma_j \tag{4.22}$$

These $\boldsymbol{PF}$ functions evaluate the quality of the offspring populations produced by each mechanism in the generation i and recalculate the amount of individuals that each of these techniques will produce in the next generation.

$$\mathcal{P}(\mathbf{S}) \xrightarrow{qual} \mathbb{R}$$
$$O_i^{(j)} \longrightarrow \boldsymbol{qual}(O_i^{(j)}) = \theta \tag{4.23}$$

Other dynamic adaptive methods in the literature have used the evolutionary pressure to adjust dynamic parameters of the algorithms. In [39] the effective representation or in [45] the recombination and mutation operators are parameterised. This means that the parameter is included in the genome as one additional gene and it is evolved and selected according to the fitness of the individual it belongs to.

To measure the quality of a population several strategies can be used, depending on the characteristic to focus on. One possibility is to consider that one population is better than another if its average fitness value is better than the other's.

$$\boldsymbol{qual}(O_i^{(j)}) = \boldsymbol{AvgFit}(O_i^{(j)}, \gamma) \tag{4.24}$$

Where γ is the top percentage of the population to be considered to calculate the average fitness value.

A special case: two different optimisation techniques

All previous equations deal with the case of n different optimisation techniques. Nevertheless, the experiments performed for this chapter have focused on the case of two competing techniques. Therefore, the next equations will be presented for this particular situation to facilitate their comprehension.

Different functions have already been proposed in other scenarios by [37]. In this contribution the following dynamic function is used:

$$\boldsymbol{PF}_{dy}(O_i^{(1)}, O_i^{(2)}) = \begin{cases} (|O_i^{(1)}| + \delta, |O_i^{(2)}| - \delta) & \text{if } \boldsymbol{qual}(O_i^{(1)}) > \boldsymbol{qual}(O_i^{(2)}1), \\ (|O_i^{(1)}| - \delta, |O_i^{(2)}| + \delta) & \text{otherwise} \end{cases} \tag{4.25}$$

Where δ is a trade-off factor that represents the relative difference between the fitness of the best and the worst offspring populations.

$$\delta = \alpha \frac{\boldsymbol{qual}(\boldsymbol{best}(O_i^{(1)}, O_i^{(2)})) - \boldsymbol{qual}(\boldsymbol{worst}(O_i^{(1)}, O_i^{(2)}))}{\boldsymbol{qual}(\boldsymbol{best}(O_i^{(1)}, O_i^{(2)})) - \beta} |\boldsymbol{worst}(O_i^{(1)}, O_i^{(2)})| \tag{4.26}$$

where α is the factor from the ratio transferred from one offspring to the other (usually 0.05), and β is a reduction factor obtained from the fitness value of the worst individual in the first population.

best and **worst** are functions that compare the quality of two different populations selecting the best (or worst, respectively) according to the specified comparison criterion (maximisation or minimisation), as shown in equation 4.27. **worst** is defined in an analogous way.

$$\mathcal{P}(\mathbf{S}) \times \mathcal{P}(\mathbf{S}) \xrightarrow{best} \mathcal{P}(\mathbf{S}) \\ (O_i^{(1)}, O_i^{(2)}) \longrightarrow \textbf{best}(O_i^{(1)}, O_i^{(2)}) = \overline{O_i} \tag{4.27}$$

$$\textbf{best}(O_i^{(1)}, O_i^{(2)}) = \begin{cases} O_i^{(1)} & \text{if } \textbf{qual}(O_i^{(1)}) > \textbf{qual}(O_i^{(2)}), \\ O_i^{(2)} & \text{otherwise} \end{cases} \tag{4.28}$$

For example, let the sizes of the offspring population of two given techniques, in a certain iteration, be $|O_i^{(1)}| = 89$ and $|O_i^{(2)}| = 111$. If it is considered that the quality of an offspring population is the average fitness of the top 25% of the individuals in this population, it could be for example:

$$\textbf{qual}(O_i^{(1)}) = 0.045 \tag{4.29}$$

$$\textbf{qual}(O_i^{(2)}) = 0.027 \tag{4.30}$$

If the following values are defined: $\alpha = 0.05$ and $\beta = 0.010$ (the best fitness value obtained in the first iteration) the value of δ is computed as:

$$\delta = \alpha \frac{\textbf{qual}(\textbf{best}(O_i^{(1)}, O_i^{(2)})) - \textbf{qual}(\textbf{worst}(O_i^{(1)}, O_i^{(2)}))}{\textbf{qual}(\textbf{best}(O_i^{(1)}, O_i^{(2)})) - \beta} |\textbf{worst}(O_i^{(1)}, O_i^{(2)})| \tag{4.31}$$

$$= 0.05 \frac{0.045 - 0.027}{0.045 - 0.010} \cdot 111 \tag{4.32}$$

$$= 0.0257 \cdot 111 = 2.85 \simeq 3 \tag{4.33}$$

Then, the participation function for the next iteration is:

$$\textbf{PF}_{dy}(O_i^{(1)}, O_i^{(2)}) = (|O_i^{(1)}| + \delta, |O_i^{(2)}| - \delta) \tag{4.34}$$

$$= (89 + 3, 111 - 3) \tag{4.35}$$

4.6 Experiments

In this section, we describe the experimental scenario and a comparison of the results obtained by classical methods (FCFS, Backfilling, Backfilling with reservations, SJF and LJF), single genetic algorithms and a hybrid evolutionary technique (MOS).

This experimentation tries to optimise the scheduling policy of the *Magerit* cluster, located at the CeSViMa (*Centro de Supercomputación y Visualización de Madrid*)[6]. This system consists of 1080 eServer BladeCenter JS20, each of them with 2 Power970 2.2 GHz processors and 4 GB of shared RAM. All the results presented hereafter have been obtained for this machine configuration.

4.6.1 Evolutionary Techniques for Supercomputer Scheduling

Solutions generated by the algorithms represent ordered sequences of jobs to be dispatched by the supercomputer. In this sense, these solutions are actually permutations of the jobs. Thus, any of the possible codings that represent permutations could be applied.

Once the supercomputer receives the queue of jobs to be executed it uses a deterministic policy to process it. If the supercomputer has no available resources to run the next job from the queue, it will wait until the end of a running job. When one job terminates all the assigned resources are released. Therefore the waiting job has a new chance to be able to run.

The following scenario is played on the Magerit system. In this scenario, shown in Fig. 4.2, the first two jobs in the queue are dispatched to the supercomputer simultaneously (see Fig. 2(a)). Job3 can not be executed because there are not enough available processors, so it must wait until both jobs have finished (see Fig. 2(b)). Once the number of processors required by Job3 are available it is submitted to the computer (see Fig. 2(c)). At this moment Job4 can not be sent to the supercomputer, in this case due to the lack of available memory. This job will run once the previous job has released its resources (see 2(d)).

4.6.2 First Experimental Scenario

We have first tested our approach with three datasets with sizes ranging from sixty to one hundred and twenty jobs. Each job is described by the required amount of CPUs, memory and execution time. These datasets can be freely downloaded from our homepage[7].

The experiments were executed on the aforementioned Magerit system, making use of 2 of its 2160 processors. They have been coded using a parallel asynchronous genetic algorithm implemented in GAEDALib coded by [12] with the configuration described in Table 4.1.

For this experiment, MOS algorithm uses two different codings: path and order representation. The first one uses integer numbers to represent the permutation of the execution order of jobs. The second one uses real numbers to code the individuals. The real values are sorted to obtain the execution order of the jobs (which are implicitly coded as the position in the real vector).

---

[6] http://www.cesvima.upm.es
[7] http://laurel.datsi.fi.upm.es/research/mos/scs

4 Supercomputer Scheduling with Combined Evolutionary Techniques   113

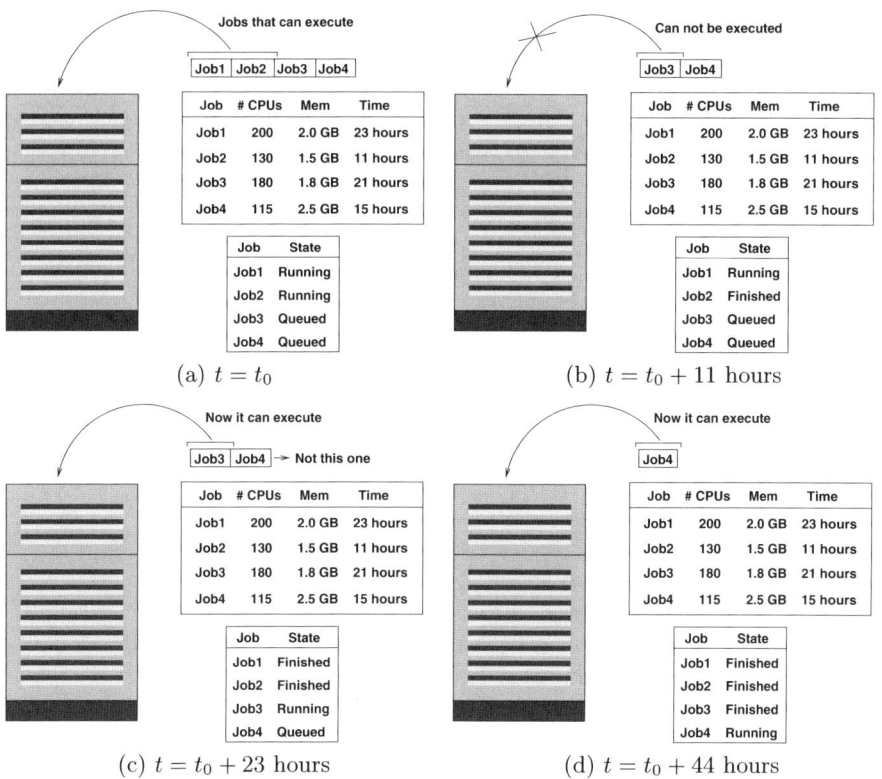

Fig. 4.2. Scheduler description

This approach faces the problem of mixing individuals of both types. This issue has been overcome by implementing two conversion functions for converting from integer to real coding and vice versa. The first function sorts the real vector contained in the genome and takes as integer value the position of each gene in the ordered vector. The second function simply generates for each gene a random real number within an interval bounded by two values proportional to the integer value of the gene.

Different genetic operators have been proposed for the codings used in this experiment. For the real-valued coding we have used the classical one-point crossover and uniform mutator operators [15]. For the integer-valued coding we selected the Order Crossover operator [10] and the Exchange Mutation operator [33]. Another crossover operator was tested for the integer coding (Cycle Cross operator [33]) but with poorer results, as previously stated [28].

Finally, the fitness of each individual is calculated as the percentage of processor time the system is busy (% of CPU usage):

Table 4.1. Experimental scenario

(a) GA configuration

(Global) Pop. size	100
Termination	Pop. convergence
Convergence %	99 %
Individuals selection	Roulette wheel
Crossover %	90 %
Mutation %	1 %

(b) Parallel configuration

Paradigm	islands model
Model	asynchronous
Topology	mesh
Migration rate	10 gens.
Migration pop.	Top 20 %
Nodes	2

$$fitness = \frac{total\_processor\_time}{scheduled\_time * cpus} \quad (4.36)$$

In addition, in order to evaluate the performance of MOS approach, the same datasets would be scheduled using some classical non-combinatorial techniques (described in subsection 4.4.1). These techniques are:

❶ *FCFS*,
❷ *Backfilling* without reservations,
❸ *Backfilling* with one reservation,
❹ two variants of *SJF* (shortest job first), considering the required amount of processors and the expected execution time as the criterion to sort the queue of jobs, and
❺ two variants of *LJF* (longest job first), which are similar to *SJF* but in reverse order, both in number of processors and in the expected execution time.

4.6.3 Results and Discussion of First Experiment

A summary of the obtained results can be found in Table 4.2. We can see that MOS performs as well as the best other technique for the first dataset and that it clearly outperforms classical approaches solving the two bigger datasets. All the problems have been executed ten times and the results are the average of all the executions, except for the deterministic methods (classical scheduling models) that were executed only once.

The result on the first dataset is the optimal overall value. The hybrid evolutionary technique, as well as three of the other techniques, are able to reach this value. This circumstance does not happen on the more complex scheduling problems where hybrid evolutionary techniques are able to improve the results of other techniques. These problems are too difficult for most of the other techniques.

It is also interesting to check that the standard deviation of the evolutionary techniques, although they are heuristic/stochastic methods, is very low. This makes this approach stable and reliable for real-world applications.

The convergence to solutions with the same fitness does not mean that the actual same job order is obtained. Many different job combinations would lead

Table 4.2. Results summary of first experiment

	60 Jobs	80 Jobs	120 Jobs
MOS	0.5321 ± 0.0000	0.9821 ± 0.0318	1 ± 0.0000
FCFS	0.3674	0.6499	0.7068
Backfilling	0.5321	0.7029	0.8065
Backfilling Res.	0.3582	0.7398	0.7503
SJF Procs.	0.5321	0.6564	0.7022
LJF Procs.	0.4050	0.6305	0.6957
SJF Time	0.4349	0.6766	0.6483
LJF Time	0.5321	0.7779	0.9047

to a similar, or even equal, performance (CPU usage). This feature is explained because it is more important to keep groups of jobs together, that are able to fit with the highest CPU usage in the parallel system. These groups of jobs may be swapped among them. Also the jobs belonging to one of these groups may be swapped within the group context.

These two kinds of organisations are preserved by the two evolutionary techniques combined by MOS. Integer-based encoding crossover operator keeps large portions of the parents' orders (preserving job groups). Real-based mutation would move jobs inside the same job group.

Focusing on the classical techniques, we can see that *LJF*, when dealing with expected execution time, behaves the best compared with the other non-combinatorial methods. This performance is due to the fact that the longer the job is the earlier it is scheduled. This makes shortest jobs be submitted at the end of the execution time, filling the gaps in the last part of the execution.

It can be noted that the hardness of the scheduling problem is not directly proportional to the number of jobs. Although the number of possible permutations is bigger, there could be more equivalent solutions with the best performance. The number of best solutions (with different ordering) depends on other characteristics of the problem, rather than the number of jobs. Sometimes, with a reduced number of jobs, the possible combinations are so limited that there is no job ordering schema with more than 60% of the CPU usage, similar to the figures obtained by classical methods.

4.6.4 Second Experimental Scenario

For this second experiment we have used a bigger dataset of 248 jobs, described in the same way (#CPUs, memory and execution time). The same configuration for the genetic algorithm has been used (see Table 4.1).

In this case, instead of using two different codings we have focused our attention at two different mutation operators for the same genetic representation (integer-valued representation). These mutation operators are the aforementioned Exchange Mutation operator [33] and the Simple Inversion Mutation operator [15].

Again, the performance of the hybrid genetic algorithm is compared against the non-combinatorial techniques but also with the single genetic algorithm each of them using a different mutation operator.

Finally, the fitness of each individual is calculated in the same way as it was done in the previous experiment.

4.6.5 Results and Discussion of Second Experiment

The proposed dataset, even if it doubles in size the biggest dataset used in the first experiment, presents the same properties as those seen before, i.e., the hardness of the problem does not increase proportionally to the problem size. These problems can be solved by multiple different solutions with the same fitness. Nevertheless, we can observe that even under those circumstances, MOS is able to obtain better results in terms of average fitness and standard deviation than any other technique. The whole result list is provided in Table 4.3.

Table 4.3. Results summary of second experiment

	248 Jobs
MOS	0.9632 ± 0.0002
GA-SIM	0.9237 ± 0.0137
GA-EM	0.9451 ± 0.0015
FCFS	0.7399
Backfilling	0.7819
Backfilling Res.	0.7783
SJF Procs.	0.7861
LJF Procs.	0.7596
SJF Time	0.7529
LJF Time	0.8626

Finally, the non-parametric Wilcoxon test was applied to MOS and the single genetic algorithm (using different mutation operators) to test the null hypothesis of both algorithms (MOS and single GAs) having the same distribution. The p-values obtained were $p = 0.0002$ and $p = 0.02$ for GA-SIM and GA-EM respectively, which let us reject the null hypothesis and state that there is statistical significance. Thus, MOS outperforms each single genetic algorithms.

4.7 Conclusions

In this chapter a new definition for a scheduling problem has been presented. Supercomputer scheduling (SCS) has been solved as a particular case of permutation ordering problems. To illustrate this problem the sample scenario of a supercomputer called Magerit hosted at the CeSViMa (*Centro de Supercomputación y Visualización de Madrid*)[8] has been used.

---
[8] http://www.cesvima.upm.es

Alternative coding models and genetic operators have been proposed for a large number of optimisation problems, but in all the cases the studies were based on comparing the whole search process for two or more encoding formats (or recombination operators). We propose a dynamic alternative to combine the benefits of different encodings. This approach has been developed under the formalism of Multiple Offspring Sampling (MOS) also introduced in this contribution.

MOS approach is, somehow, close to the work started by [39], but it differs in the following key aspects:

- This work considers permutation-based problems extending any preliminary work of benchmark functions retrieved from the literature.
- In MOS approach, the creation of an offspring from a given population is the central element when multiple alternatives are handled. This offspring creation includes not only coding formats or operators, but also selection operators and other evolutionary aspects.
- MOS would perform the selection of the most appropriate features using different criteria. Evolutionary pressure is one option, but other alternatives to evaluate the quality of populations might be more interesting. These new alternatives might consider diversity, coverage and overall fitness of the population. The reason to use the selection pressure to select this parameter is not clear. The GA evolves to optimise the fitness function. Thus, coding additional information in the individual's genotype represents that the algorithm is able to optimise both objectives and they are not contradictory. There is no clear assumption under which this might be always true, and in some cases, it would mislead the optimisation process. MOS is open to define both adaptive techniques (based on selection pressure or using any population quality measure).
- MOS also introduces a formal framework to represent multiple individual creation techniques. This mathematical model, together with the proposed taxonomy, open very interesting issues in a future work. Many of the previous works on dynamic and self-adaptive evolutionary techniques can be translated into MOS formalism.

In addition, experimental results have been tested for statistical significance, including a more detailed discussion about the performance of the combined techniques.

The work presented on this chapter has dealt with CPU usage, which is a key factor in the resource management of expensive supercomputing facilities. However, the user has a different perception of the system: for him, it is more important the waiting time of the submitted jobs. More complex approaches would take into account both aspects (or even more). In these cases, classical heuristic methods are not able to adapt properly. Instead, evolutionary techniques are able to deal with them, once a well-balanced fitness function is provided to evaluate all the desired aspects of the scheduling.

This chapter shows that combined meta-heuristic methods are able to outperform traditional approaches. MOS provides also the possibility of combining different meta-heuristic methods and selecting the most appropriate to optimise the given problem. This abstraction is quite interesting when the behaviour of different techniques is previously unknown.

References

1. Agarwal, A., Colak, S., Eryarsoy, E.: Improvement heuristic for the flow-shop scheduling problem: An adaptive-learning approach. European Journal of Operational Research 169, 801–815 (2006)
2. Auyeung, A., Gondra, I., Dai, H.K.: Multi-heuristic list scheduling genetic algorithm for task scheduling. In: Proceedings of the 18th Annual ACM Symposium on Applied Computing, pp. 721–724. ACM Press, New York (2003)
3. Balas, E., Vazacopoulos, A.: Guided local search with shifting bottleneck for job-shop scheduling. Management Science 44(2), 262–275 (1998)
4. Bertel, S., Billaut, J.C.: A genetic algorithm for an industrial multiprocessor flow shop scheduling problem with recirculation. European Journal of Operational Research 159(3), 651–662 (2004)
5. Bierwirth, C.: A generalized permutation approach to job shop scheduling with genetic algorithms. OR Spectrum 17, 87–92 (1995)
6. Bierwirth, C., Mattfeld, D.C.: Production scheduling and reschedunling with genetic algorithms. Evolutinary Computation 7(1), 1–17 (1999)
7. Bryant, K.: Genetic algorithms and the traveling salesman problem. Master's thesis, Harvey Mudd College, Department of Mathematics (December 2000)
8. Chiang, S.-H., Arpaci-Dusseau, A., Vernon, M.K.: The impact of more accurate requested runtimes on production job scheduling performance. In: Feitelson, D.G., Rudolph, L., Schwiegelshohn, U. (eds.) JSSPP 2002. LNCS, vol. 2537, pp. 103–127. Springer, Heidelberg (2002)
9. Coli, M., Palazzari, P.: Searching for the optimal coding in genetic algorithms. In: Proceedings of the IEEE International Conference on Evolutionary Computation, Perth, WA, Australia, vol. 1, pp. 92–96 (1995)
10. Davis, L.: Applying adaptive algorithms to epistatic domains. In: Proceedings of the 9th IJCAI, pp. 162–164 (1985)
11. Davis, L.: Job shop scheduling with genetic algorithms. In: Grefenstette, J.J. (ed.) Proceedings of the first International Conference on Genetic Algorithms and their Applications, pp. 136–140 (1985)
12. Díaz, P.: Diseño e implementación de una librería de algoritmos evolutivos paralelos. Master's thesis, Facultad de Informática, Universidad Politécnica de Madrid (November 2005)
13. Feitelson, D., Rudolph, L., Schwiegelshohn, U.: Parallel job scheduling – a status report. In: Proceedings of the 10th Workshop on Job Scheduling Strategies for Parallel Processing, New York, NY, pp. 1–16 (2004)
14. Grabowski, J., Wodecki, M.: A very fast tabu search algorithm for the permutation flow shop problem with makespan criterion. Technical Report PRE 64/2002, Institute of Engineering Cybernetics, Technical University of Wroclaw, Wroclaw, Poland (2002)
15. Holland, J.H.: Adaptation in natural and artificial systems. University of Michigan Press (1975)

16. Jette, M., Grondona, M.: SLURM: Simple linux utility for resource management. Technical Report UCRL-MA-147996-REV, Lawrence Livermore National Laboratory (2002)
17. Jones, T., Forrest, S.: Fitness distance correlation as a measure of problem difficulty for genetic algorithms. In: Eshelman, L. (ed.) Proceedings of the Sixth International Conference on Genetic Algorithms, pp. 184–192. Morgan Kaufmann, San Francisco (1995)
18. Jonikow, C.Z., Michalewicz, Z.: An experimental comparison of binary and floating point representations in genetic algorithms. In: International Conference on Genetic Algorithms, pp. 31–38 (1991)
19. Kannan, S., Roberts, M., Mayers, P., Brelsford, D., Skovira, J.F.: Workload Management with LoadLeveler. IBM Red Books (2001)
20. Korkmaz, E.E., Du, J., Alhajj, R., Barker, K.: Combining advantages of new chromosome representation scheme and multi-objective genetic algorithms for better clustering. Intelligent Data Analysis Journal 10(2), 163–182 (2006)
21. Kwok, Y., Ahmad, I.: Static Scheduling Algorithms for Allocating Directed Graphs to Multiprocessors. ACM Computing Surveys 31(4), 406–471 (1999)
22. Larrañaga, P., Kuijpers, C.M.H., Murga, R.H., Inza, I., Dizdarevic, S.: Genetic algorithms for the travelling salesman problem: A review of representations and operators. Articial Intelligence Review 13, 129–170 (1999)
23. Larrañaga, P., Lozano, J.A.: Estimation of Distribution Algorithms. A New Tool for Evolutionary Computation. Kluwer Academic Publishers, Dordrecht (2002)
24. Lee, Y.H., Chen, C.: A modified genetic algorithm for task scheduling in multiprocessor systems. In: The Ninth Workshop on Compiler Techniques for High-performance Computing (2003)
25. Lifka, D.: The ANL/IBM SP scheduling system. In: Job Scheduling Strategies for Parallel Processing. LNCS, pp. 295–303. Springer, Heidelberg (1995)
26. Mattfeld, D.C.: Evolutionary Search and the Job Shop. Investigations on GeneticAlgorithms for Production Scheduling. Springer, Heidelberg (1995)
27. Mattfeld, D.C.: Evolutionary Search and the Job Shop. Production and Logistics. Physica-Verlag, Heidelberg (1996)
28. Mernik, M., Crepinsek, M., Zumer, V.: A metaevolutionary approach in searching of the best combination of crossover operators for the tsp. In: Proceedings of the IASTED ICNN, Pittsburg, Pennsylvania, pp. 32–36. IASTED/ACTA Press (2000)
29. Mladenović, N., Hansen, P.: Variable neighborhood search. Comps. in Opns. Res. 24, 1097–1100 (1997)
30. M'ualem, A.W., Feitelson, D.G.: Utilization, predictability, workloads, and user runtime estimates in scheduling the IBM SP2 with backfilling. IEEE Trans. Parallel & Distributed Syst. 12(6), 529–543 (2001)
31. Nowicki, E., Smutnicki, C.: A fast taboo search algorithm for the job shop problem. Management Science 42(6), 797–813 (1996)
32. Ogbu, F.A., Smith, D.K.: The application of the simulated annealing algorithm to the solution of the n/m/Cmax flow shop problem. Computers and Operations Research 17(3), 243–253 (1990)
33. Oliver, I.M., Smith, D.J., Holland, J.R.C.: A study of permutation crossover operators on the traveling salesman problem. In: Proceedings of the Second International Conference on Genetic Algorithms on Genetic algorithms and their application, pp. 224–230. Lawrence Erlbaum Associates, Inc., Mahwah (1987)
34. Pisinger, D.: Algorithms for Knapsack Problems. PhD thesis, Department of Computer Science, University of Aarhus DenmarkDIKU (1995)

35. Schaffer, J.D., Caruna, R.A.: Representation of hidden bias: Gray vs. binary coding for genetic algorithms. In: Fifth International Conference on Machine Learning, pp. 152–161 (1988)
36. Reeves, C.R., Yamada, T.: Genetic algorithms, path relinking, and flow shop problem. Evolutionary Computation 6, 45–60 (1998)
37. Robles, V., Peña, J.M., Larrañaga, P., Pérez, M.S., Herves, V.: Towards a New Evolutionary Computation. Advances in Estimation of Distribution Algorithms. In: GAEDA: A New Hybrid Cooperative Search Evolutionary Algorithm. Studies in Fuzziness and Soft Computing, vol. 192, pp. 187–220. Springer, Heidelberg (2006)
38. Salomon, R.: The influence of different coding schemes on the computational complexity of genetic algorithms in function optimization. In: Ebeling, W., Rechenberg, I., Voigt, H.-M., Schwefel, H.-P. (eds.) PPSN 1996. LNCS, vol. 1141, pp. 227–235. Springer, Heidelberg (1996)
39. Schnier, T., Yao, X.: Using multiple representations in evolutionary algorithms. In: Proceedings of the 2000 Congress on Evolutionary Computation, pp. 479–486. IEEE Press, Los Alamitos (2000)
40. Schwiegelshohn, U., Yahyapour, R.: Analysis of first-come-first-serve parallel job scheduling. In: Proceedings of the 9th SIAM Symposium on Discrete Algorithms, pp. 629–638 (1998)
41. Sengoku, H., Yoshihara, I.: A fast tsp solver using ga on java. In: Proceedings of the 3rd International Symposium on Artificial Life and Robotics, AROB 1998, vol. 1 (1998)
42. Shenassa, M.H., Mahmoodi, M.: A novel intelligent method for task scheduling in multiprocessor systems using genetic algorithmg. Journal of the Franklin Institute (in press, 2006)
43. Shi, L., Pan, Y.: An efficient search method for job-shop scheduling problems. IEEE Transactions on Automation Science and Engineering 2(1), 73–77 (2005)
44. ElGhazawi, T.A., Alaoui, S.M., Frieder, O.: A parallel genetic algorithm for task mapping on parallel machines. In: Rolim, J.D.P. (ed.) IPPS-WS 1999 and SPDP-WS 1999. LNCS, vol. 1586, pp. 201–209. Springer, Heidelberg (1999)
45. Smith, J.E., Fogarty, T.C.: Adaptative parametrised evolutionary algorithms: Self adaptive recombination and mutation in a genetic algorithm. In: Ebeling, W., Rechenberg, I., Voigt, H.-M., Schwefel, H.-P. (eds.) PPSN 1996. LNCS, vol. 1141, pp. 441–451. Springer, Heidelberg (1996)
46. Tamaki, H., Kita, H., Shimizu, N., Maekawa, K., Nishikawa, Y.: A comparison study of genetic codings for the travelling sallesman problem. In: Proceedings of the First IEEE Conference on Computational Intelligence, vol. 1, pp. 1–6 (1994)
47. Watson, J.-P., Beck, J.C., Howe, A.E., Whitley, L.D.: Problem difficulty for tabu search in job-shop scheduling 143, 189–217 (2003)
48. Wu, A.S., Yu, H., Jin, S., Lin, K., Schiavone, G.: An incremental genetic algorithm approach to multiprocessor scheduling. IEEE Transactions on Parallel and Distributed Systems 15(9), 824–832 (2004)
49. Yamada, T., Nakano, R.: A genetic algorithm with multi-step crossover for job-shopscheduling problems. In: First International Conference on Genetic Algorithms in Engineering Systems: Innovations and Applications, pp. 146–151. IEE Press (1995)
50. Ying, K.C., Liao, C.J.: An ant-colony system for permutation flow shop sequencing. Computers and Operations Research 31(5), 791–801 (2004)

5

Adapting Iterative-Improvement Heuristics for Scheduling File-Sharing Tasks on Heterogeneous Platforms

Kamer Kaya[1], Bora Uçar[2], and Cevdet Aykanat[3]

[1] Department of Computer Engineering, Bilkent University, Ankara, Turkey
kamer@cs.bilkent.edu.tr
[2] CERFACS, 42 Av. Gaspard Coriolis, Toulouse, Cedex 1, 31057 France
ubora@cerfacs.fr
[3] Department of Computer Engineering, Bilkent University, Ankara, Turkey
aykanat@cs.bilkent.edu.tr

Summary. We consider the problem of scheduling an application on a computing system consisting of heterogeneous processors and one or more file repositories. The application consists of a large number of file-sharing, otherwise independent tasks. The files initially reside on the repositories. The interconnection network is heterogeneous. We focus on two disjoint problem cases. In the first case, there is only one file repository which is called as the master processor. In the second case, there are two or more repositories, each holding a distinct set of files. The problem is to assign the tasks to the processors, to schedule the file transfers from the repositories, and to order the executions of tasks on each processor in such a way that the turnaround time is minimized.

This chapter surveys several solution techniques; but the stress is on our two recent works [22, 23]. At the first glance, iterative-improvement-based heuristics do not seem to be suitable for the aforementioned scheduling problems. This is because their immediate application suggests iteratively improving a complete schedule, and hence building and exploring a complex neighborhood around the current schedule. Such complex neighborhood structures usually render the heuristics time-consuming and make them stuck to a part of the search space. However, in both of the our recent works, we show that these issues can be solved by using a three-phase approach: initial task assignment, refinement, and execution ordering. The main thrust of these two works is that iterative-improve-based heuristics can efficiently deliver effective solutions, implying that iterative-improve-based heuristics can provide highly competitive solutions to the similar scheduling problems.

Keywords: Scheduling File-Sharing Tasks, Iterative-Improvement Heuristics, Heterogeneous Platforms, Neighborhood exploration.

5.1 Introduction

Task scheduling in heterogeneous systems is an important problem for today's computational Grid environments [9], as heterogeneous systems become more and more prevalent. There are important Grid applications [10] which are typically composed of a large number of independent but file-sharing tasks. Therefore, the problem of scheduling a large number of independent but file-sharing tasks on heterogeneous platforms has recently attracted much attention, see for example [12, 13, 17, 18, 19, 20, 22, 23, 25] and the references therein. By file sharing, we mean that a file may be requested by a number of tasks. The computing system consists of heterogeneous processors and one or more repositories that store input files. The files are not replicated, i.e., if there are two or more repositories, each one stores a distinct set of files. The repositories are decoupled from the processors. The processors and the repositories are connected through a heterogeneous interconnection network. The problem is to schedule the task executions on processors and to schedule the input file transfers in such a way that the turnaround time, i.e., the completion time of the application is minimized.

Once the tasks are assigned to the processors, the files should be transferred from the repositories to the processors. A task execution can start only after its input files are delivered to the respective processor. Once a file is transferred to a processor, it can be used by all tasks assigned to the same processor without any additional cost. Since the interconnection network is heterogeneous, the costs of transferring a certain file between different source and destination pairs are not necessarily equal. We assume the one-port communication model in which a data repository or a processor can, respectively, send or receive at most one file at a given time. In order to minimize the turnaround time, the scheduler must decide the task-to-processor assignment, the order of file transfers, and the order of task executions on each processor.

Task scheduling for heterogeneous environments is harder than task scheduling for homogeneous ones, since in a heterogeneous environment, different tasks which need the same files might have different favorite processors. Therefore, it may not be feasible to assign them to the same processor on the grounds of efficient resource utilization. Even if such tasks may have the same favorite processor, that processor might have relatively low bandwidth so that assigning these tasks to that processor can increase the file transfer time while decreasing the file transfer amount.

The application and computing models and the objective function which characterize the scheduling problems at hand are introduced formally in Sect. 5.2. In Sect. 5.3, we discuss the single repository case and review the heuristics from the works [12, 13, 17, 20, 22]. Then, in Sect. 5.4, we discuss the multiple repository case and review the heuristics from the works [18, 19, 23, 25].

5.2 Framework

5.2.1 Application Model

The application is defined as a two tuple $\mathcal{A} = (\mathcal{T}, \mathcal{F})$, where $\mathcal{T} = \{1, 2, \ldots, T\}$ denotes the set of T tasks, and $\mathcal{F} = \{1, 2, \ldots, F\}$ denotes the set of F input files. Each task t depends on a subset of files denoted by files(t); these files should be delivered to the processor that will execute the task t. We extend the operator

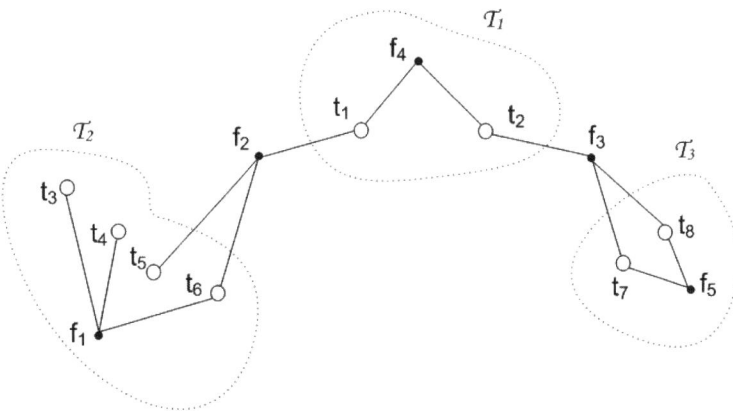

Fig. 5.1. Hypergraph model $\mathcal{H}_\mathcal{A} = (\mathcal{T},\mathcal{F})$ for an application with a set of 8 tasks $\mathcal{T} = \{1,2,\ldots,8\}$ and a set of 5 files $\mathcal{F} = \{1,2,\ldots,5\}$. Vertices are shown with empty circles and correspond to the tasks; nets are shown with filled circles and correspond to the files. File requests are shown with lines connecting vertices and nets. For example, task t_6 needs files f_1 and f_2 and hence vertex t_6 is in the nets f_1 and f_2. A 3-way partition on the vertices of the hypergraph is shown with dashed curves encompassing the vertices.

files($\cdot$) to a subset of tasks $\mathcal{S} \subseteq \mathcal{T}$ such that files($\mathcal{S}$) = $\bigcup_{t \in \mathcal{S}}$ files(t) denotes the set of files that the set $\mathcal{S}$ of tasks depend on. Apart from sharing the input files, there are no dependencies and interactions among the tasks. The size of a file f is denoted by $w(f)$. We extend the operator $w(\cdot)$ to a subset $\mathcal{E} \subseteq \mathcal{F}$ of files such that $w(\mathcal{E})$ denotes the total size of the files in $\mathcal{E}$, i.e., $w(\mathcal{E}) = \sum_{f \in \mathcal{E}} w(f)$. We use $|\mathcal{A}|$ to denote the total number of file requests in the application, i.e., $|\mathcal{A}| = \sum_{t \in \mathcal{T}} |\text{files}(t)|$.

It seems natural to use a hypergraph $\mathcal{H}_\mathcal{A} = (\mathcal{T},\mathcal{F})$ to model the application $\mathcal{A} = (\mathcal{T},\mathcal{F})$, see [22, 23]. Recall that a hypergraph is defined as a set of vertices and a set of hyperedges (nets) each of which contains a subset of vertices [8]. We use $\mathcal{T}$ and $\mathcal{F}$ to denote, respectively, the vertex and net sets of the hypergraph. In this setting, the net corresponding to the file f contains the vertices that correspond to the tasks depending on f. Fig. 5.1 contains an example hypergraph model.

5.2.2 Computing Model

The tasks are to be executed on a heterogeneous system consisting of a set $\mathcal{P} = \{1,2,\ldots,P\}$ of P computing resources, and a set $\mathcal{R} = \{1,2,\ldots,R\}$ of R repositories. Each computing resource can be any computing system ranging from a single processor workstation to a parallel computer. Throughout this chapter we use "processor" to refer to any type of computing resource. The set of files stored on a repository r is denoted as $\mathcal{F}(r)$. We assume that the files are not duplicated, i.e., $\mathcal{F}(r) \cap \mathcal{F}(s) = \emptyset$ for distinct repositories r and s. We use store(f) to denote the repository which holds the file f.

We use $\Pi = \{\mathcal{T}_1, \mathcal{T}_2, \ldots, \mathcal{T}_P\}$ to denote a partition on the vertices of the hypergraph $\mathcal{H}_\mathcal{A}$ and hence an assignment of the tasks to the processors. In other words, we denote the set of tasks assigned to processor p as $\mathcal{T}_p$. Given a task assignment, we use Λ_f to denote the set of processors to which the file f is to be transfered, i.e., $\Lambda_f = \{p \mid f \in \text{files}(\mathcal{T}_p)\}$. The three dashed curves encompassing the vertices in Fig. 5.1 show a partition on the vertices of the hypergraph, and hence an assignment of tasks to processors. For example, the tasks t_1 and t_2 are assigned to the processor 1 since the vertices t_1 and t_2 are in $\mathcal{T}_1$.

The authors of [12,13,17,20,22,23] assume the one-port communication model for the file transfers from the repositories to the processors. In this model, a processor can receive at most one file, and a repository can send at most one file at a given time. This model is deemed to be realistic [5,7,30] and it is prevalent in the scheduling for Grid computing literature, however, alternatives exist (see [4,11]). Task executions and file transfers can overlap at a processor. That is, a processor can execute a task while it is downloading a file for other tasks. The file transfer operations take place only between a repository and a processor. The congestion in the communication network during the file transfers is ignored. In other words, each processor is assumed to be connected to all repositories through direct communication links. Note that the resulting topology is a complete bipartite graph ($K_{P \times R}$). Computing platforms of this topology are called heterogeneous fork-graphs [17,20] when $R = 1$. Such complete graph models are used to abstract wide-area networking infrastructures [11]. The network heterogeneity is modeled by assigning different bandwidth values to the links between the repositories and the processors. We use b_{rp} to represent the bandwidth from the repository r to the processor p. The heuristics in the literature generally use

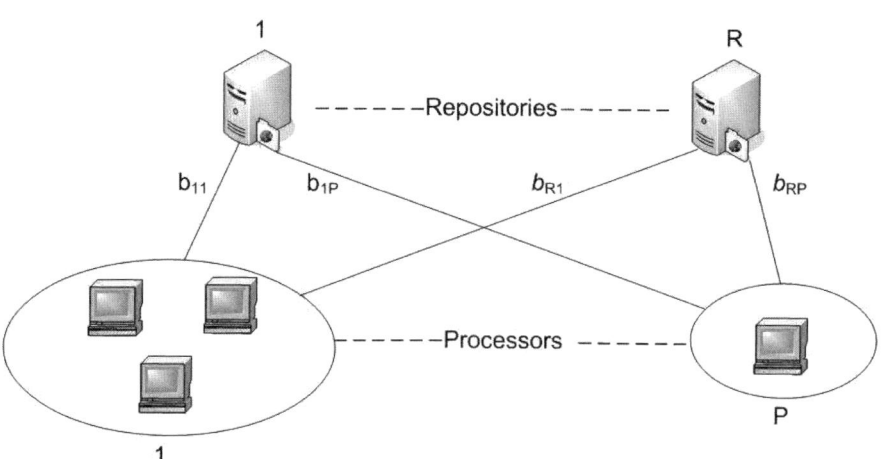

Fig. 5.2. Computing system

the linear cost model [6,11] for file transfers, i.e., transferring the file f from the repository r to the processor p takes $w(f)/b_{rp}$ time units. Fig. 5.2 displays the essential properties of the computing system described.

The task and processor heterogeneity are modeled by incorporating different execution costs for each task on different processors. The execution-time values of the tasks are stored in a $T \times P$ expected-time-to-compute (ETC) matrix. We use x_{tp} to denote the execution time of the task t on the processor p. The ETC matrices are classified into two categories [1]. In the *consistent* ETC matrices, there is a special structure which implies that if a machine has a lower execution time than another machine for some task, then the same is true for the other tasks. The *inconsistent* ETC matrices have no such special structure. In general, the inconsistent ETC matrices are more realistic for heterogeneous computing environments, since they can model a variety of computing systems and applications that arise in Grid environments.

5.2.3 Objective Function

The cost of a schedule is the turnaround time, i.e., the length of the time interval whose start and end points are defined by the start of the first file transfer operation and the completion of the last task execution, respectively. Therefore, the objective of the scheduling problem is to assign the tasks to processors, to determine the order in which the files are transfered from the repositories to the processors, and to determine the task execution order on each processor in order to minimize the turnaround time. Scheduling file-sharing tasks on heterogeneous systems with $R = 1$ repository is NP complete [17]. The NP completeness of the multiple repositories case, i.e., $R > 1$ case, follows easily.

5.3 Scheduling File-Sharing Tasks with Single Repository

In this section, we survey the heuristics proposed for the scheduling problem on heterogeneous systems with $R = 1$ repository, e.g., heterogeneous master-slave environments where the master processor stores all files. This framework has been studied in [10,12,13,17,20,22] for adaptive scheduling of parameter-sweep-like applications in Grid environments. Such applications arise in the *Application Level Scheduling* (AppLeS) project [10].

For the single-repository case, Casanova et al. [12,13] extend three heuristics, namely *MinMin*, *MaxMin* and *Sufferage*, which are initially proposed in [28] for scheduling independent tasks. They use these extended heuristics in the *AppLeS Parameter Sweep Template* (APST) project [10]. They also proposed a new heuristic *XSufferage* exclusively for APST. After this work, Giersch et al. [17,20] proposed several different heuristics which reduce the time complexity while preserving the quality of schedules.

The heuristics in [12,13,17,20] are based on the greedy choices that depend on the momentary completion time values of tasks. Kaya and Aykanat claim that this greedy decision criterion cannot use the file sharing information effectively, since the completion time values are not sufficient to extract the global view of the interaction among the tasks [22]. Instead of a direct construction of schedules, Kaya and Aykanat propose a three-phase scheduling approach which involves initial task assignment, refinement and execution ordering phases.

Kaya and Aykanat argue in [22] that an iterative-improvement-based method which uses task reassignments to improve the actual length of the schedule, i.e., the turnaround time, have a global perturbation on the given schedule. However, the effectiveness and efficiency of the iterative-improvement-based heuristics, which are widely and successfully used for hypergraph partitioning, depend on the perturbations being local [2]. When the perturbations are local, the objective functions become *smooth* over the search space, and the iterative-improvement-based heuristics explore a relatively large part of the search space in relatively small time.

In the refinement phase of the proposed three phase approach, Kaya and Aykanat use two novel smooth objective functions in a hypergraph-partitioning-like formulation to refine task-to-processor assignments. The first objective function represents an upper bound while the second one represents a lower bound for the turnaround time of a schedule by considering only the task-to-processor assignments. The first and the second objective functions relate, respectively, to a pessimistic and an optimistic view of the execution time of an application. In the rest of this section, we will investigate the heuristics in detail.

The notation described in Sect. 5.2 is slightly modified for the master-slave case. In this section, we will omit the notation for the repositories since in this framework there is a single repository. As an example, the bandwidth of a processor p will be denoted as b_p instead of b_{rp}. Similarly, for a file f the notation store(f) is not used.

5.3.1 Greedy Constructive Scheduling Heuristics

Algorithm 5.1 shows the structure of the heuristics used by Casanova et al. [12, 13]. In Alg. 5.1, the completion time $CT(t,p)$ of task t on processor p is computed

---

Algorithm 5.1. Structure of heuristics by Casanova et al. [12, 13]

1: **while** there remains a task to schedule **do**
2:   **for** each unscheduled task t **do**
3:     **for** each processor p **do**
4:       Evaluate completion time $CT(t,p)$ of t on p
5:     **end for**
6:     Evaluate schedule cost $g(CT(t,p_1), \ldots, CT(t,p_P))$ for t
7:   **end for**
8:   Choose task t_b with the "best" schedule cost
9:   Pick the best processor p_b for t_b with min. completion time
10:   Schedule t_b on p_b and its file transfers
11:   Mark t_b as scheduled
12: **end while**

by taking the previously scheduled tasks into account. That is, the file transfers for unscheduled tasks cannot be initialized before the file transfers for scheduled tasks, and the executions of unscheduled tasks on a candidate processor cannot be initialized before the completion of the scheduled tasks on the same processor. The scheduling objective function g and the meaning of the "best" characterize these heuristics as shown in Table 5.1. As seen in Alg. 5.1, computing the completion times for all task-processor pairs takes $O(TP + P|\mathcal{A}|)$ time for each scheduling decision. As this decision is made once for each task, the total time complexity of these heuristics is $O(T^2P + TP|\mathcal{A}|)$.

Table 5.1. Definitions for the heuristics proposed by Casanova et al. [12,13]

Heuristics	Function g	best
MinMin	minimum of all $CT(t,p)$ values	minimum
MaxMin	minimum of all $CT(t,p)$ values	maximum
Sufferage	difference between 2nd minimum and minimum of all $CT(t,p)$ values	maximum

After Casanova et al. [12,13], Giersch et al. [17,20] proposed several different heuristics. These heuristics have better time complexity and their solution quality is comparable with those of the previous heuristics. Algorithm 5.2 shows the structure of these heuristics. Table 5.2 displays the objective functions proposed by Giersch et al. [17,20] for a task-processor pair (t,p) based on the computation time $\text{Comp}(t,p) = x_{tp}$ and communication time $\text{Comm}(t,p) = w(\text{files}(t))/b_p$ values of the task t when it is executed on the processor p. The additional policies

---

Algorithm 5.2. Structure of heuristics by Giersch et al. [17,20]

1: **for** each processor p **do**
2:   **for** each task t **do**
3:     Evaluate $OBJECTIVE(t,p)$
4:   **end for**
5:   Build the list $L(p)$ of the tasks sorted according according to the value of $OBJECTIVE(t,p)$
6: **end for**
7: **while** there remains a task to schedule **do**
8:   **for** each processor p **do**
9:     Let t be the first unscheduled task in $L(p)$
10:     Evaluate completion time $CT(t,p)$ of t at p
11:   **end for**
12:   Pick a task-processor pair (t_b, p_b) with minimum completion time
13:   Schedule t_b on p_b and its file transfers
14:   Mark t_b as scheduled
15: **end while**

Table 5.2. Definitions for the heuristics proposed by Giersch et al. [17, 20]

Heuristic	Objective Function	Task Selection Order w.r.t. Objective Func.
Computation	$\text{Comp}(t,p)$	increasing
Communication	$\text{Comm}(t,p)$	increasing
Duration	$\text{Comp}(t,p) + \text{Comm}(t,p)$	increasing
Payoff	$\text{Comp}(t,p) / \text{Comm}(t,p)$	decreasing
Advance	$\text{Comp}(t,p) - \text{Comm}(t,p)$	decreasing

Additional Policy	Explanation
Readiness	Selects a ready task for a processor if one exists. A task is called ready for processor p if the transfers of all input files of the task to p are previously scheduled.
Shared	While calculating $w(\text{files}(t))$, scaled versions of file sizes are used. The scaled size of a file is calculated by dividing its original size to the number of tasks that need this file as an input. This policy is redundant with the Computation objective function
Locality	To reduce the file transfer amount, locality tries to avoid assigning a task to a processor if some files used by the task were already scheduled to be transferred to another processor.

readiness, *shared* and *locality* proposed by Giersch et al. [17, 20] are also explained in Table 5.2. As seen in Alg. 5.2, the heuristics construct a task list for each processor. These lists are sorted with respect to various objective values in step 4. For an efficient implementation, we compute the total file sizes for all tasks, i.e., $w(\text{files}(t))$ values, in $\Theta(|\mathcal{A}|)$ time in a preprocessing step. In this way, the objective value computations for all task-processor pairs take $\Theta(TP + |\mathcal{A}|)$ time, so the construction of all sorted lists takes $O(TP \log T + |\mathcal{A}|)$ time. The while loop for scheduling tasks in step 5 takes $O(TP|\mathcal{A}|)$ time. Therefore, the overall time complexity becomes $O(TP \log T + TP|\mathcal{A}|)$.

Flaws of the Greedy Heuristics

The task-processor pair selection according to the momentary completion time values is the greedy decision criterion commonly used in all existing constructive heuristics. Kaya and Aykanat show that this criterion suffers from ineffective use of information about file sharing among the tasks [22]. This flaw is likely to increase with the increasing amount of file sharing and can incur extra file transfers in the resulting schedule. Since the amount of the total file transfers from the server is a bottleneck under the one-port communication model, extra

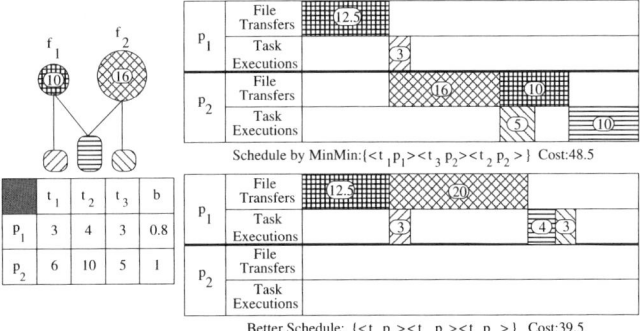

Fig. 5.3. A flaw of the greedy constructive approach for communication-intensive tasks

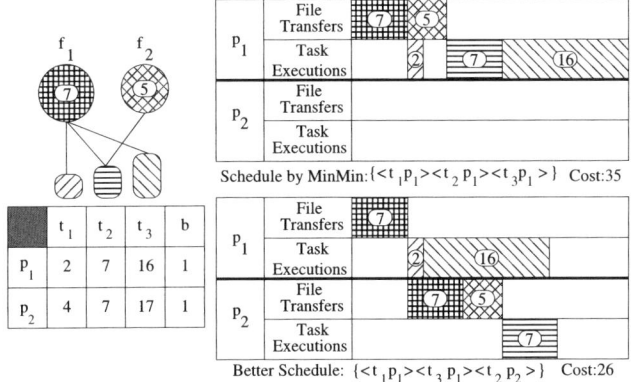

Fig. 5.4. Another flaw of the greedy constructive approach

file transfers can deteriorate the quality of the schedule. This effect is amplified for communication-intensive tasks where the cost of file transfers is considerably higher than the cost of task executions.

Fig. 5.3 displays a sample communication-intensive application with three tasks and two large files. As seen in the figure, *MinMin* schedules t_3 on p_2 after scheduling t_1 on p_1 ignoring the fact that t_2 needs both files. This greedy choice incurs an extra transfer of file f_1. However, there is another schedule without this extra file transfer and with much less turnaround time as shown in Fig. 5.3.

Although extra file transfers constitute crucial bottleneck, it is stated in [22] that they can also be necessary for efficient utilization of computational resources, especially when tasks have comparable computation and communication times. However, if initial scheduling decisions create a computational imbalance, the following greedy decisions may aggravate this problem. The processors that are computationally overloaded due to the previous scheduling decisions are likely to be more favorable for future task assignments since in addition to being already favorable, they have lots of file transfers already scheduled.

Fig. 5.4 illustrates a sample application with three tasks and two small files. As seen in the figure, *MinMin* schedules t_2 on p_1 after scheduling t_1 on p_1 because of the cost of the extra transfer of file f_1 in case of scheduling t_2 on p_2. However, *MinMin* ignores the fact that scheduling t_3 on p_1 does not require any extra file transfer. After faster processor p_1 is overloaded by these two scheduling decisions, it becomes more favorable since both f_1 and f_2 are already transferred to p_1. Finally, *MinMin* schedules t_3 on the overloaded processor p_1 because of the extra transfer of file f_1 required for the other choice of scheduling t_3 on the empty processor p_2. However, there is a much better schedule that utilizes both processors as shown in Fig. 5.4.

5.3.2 Iterative-Improvement-Based Scheduling Heuristics

In [22], Kaya and Aykanat propose an iterative-improvement-based heuristic for scheduling file-sharing tasks on a heterogeneous framework with a single repository. They propose a three-phase scheduling approach which involves initial task assignment, refinement and execution ordering phases. For the refinement phase, they model the target application as a hypergraph and with a hypergraph-partitioning-like formulation, they propose iterative-improvement-based heuristics for refining the task assignments according to two novel objective functions. Unlike the turnaround time, which is the actual schedule cost, the smoothness of proposed objective functions enables the use of iterative-improvement-based heuristics successfully.

Before a detailed analysis of the heuristics in [22], we first give the background material on hypergraph partitioning and iterative-improvement heuristics which are exploited in the scheduling approach.

Hypergraph Partitioning Problem

A hypergraph $\mathcal{H} = (\mathcal{V}, \mathcal{N})$ is defined as a set of vertices $\mathcal{V}$ and a set of nets (hyperedges) $\mathcal{N}$ among these vertices [8]. Every net n in $\mathcal{N}$ is a subset of vertices, i.e. $n \subseteq \mathcal{V}$. The vertices in a net n are called its pins. The set of nets that contain vertex v is denoted as $nets(v)$. The total number of pins denotes the size of the hypergraph. Weights can be associated with vertices and nets. Graph is a special instance of hypergraph such that each net has exactly two pins.

$\Pi = \{\mathcal{V}_1, \mathcal{V}_2, \ldots, \mathcal{V}_K\}$ is a K-way vertex partition of $\mathcal{H}$ if each part $\mathcal{V}_k$ is nonempty, parts are pairwise disjoint and the union of parts gives $\mathcal{V}$. In Π, a net is said to connect a part if it has at least one pin in that part. The connectivity set Λ_n of a net n is the set of parts that n connects and the connectivity $\lambda_n = |\Lambda_n|$ of n is the number of parts it connects. In Π, the weight of a part is the sum of the weights of the vertices in that part.

The K-way hypergraph partitioning (HP) problem is defined as finding a K-way vertex partition that optimizes a given objective function while preserving a given partitioning constraint. The *connectivity*-1 metric is frequently used in hypergraph partitioning [26]. The partitioning objective in this metric is the minimization of CutSize(Π) which is given as:

$$\text{CutSize}(\Pi) = \sum_{n \in \mathcal{N}} w(n)(\lambda_n - 1), \tag{5.1}$$

where $w(n)$ denotes the weight of net n. The partitioning constraint is to maintain a balance on the part weights, i.e.,

$$(W_{max} - W_{avg})/W_{avg} \leq \epsilon, \tag{5.2}$$

where W_{max} is the weight of the part with the maximum weight, W_{avg} is the average part weight, and ϵ is a predetermined imbalance ratio.

Iterative-Improvement Heuristics

The refinement heuristics proposed by Kaya et al. [22, 23] are based on the iterative-improvement heuristics introduced by Kernighan-Lin (KL) [24] and Fidducia-Mattheyses (FM) [16] for graph/hypergraph partitioning. Both KL and FM are move-based approaches with the neighborhood operator of swapping a pair of vertices between parts and shifting a vertex from one part to another, respectively. These heuristics have been widely used for graph/hypergraph partitioning by the VLSI [26] and scientific computing [3, 14, 15, 21, 32] communities because of their effectiveness with good-quality results and efficiency with short run times.

The FM algorithm, starting from an initial bipartition, performs a number of passes until it finds a locally-optimal partition, where each pass contains a sequence of vertex moves. The fundamental idea is the notion of *gain*, which is the decrease in the cost of a bipartition by moving a vertex to the other part. Several FM variants are proposed for the generalization of the approach to the K-way refinement [31].

Iterative-Improvement-Based Refinement Approach

Both effectiveness and efficiency of FM-based heuristics depend on "the smoothness" of the objective function over the neighborhood structure [2], i.e., the neighborhood operator should be small and local. However, a direct generalization of FM-based heuristics to the task scheduling problem suffers from disturbing this smoothness criterion. Removing a task from a processor and scheduling it among previously scheduled tasks of another processor incurs a global perturbation in the schedule, because previously scheduled tasks affect the initialization and completion times of executions of the waiting tasks. Due to this global effect of a task move, computing the gain, which is the change in the turnaround time, is a time consuming work and its time complexity is as high as computing the turnaround time of a given schedule.

In order to alleviate the above problem, Kaya and Aykanat [22] consider the task scheduling problem as involving two consecutive processes: task assignment process which determines the task-to-processor assignment, and execution-ordering process which determines the order of inter- and intra-processor task executions. This view enables the use of FM-based heuristics effectively and efficiently in the task-assignment process by proposing smooth assignment objective functions that are closely related to the turnaround time of a schedule. This refined task-to-processor assignment can then be used to generate better schedules during execution-ordering process.

HP Models for Task Assignment in Heterogeneous Environments:

Kaya and Aykanat use the hypergraph model $\mathcal{H}_A = (\mathcal{T}, \mathcal{F})$ described in Sect. 5.2.1 to represent the interaction among the tasks of the target application $\mathcal{A} = (\mathcal{T}, \mathcal{F})$. Recall that in this model, the vertices of the hypergraph represent the tasks and the nets represent the files. The pins of a net correspond to the tasks that use the respective file. Because of this natural correspondence between a target application and a hypergraph, we describe the heuristics using the problem-specific notation of Sect. 5.2 instead of hypergraph-specific notation, as much as possible, for clarity of presentation. For example, we will use files(t) instead of nets(t). The size of a file f is the weight of the corresponding net. Recall also from Sect. 5.2.2 that a P-way vertex partition $\Pi = \{\mathcal{T}_1, \mathcal{T}_2, \ldots, \mathcal{T}_P\}$ of $\mathcal{H}_A$ is decoded as inducing a task-to-processor assignment for a target schedule. That is, all tasks in a part $\mathcal{T}_p$ will be executed by processor p in the target schedule.

Successful hypergraph partitioning formulations have been recently proposed for solving the task-to-processor assignment problem arising in the parallelization of several applications on homogeneous platforms [3,14,15,32]. If the master-slave platform is homogeneous, i.e., processors are identical and server-to-processor bandwidth values are equal, the partitioning objective given in (5.1) and the load balancing constraint given in (5.2) can be used effectively and efficiently for the refinement. However, the heterogeneity of the environment brings difficulties to the formulation of the task assignment problem. For this reason, Kaya and Aykanat propose new assignment objectives, which can be generalized as partitioning objectives of the hypergraph partitioning problem for heterogeneous environments.

In a given task-to-processor assignment Π, each file will be transferred at least once since it is used by at least one task. Consider a cut net n with connectivity λ_n in Π. Let f_n be the corresponding file for n. is clear that $\lambda_n - 1$ denotes the number of additional transfers of file f_n incurred by Π. Hence $w(f_n)(\lambda_n - 1)$ represents the additional transfer volume, whereas $w(f_n)\lambda_n$ denotes the total transfer volume for file f_n. That is, the *connectivity* metric is the correct metric, rather than the *connectivity*-1 metric, for encoding the total file transfer volume in a given task-to-processor assignment as shown below:

$$\mathrm{CommVol}(\Pi) = \sum_{f_n \in \mathcal{F}} w(f_n)\lambda_n. \tag{5.3}$$

Note that minimizing $\mathrm{CommVol}(\Pi)$ is equal to minimizing $\mathrm{CutSize}(\Pi)$ since $\mathrm{CommVol}(\Pi) = \mathrm{CutSize}(\Pi) + \sum_{f \in \mathcal{F}} w(f)$ and the second term is only a constant factor.

Equation (5.2) can also be used to represent the total file transfer time if the network is homogeneous by normalizing file sizes with respect to the bandwidth value. That is, minimization of the total file transfer volume and the total file transfer time are equivalent in the homogeneous case. To encapsulate the network heterogeneity of the target master-slave platform, we need to modify the conventional definition of the connectivity λ_n of a net n in which different parts connected by n make equal contribution to λ_n. Since we want to formulate the total file transfer time as the real communication cost and bandwidth values of the links are different, Kaya and Aykanat define a *heterogeneous connectivity* λ'_f of a file f as:

$$\lambda'_f = \sum_{p \in \Lambda_f} \frac{1}{b_p}, \tag{5.4}$$

where Λ_f denotes the set of processors that have at least one task needing f as input. Then the total communication time, i.e., the total file transfer time, for the single-repository case can be defined as:

$$\mathrm{CommTime}(\Pi) = \sum_{f_k \in \mathcal{F}} w(f_k)\lambda'_k. \tag{5.5}$$

The computational cost of a task-to-processor assignment Π to the environment is the load of the maximally loaded processor since computations are done in parallel. That is,

$$\mathrm{CompTime}(\Pi) = \max_p \left(\sum_{t \in \mathcal{T}_p} x_{tp} \right). \tag{5.6}$$

Since the assignment Π is clear from the context, we drop Π while referring to CompTime and CommTime in the following text. The processor heterogeneity creates difficulties in modeling the computational cost of a task-to-processor assignment Π. In homogeneous environments, the average part weight – W_{avg} in (5.2) – can be considered as a lower bound for CompTime if a vertex weight represents a computational cost. Similarly, W_{max} can be considered as CompTime which is the exact parallel computational cost of the partition. Therefore in homogeneous environments, the load balancing constraint given in (5.2) can be used for minimizing CompTime. However, in heterogeneous environments, since the same task incurs different computational costs to different processors, a lower bound for parallel computational cost of Π cannot be treated as a balancing constraint as in the hypergraph partitioning formulation for homogeneous environments. Therefore, CompTime should be explicitly included in the assignment objective function as well as CommTime.

By using CompTime and CommTime, Kaya and Aykanat propose two novel objective functions. The first one represents an upper bound for the turnaround time of a schedule with a pessimistic view that assumes no overlap between communication and computation. It is a pessimistic view since it excludes the possibility of communication-computation overlap between different processors as well as on the same processor. For example, a schedule, in which all task executions commence only after the completion of all file transfers from the server, constitutes a typical schedule for this pessimistic view. Under this pessimistic view, the turnaround times of all possible schedules that can be derived from a given task-to-processor assignment Π are bounded above by

$$\text{UBTime} = \text{CommTime} + \text{CompTime}. \tag{5.7}$$

Note that this upper bound is independent of the order of task executions for a given task-to-processor assignment Π.

The second assignment objective function represents a lower bound for the turnaround time of a schedule. As mentioned in Sect. 5.2, a processor can execute a task while that or another processor is transferring a file from the server, i.e., computation and communication can overlap. Even with an optimistic view that assumes complete overlap between communication and computation, the turnaround times of all possible schedules that can be derived from a given task-to-processor assignment Π are bounded below by:

$$\text{LBTime} = \max\{\text{CommTime}, \text{CompTime}\}. \tag{5.8}$$

Note that this lower bound is also independent of the order of task executions for a given task-to-processor assignment Π. This bound is unreachable because of the non-overlapping cases at the very beginning and the end of a schedule. A schedule must begin with a file transfer, and the respective task execution cannot be initialized until the completion of this file transfer. A schedule must end with a task execution on the bottleneck processor. All file transfers from the server to all processors should be completed before the completion of the execution of this task. The length of these non-overlapping intervals are negligible compared to the turnaround time of a schedule due to the large number of tasks.

These two assignment objectives are closely related to the turnaround time of a schedule, and their minimization can generate good task-to-processor assignments. The resulting task-to-processor assignments can be used to obtain schedules with better turnaround times. Instead of one objective as in the hypergraph partitioning problem, we have two assignment objectives and there are various options to improve them. The details of the iterative-improvement-based approach are given in the following subsection.

Structure of the Refinement Heuristics

It is clear that the effectiveness of the refinement phase depends on considering both objective functions simultaneously. Since the objective functions represent upper and lower bounds for the turnaround time, the overall objective should be closing the gap between these two objective functions while minimizing both of them. For this purpose, Kaya and Aykanat propose to use an alternating refinement scheme in which refinement according to one objective function follows the refinement according to the other one in a repeated pattern. The refinement of a task-to-processor assignment Π according to UBTime or LBTime is referred to here as *UB-Refinement* or *LB-Refinement* stage, respectively.

Kaya and Aykanat state that using FM-based heuristics separately and independently for the minimization of the respective objective function is only a partial remedy for satisfying the overall objective. While choosing the best move according to one objective function, the effect of the move according to the other one should also be considered indirectly since the minimization of one objective function may degrade the value of the other one. For this purpose, the authors propose to modify the move selection policy of FM-based approach accordingly in the LB-Refinement stage and/or in the UB-Refinement stage.

In the general FM-based approach, the *best move* associated with a task corresponds to reassigning the task to another processor that incurs maximum decrease in the respective objective function. In the proposed modification, a two-level gain scheme is applied to determine the best move associated with a task through considering the respective objective function as the primary one while considering the other objective function as the secondary one. For the first level, a *good move* concept is introduced, which selects the moves that decrease the primary objective function. In the second level, the best move associated with that vertex is selected among these good moves that incurs the minimum increase to the secondary objective function.

In [22], the proposed two-level gain computation scheme is used in the LB-Refinement stage. The rationale behind this decision is explained as follows: First, the variations in the task-move gains are expected to be larger in UBTime compared to LBTime. Second, UBTime is a relatively loose bound compared to LBTime. Therefore, providing more freedom in the minimization of the loose upper bound while incorporating the constraint to the minimization of the relatively tight lower bound is expected to be more effective for reducing the gap between these two bounds. Based on these two reasons, they also recommend to start the alternating refinement sequence with UB-Refinement stage.

In [22], both UB- and LB-Refinement stages contain multiple FM-like passes. In each pass, all tasks are visited in random order. The best move associated with each visited task is computed according to the adopted gain computation scheme, and this move is realized if it incurs a positive gain according to the respective objective function. Note that each task is visited exactly once in a pass and these passes are repeated until a stopping criterion is met. Algorithms 5.3 and 5.4 show the general structures of UB- and LB-Refinement stages, respectively. In these

Algorithm 5.3. UB-Refinement(Π)

1: **while** a stopping criterion is not met **do**
2:     Create a random visit order of tasks
3:     **for** each task t in this random order **do**
4:         $leaveGain \leftarrow$ **UB-ComputeLeaveGain**(t)
5:         **if** $leaveGain > 0$ **then**
6:             $p_b \leftarrow$ **UB-SelectBestMove**($t, leaveGain$)
7:             **if** p_b is not equal to $Map(t)$ **then**
8:                 **UpdateGlobalData**(t, p_b)
9:                 $Map(t) \leftarrow p_b$
10:            **end if**
11:        **end if**
12:    **end for**
13: **end while**

Algorithm 5.4. LB-Refinement(Π)

1: **while** a stopping criterion is not met **do**
2:     Create a random visit order of tasks
3:     **for** each task t in this random order **do**
4:         $\{commLeaveGain, compLeaveGain\} \leftarrow$ textbfLB-ComputeLeaveGain(t)
5:         **if** ($CommCost(\Pi) > CompCost(\Pi)$ **and** $commLeaveGain > 0$) **or**
             ($CompCost(\Pi) > CommCost(\Pi)$ **and** $compLeaveGain > 0$) **then**
6:             $\{p_b, bestCommGain, bestCompGain\} \leftarrow$
                 LB-SelectBestMove($t, commLeaveGain, compLeaveGain$)
7:             **if** p_b is not equal to $Map(t)$ **then**
8:                 **UpdateGlobalData**(t, p_b)
9:                 $CommCost(\Pi) \leftarrow CommCost(\Pi) - bestCommGain$
10:                $CompCost(\Pi) \leftarrow CompCost(\Pi) - bestCompGain$
11:                $Map(t) \leftarrow p_b$
12:            **end if**
13:        **end if**
14:    **end for**
15: **end while**

figures, $Map(t)$ denotes the processor to which task t is currently assigned. For a more detailed structure of the refinement phase, we refer the reader to [22].

The Three-Phase Approach

In the first phase, initial task-to-processor assignments are derived from the schedules created by some of the existing constructive scheduling heuristics. Kaya and Aykanat prefer this approach to a direct task-to-processor assignment heuristic, because the proposed refinement heuristics are developed by taking the flaws of existing constructive scheduling heuristics into account. They use the heuristics proposed by Giersch et al. [17,20] because of their short execution times. The additional policies are not used, but all of the five heuristics, each having a different objective function, are used since their relative performances vary with the characteristics of applications, e.g., with the number of tasks and files, the average execution time of the tasks, and the average transfer time of the files. Each one of the five initial task-to-processor assignments obtained in this way is fed to the next two phases to obtain five schedules. At the end, the best schedule in terms of the turnaround time is taken as the schedule for the target application.

After the initial task assignment phase, these task assignments are refined with respect to the UBTime(Π) and LBTime(Π), the two proposed objective functions. The authors state that the main improvement in the turnaround time of a schedule can be obtained within only a few passes, whereas the following passes incur negligible improvement. Likewise, the main improvement in the turnaround time of a schedule can be obtained within the first two alternating sequences of UB- and LB-Refinement stages, whereas the following alternating sequences incur negligible improvement. For this reason, a constant number of alternating sequences of UB- and LB-Refinement stages is allowed in the implementation.

In the execution ordering phase, each task-to-processor assignment Π obtained in the refinement phase is preserved while determining the inter- and intra-processor ordering of the task executions. Note that CommTime, CompTime and hence the improved values of both objective functions remain the same as determined in the refinement phase. The structure of the execution ordering heuristic is similar to the scheduling heuristics proposed by Giersch et al. [17, 20]. However, the execution ordering heuristic is asymptotically faster since the same task-to-processor assignment Π is used during the course of the heuristic. For each Π, the execution ordering heuristic is run five times by using each one of the five objective functions proposed by Giersch et al. [17, 20] and the best schedule is selected for this Π.

We omit the details of the subroutines used in LB- and UB-Refinement stages. Slightly different versions of some of them will be explained in detail for the general framework with multiple repositories. The time complexity of the iterative-improvement-based scheduling heuristic is $O(TP\log T + TP|\mathcal{A}|)$. A detailed explanation of the heuristics and complexity analysis can be found in [22].

Experimental Analysis

Kaya and Aykanat give various experimental results for the assessment of the proposed iterative-improvement-based approach. To make the section self-contained, we give the details of the experimental framework in [22] and restate some important results to show the effectiveness of the proposed heuristic. A detailed and complete list of the experiments conducted to analyze the performance of the heuristics can be found in [22].

Kaya and Aykanat demonstrate the performance of the proposed heuristic in comparison with the existing constructive heuristics. They simulate a total of 250 applications, each consisting of $T = 2000$ tasks and $F = 2000$ files. Each task in an application uses a random number of files between 1 and 10. The file sizes are randomly selected to vary between 100 Mbytes and

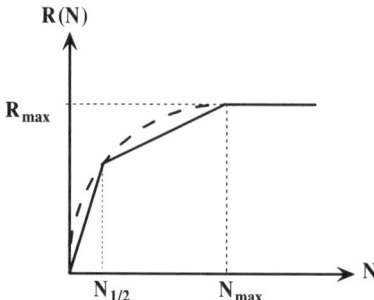

Fig. 5.5. Piecewise linear approximation for task-execution time estimation

200 Gbytes. The experiments vary with the computation-to-communication ratio $\rho = Comp_{avg}/Comm_{avg}$ of the target application, where $Comp_{avg} = (1/P)\sum_{t\in\mathcal{T}}\sum_{p\in\mathcal{P}} x_{tp}$, and $Comm_{avg} = (1/b_{avg})\sum_{i=1}^{n} w(files(t_i))$. Note that $b_{avg}=(1/P)\sum_{p\in\mathcal{P}} b_p$ denote the average server-to-processor bandwidth. They show results with five different ratios $\rho = 10.0, 5.0, 1.0, 0.2$, and 0.1, where for each ρ value there are 50 randomly created applications; thus totaling 250 applications. These choices of ρ characterize a range of applications containing including computation intensive ($\rho = 10$) and communication intensive ($\rho = 0.1$) ones.

Kaya and Aykanat use the GridG topology generator [27] for creating a heterogeneous master-slave platform with $P=32$ processors. The network contains communication links with bandwidth values varying between 20 Mbit/s and 1 Gbit/s.

The Top500 supercomputer list maintained by Dongarra et al. [29] is used to generate the task execution times. Since the Top500 list depends on the LIN-PACK benchmark, the individual tasks are instances of the same problem approximately incurring $(2/3)N^3$ floating point operations for an instance size N. The benchmark values R_{max}, N_{max} and $N_{1/2}$, provided in [29] for each supercomputer, are used to make realistic approximations (inconsistent ETC matrices) for task execution times in a heterogeneous Grid system. Here, R_{max} denotes the maximum processor performance (in terms of FLOPS) that can be achieved for a task with an instance size greater than or equal to N_{max}. Here, $N_{1/2}$ represents the instance size for which half of the R_{max} is achieved. For specific ρ value, the instance sizes for the tasks are uniformly distributed on an interval which is selected judiciously to achieve ρ. Therefore, the performance variation of a task with instance size N can be represented approximately with a piecewise linear function $R(N)$ as shown in Fig. 5.5. The execution time of a task t with instance size N on a processor p is estimated as $x_{tp}=(2/3)N^3/R_p(N)$.

Table 5.3 summarizes the results of the experiments conducted to validate the relation between the proposed assignment objective functions and the actual schedule cost which is the turnaround time of a schedule. The values in the table are derived by using scheduling heuristics individually in the initial task assignment phase as follows: For each heuristic used, the amount of decrease achieved in both UBTime and LBTime during the refinement phase are normalized with respect to the amount of the resulting decrease in the actual schedule cost. That is, these values display the amount of improvements needed in UBTime and LBTime simultaneously to attain one time unit of improvement in the actual schedule cost. Note that performance results are also given for *MinMin* and *Sufferage*, which are not adopted in IIS, in the last two rows of the table. As seen in Table 5.3, close to one time-unit (between 0.91 and 1.00) of improvements are needed in LBTime which is a rather tight bound, whereas a large variation (between 0.16 and 1.95) can be seen for the improvements needed in UBTime which is a loose bound.

Table 5.3. Effectiveness of the objective functions

Heuristic in	Min		Max		Avg	
the first phase	UB	LB	UB	LB	UB	LB
Communication	0.703	0.955	1.879	0.996	1.281	0.980
Computation	0.331	0.928	1.718	0.993	0.989	0.966
Duration	0.570	0.905	1.647	0.997	1.049	0.964
Payoff	0.746	0.988	1.790	1.000	1.291	0.994
Advance	0.747	0.975	1.470	0.999	1.378	0.992
MinMin	0.266	0.923	1.759	0.986	0.923	0.958
Sufferage	0.160	0.993	1.951	0.999	1.128	0.995

The amount of improvements in LBTime and UBTime objective values required to obtain one unit of improvement in the turnaround time, i.e., Δ(LBTime)/Δ(TurnaroundTime) and Δ(UBTime)/Δ(TurnaroundTime), respectively. Here, Δ(Obj) is the difference between Obj values after the first and the third phases of the proposed heuristic.

Table 5.4. Relative performances of the heuristics for the single-repository case

Heuristic	Cost	Execution Time
Iterative-Improvement-Based Heu.	1.000	46.5
Sufferage	1.251	606.9
MinMin	1.303	655.6
Computation+Readiness	1.415	3.9
Communication+Shared+Readiness	1.418	1.3
Computation+Shared	1.426	1.1
Computation	1.435	3.6
Advance+Shared+Readiness	1.439	4.6
Communication+Readiness	1.455	1.3
Communication	1.468	1.0

Table shows the averages of the relative performances of good heuristics normalized with respect to the best/fastest heuristic for each scheduling instance.

Table 5.4 summarizes the results of the experiments conducted to compare the performance of the proposed iterative-improvement-based approach with the best greedy constructive heuristics. The last column of the table also shows the relative runtime performances of these heuristics. For each scheduling instance, the relative runtime performance of every heuristic is calculated by dividing the execution time of the heuristic to that of the fastest heuristic. As seen in Table 5.4, the iterative-improvement-based heuristic performs significantly better than all existing heuristics on the average. For example, *Sufferage*, which is the second best heuristics for the single-repository case, produces 25.1% worse schedules than the iterative-improvement-based heuristic on the average.

5.3.3 An Extension: Clustered Platform

In [12, 20, 22], a slightly different version of the basic platform, a clustered platform, is also considered as the target computing environment. The clustered platform also has a single-repository but differs from the above-mentioned basic one in the following aspects: Each processor node of the basic master-slave platform effectively becomes a cluster of processors, which is served by a local file storage unit for that cluster. That is, we have a set $\mathcal{CL} = \{cl_1, cl_2, \ldots, cl_c\}$ of c clusters and a set $\mathcal{FS} = \{fs_1, fs_2, \ldots, fs_c\}$ of c local file storage units, where fs_i is the file storage unit of cluster cl_i. fs_i is responsible for storing the files, that are transferred to cluster cl_i, until the end of the schedule. The network heterogeneity is modeled by assigning different bandwidth values to the links between the server and the file storage units of the clusters. The intra-cluster communication costs due to the local file transfers from a file storage unit are not considered, because intra-cluster file transfers are assumed to be much faster than the file transfers from the server.

Table 5.5. Relative performances of the heuristics for the single-repository case: clustered platform

Heuristic	Cost	Execution Time
Iterative-Improvement-Based Heu.	1.000	22.4
XSufferage	1.164	280.8
MinMin	1.193	263.0
Sufferage	1.236	263.5
Computation+Readiness	1.270	3.6
Computation	1.275	3.5
Duration+Readiness	1.358	3.7
Duration	1.370	3.7
Communication+Shared	1.445	1.0
Communication+Shared+Readiness	1.446	1.1

Table shows the averages of the relative performances of every heuristic normalized with respect to the best/fastest heuristic for each scheduling instance.

The greedy constructive heuristics [12, 13, 17, 20] and the iterative-improvement-based heuristic [22] can be easily extended for the clustered platform. In addition to the heuristics given in Table 5.1, Casanova et al. [12] also propose a new heuristic called *XSufferage* for the clustered master-slave platforms. Unlike other three scheduling heuristics, *XSufferage* computes cluster-based minimum completion times for each task t from $CT(t, p)$ values. The function g is defined as the difference between the second minimum and the minimum of these minimum completion times and "best" is defined as maximum. For the case of the heuristics by Giersch et al. [17, 20], to adapt the readiness policy, a task is called ready for a cluster if all of its input files are available at that cluster. Similarly for adapting the locality policy, assignment of a task to a processor of a cluster is avoided if some of the input files of that task were already transferred to another cluster. Experiments in [22] show that the iterative-improvement-based approach performs better than all other heuristics. The results are summarized in Table 5.5.

5.4 Scheduling with Multiple Repositories

For the multiple-repository case, Giersch et al. [18, 19] assume a fully decentralized system composed of servers linked through an interconnection network. Each server acts both as a file repository and as a computing node consisting of a cluster of heterogeneous processors. This system is slightly different from the framework given in Sect. 5.2. In [18, 19], files can be replicated and they are initially assumed to be stored at one or more repositories. In addition to the objectives stated above for the single-repository case, the scheduler has to decide how to route the files from repositories to other servers. The paper [18] establishes NP-completeness results for this instance of the scheduling problem and proposes several practical heuristics. The proposed heuristics include extensions of the *MinMin* heuristic, *Sufferage* heuristic, and the heuristics presented in the previous works of the authors [17, 20]. The structure of the extended *MinMin*, *MaxMin*, and *Sufferage* and the heuristics by Giersch et al. [17, 20] is similar to Algs. 5.1 and 5.2. We refer the reader to [18, 19, 23] for a detailed explanation and analysis of the extended heuristics.

Khanna et al. [25] deal with a scheduling problem for a slightly different computing system. They assume a decoupled system consisting of processors and storage nodes (repositories) connected in a local area network. As in the works discussed above, the application consists of file-sharing otherwise independent tasks. They assume that the computation time of a task is a linear function of the total size of the requested files, and hence the expected execution time of a task can be calculated as a constant multiple of the total size of the requested files. This execution time model incorporates the local disk access costs in addition to the file transfer and processing costs. Under these assumptions, the problem addressed in [25] can be specified as scheduling file-sharing tasks on a set of homogeneous processors connected to a set of storage nodes through a uniform (homogeneous) network. Khanna et al. also use a hypergraph to model the application. They propose a two-stage strategy for scheduling task executions and file transfers. In the first stage, they partition the tasks into groups—one group to be assigned to a processor—using a hypergraph partitioning tool. In the second stage, they order the tasks in each group and file transfers from the storage nodes. Due to the homogeneous processors and network assumptions, hypergraph partitioning objective and constraint correspond, respectively, to minimizing total volume of file transfers (excluding local access) and maintaining a balance on the loads (including I/O) of the processors. Khanna et al. report better performance than some existing heuristics, including *MinMin*, *MaxMin*, and *Sufferage*, on two real world applications.

Kaya et al. [23] extend the approach in [22] for the multi-repository case and propose a similar three phase heuristic for scheduling file-sharing tasks on a heterogeneous network with multiple data repositories. They state that, the objective functions given in Sect. 5.3.2 cannot be used for the general framework because of the existence of distributed repositories. We will give the details of the new objective functions in this section. Kaya et al. also implement the *MinMin*, *MaxMin*, and *Sufferage* heuristics [12, 13] and compare the performances of the greedy constructive and iterative-improvement-based approach.

5.4.1 Iterative-Improvement-Based Scheduling Heuristics

Since we are dealing with heterogeneous environments, existing hypergraph partitioning techniques and iterative-improvement-based approaches that are used by Khanna et al. [25] are not applicable. Therefore, Kaya et al. adopt the techniques proposed in [22] and reviewed in the previous section. The objective functions proposed in [22] cannot be used when the files are stored in multiple repositories. Hence, new smooth objective functions are required to design iterative-improvement-based heuristics on heterogeneous environments with multiple repositories. Here, we give the details of the heuristics proposed in [23].

Iterative-Improvement-Based Refinement Approach

Objective Functions for Scheduling with Multiple Repositories

In an attempt to obtain bounds on the turnaround time, Kaya et al. [23] make the following observations. The computational cost, CompTime, for the single-repository case given in (5.6) is applicable as is in the multiple repositories case. However, the communication cost, CommTime, for the single repository case given in (5.5) is not applicable to the multiple repository case. Kaya et al. identify two other cost components that are associated with the turnaround time and that can be used instead of the CommTime. These are:

- UploadTime(Π): File transfer cost from the repositories' perspective. In particular, this is the maximum file transfer time spent by a single repository.
- DownloadTime(Π): File transfer cost from the processors' perspective. In particular, this is the maximum file download time spent by a single processor.

Since the assignment Π is clear from the context, we drop Π in the following text. Suppose that the file f is stored in the repository r, i.e., store$(f) = r$. Recall that Λ_f denotes the set of processors to which file f is to be uploaded. The time spent by the repository r on transferring the file f is

$$\text{Upload}(f) = w(f) \sum_{p \in \Lambda_f} \frac{1}{b_{rp}} . \tag{5.9}$$

For each repository r, the total upload time U_r is defined as the summation of Upload(f) costs over all files stored in r, i.e.,

$$U_r = \sum_{f \in \mathcal{F}(r)} \text{Upload}(f) . \tag{5.10}$$

Since the files can be transfered in parallel, with an optimistic view, the maximum upload time spent by a single repository is

$$\text{UploadTime} = \max_r \{U_r\} . \tag{5.11}$$

The time spent by the processor p on downloading the file f is

$$\text{Download}(f, p) = \frac{w(f)}{b_{\text{store}(f), p}} . \tag{5.12}$$

Recall that files$(\mathcal{T}_p) = \bigcup_{t \in \mathcal{T}_p} \text{files}(t)$ is the set of files to be transferred to processor p. For each processor p, the total download time D_p is defined as the summation of Download(f, p) costs over all files that are needed by the tasks assigned to the processor p, i.e.,

$$D_p = \sum_{f \in \text{files}(\mathcal{T}_p)} \text{Download}(f, p) . \tag{5.13}$$

Since the files can be downloaded in parallel, with an optimistic view, the maximum download time spent by a single processor is

$$\text{DownloadTime} = \max_p \{D_p\} . \tag{5.14}$$

Although the three cost components given in (5.6), (5.11), and (5.14) do not represent the turnaround time, they are closely related to it. By using these components, Kaya et al. define lower and upper bounds on the turnaround time. First, observe that the turnaround time cannot be less than any of these components. Therefore, a lower bound on the turnaround time is

$$\text{LBTimeEX} = \max \{\text{CompTime}, \text{UploadTime}, \text{DownloadTime}\} . \tag{5.15}$$

Furthermore, these components can be used to define an upper bound. A trivial upper bound is

$$\text{UBTimeEX} = \sum_{f \in \mathcal{F}} \text{Upload}(f) + \text{CompTime} \ . \tag{5.16}$$

Kaya et al. state that this bound is too pessimistic to be useful; it states that task executions start after all files have been transfered to the processors, where there are no concurrent file transfers and it is hard to define a tighter upper bound that is smooth over the search space generated by task reassignments. Therefore, they define an objective function which is estimated to be an upper bound. By assuming concurrent transfers, they obtain

$$\text{EstUBTime} = \max\{\text{UploadTime}, \text{DownloadTime}\} + \text{CompTime} \ , \tag{5.17}$$

which is likely to be an upper bound on the turnaround time. Note that this is an estimation, since it is not guaranteed to be an upper bound. This objective function is a combination of the aforementioned optimistic and pessimistic views. It expects full parallelism among the file transfers and no overlap among the task executions and file transfers.

Structure of the Refinement Heuristics

Similar to [22], for the multiple-repository case, we have two different objective functions, LBTimeEX and EstUBTime. As in [22], Kaya et al. [23] choose a similar approach and use an FM [16] based refinement heuristic to close the gap between these two bounds while minimizing both of them. For this purpose, they use an alternating refinement scheme in which first LBTimeEX and then EstUBTime are improved repeatedly until there exists no improvement in these two bounds.

Since we have two bounds to improve, a task reassignment which improves one of these functions may worsen the other one. To solve this problem, Kaya et al. use the two-level gain approach proposed in [22] which modifies the gain concept as described in Sect. 5.3.2. They adopt this modification in improving the LBTimeEX as the primary objective and refine EstUBTime without the two-level gain approach. Similar to [22], the authors state that this latter scheme gives more freedom in EstUBTime refinement and provides the future LBTimeEX refinements with a larger search space to explore.

The objective functions LBTimeEX and EstUBTime depend highly on the communication cost incurred by the file transfers. If a file f is required to be transfered to a processor p for only one task, reassigning that task from p to another processor will save the cost of transferring f to p. We call such files as *critical* to the processor p and maintain a list of such critical file and processor pairs. The critical file concept corresponds to the critical net concept in hypergraph partitioning.

Algorithm 5.5 displays the LB-RefinementEX heuristic. The heuristic first finds the values of the variables C_1, C_2, and C_3 that are used to refer to the three cost components. The variable C_1 refers to the maximum of UploadTime, DownloadTime, and CompTime, i.e., LBTimeEX $= C_1$. The variable C_2 refers to the cost component which in conjunction with C_1 defines EstUBTime $= C_1 + C_2$, e.g., if C_1

is CompTime, C_2 will be the maximum of UploadTime and DownloadTime, otherwise it will be CompTime – see (5.17). Effectively, C_1 becomes the primary objective, and $C_1 + C_2$ becomes the secondary one. The heuristics run until the cost component that defines LBTimeEX changes. If the largest cost component C_1 is the UploadTime, then a randomly permuted list of tasks that request files from the bottleneck repository is constructed. Otherwise, a randomly permuted list of tasks that are assigned to the bottleneck processor is constructed. For the sake of run time efficiency, the visit orders are constructed using only the tasks that are associated with the bottleneck repositories and processors.

Algorithm 5.5. LB-RefinementEX(Π)

1: $\langle C_1, C_2, C_3 \rangle \leftarrow$ DefBounds{UploadTime(Π), DownloadTime(Π), CompTime(Π)}
2: **while** $C_1 \geq C_2$ **and** $C_1 \geq C_3$ **do**
3:     Create a random visit order of the tasks associated with C_1
4:     **for** each task t in this random order **do**
5:        $\langle gain, q \rangle \leftarrow$**LB-ComputeGain**$(t, C_1, C_2)$
6:        **if** $gain > 0$ **then**
7:           **UpdateGlobalData**(t, q)
8:           $Assign(t) \leftarrow q$
9:           **if** $C_1 < C_2$ **or** $C_1 < C_3$ **then**
10:             **return**
11:           **end if**
12:           **if** bottleneck repository or processor is changed **then**
13:             goto 2
14:           **end if**
15:        **end if**
16:     **end for**
17: **end while**

The procedure LB-ComputeGain(t, C_1, C_2) computes the reassignment gains associated with task t and returns the reassignment with positive gain in the primary objective C_1 and the maximum gain in the secondary objective $C_1 + C_2$. If such a reassignment is found, the task is reassigned from its current owner $p = Assign(t)$ to a new processor q.

The gain computations for the cost components are performed as follows. Let $X(2)$ denote the execution time of the processor with the second maximum task execution time. Then, the gain of reassigning the task t from a bottleneck processor p to processor q is

$$g_{comp}(t,p,q) = \min \left\{ \begin{array}{c} x_{tp} \\ X_p - X(2) \\ X_p - (X_q + x_{tq}) \end{array} \right\}, \quad (5.18)$$

according to the objective CompTime. The first argument out of the three, x_{tp}, corresponds to the case in which the processor p remains to be the bottleneck processor after the reassignment. The second argument $X_p - X(2)$ corresponds to the case in which $X_q < X(2)$ and the second bottleneck processor before the reassignment becomes the bottleneck processor afterwards. The third argument $X_p - (X_q + x_{tq})$ handles the cases in which processor q becomes the bottleneck processor after the reassignment.

Let $D(2)$ denote the download cost on the processor with the second maximum file download time. Then, the gain of reassigning task t from a bottleneck processor p to processor q is

$$g_{download}(t,p,q) = \min \left\{ \begin{array}{l} \sum_{f \in \text{critical(files}(t),p)} \frac{w(f)}{b_{\text{store}(f),p}} \\ D_p - D(2) \\ D_p - \left(D_q + \sum_{f \in \text{notNeed(files}(t),q)} \frac{w(f)}{b_{\text{store}(f),q}} \right) \end{array} \right\}, \quad (5.19)$$

according to the objective DownloadTime. The first argument corresponds to the case in which the processor p remains to be the bottleneck processor after the reassignment. In this argument, the set critical(files(t), p) contains the files that are needed by task t and are critical to the processor p before the reassignment. The second argument $D_p - D(2)$ corresponds to the case in which $D_q < D(2)$ and the second bottleneck processor before the reassignment becomes the bottleneck processor afterwards. The third argument handles the cases in which processor q becomes the bottleneck processor after the reassignment. In this argument, the set notNeed(files(t), q) contains those files of task t that are not needed by any task in $\mathcal{T}_q$ before the reassignment. Note that the set of files notNeed(files(t), q) become critical to processor q after the reassignment.

Let $U(1)$ denote the upload cost of the repository with the maximum file upload time. Then, the gain of reassigning task t from the processor p to processor q is

$$g_{upload}(t,p,q) = U(1) - \max_{r} \left\{ \begin{array}{l} U_r - \sum_{f \in \text{critical(files}(t) \cap \mathcal{F}(r),p)} \frac{w(f)}{b_{rp}} \\ + \sum_{f \in \text{notNeed(files}(t) \cap \mathcal{F}(r),q)} \frac{w(f)}{b_{rq}} \end{array} \right\}, \quad (5.20)$$

according to the objective UploadTime. Here, $U(1)$ gives the bottleneck value before the reassignment. The $\max_{r}\{\cdot\}$ corresponds to the bottleneck value upon realizing the reassignment. The set files(t) $\cap \mathcal{F}(r)$ contains those files that are needed by task t and are stored in repository r. Reassigning task t changes the upload times of the repositories in which files(t) are stored. The first summation corresponds to the decrease in the upload time of the repository r due to relieving r of transferring the critical files of t to processor p. The second summation corresponds to the increase in the upload time of the repository r due to the files in the set notNeed(files(t), q).

The procedure UpdateGlobalData(t, q) computes the new loads of the repositories and the processors, and it keeps track of the changes in the cost components that define LBTimeEX and EstUBTime. It also maintains the identities

Algorithm 5.6. EstUB-Refinement(Π)

1: $\langle C_1, C_2, C_3 \rangle \leftarrow$ DefBounds{UploadTime(Π), DownloadTime(Π), CompTime(Π)}
2: **while** $C_1 \geq C_3$ **and** $C_2 \geq C_3$ **do**
3:    Create a random visit order of the tasks
4:    **for** each task t in this random order **do**
5:      $\langle gain, q \rangle \leftarrow$ **EstUB-ComputeGain**(t, C_1, C_2)
6:      **if** $gain > 0$ **then**
7:         **UpdateGlobalData**(t, q)
8:         $Assign(t) \leftarrow q$
9:         **if** $C_1 < C_3$ **or** $C_2 < C_3$ **then**
10:            **return**
11:         **end if**
12:         **if** bottleneck repository or processor is changed **then**
13:            **goto** 2
14:         **end if**
15:      **end if**
16:    **end for**
17: **end while**

of the repositories and the processors that attain the maximum and the second maximum load in terms of the three cost components.

The EstUB-Refinement heuristic is similar to the LB-RefinementEX heuristic with a few differences. This procedure visits all tasks in a random order and computes the reassignment gains of the tasks as the total gain obtained for the cost components that define EstUBTime. For example, $g_{comp}(t, p, q) + g_{download}(t, p, q)$ is computed as the gain of reassigning the task t from processor p to processor q, if EstUBTime is defined by CompTime and DownloadTime. Here, $g_{comp}(t, p, q)$ and $g_{download}(t, p, q)$ are computed according to (5.18) and (5.19), respectively. The best reassignment of a task is realized if the total gain is nonnegative. The procedure adapts itself to the cost components that define the EstUBTime and discards the tasks that are not associated with the bottleneck cost components. Observe that the CompTime is always one of the bottleneck cost components and the other may change because of the tasks reassignments throughout the execution of the EstUB-Refinement procedure.

A refinement pass consists of an LB-RefinementEX and an EstUB-Refinement. Similar to [22], Kaya et al. [23] choose to apply three refinement passes and run the LB-RefinementEX and EstUB-Refinement heuristics with at most five iterations of the while loops (see line 2 of Algs. 5.5 and 5.6).

Three-Phase Approach

Kaya et al. use the same three-phase approach described in Sect. 5.3.2. In the first phase, they use the greedy constructive heuristics [17, 20] to find an initial task-to-processor assignment. They modify these heuristics to handle multiple data repositories and implement them in such a way that their outputs are task-to-processor assignments rather than complete schedules. After refining these task assignments in the second phase by using the above-mentioned refinement approach, they use the modified versions of the greedy constructive heuristics to find the inter- and intra-processor ordering of task executions. For more details, we refer the reader to [23].

Experimental Analysis

Kaya et al. give various experimental results for the assessment of the proposed heuristics and objective functions. Here we restate some important results to show the effectiveness of the proposed heuristic. A detailed and complete list of the experiments conducted to analyze the performance of the heuristics can be found in [23].

The experimental setting of [23] is similar to the one in [22]. The authors generate five sets of applications, where the computation-to-communication ratio, ρ, is different for each application set. For each application set, they generate three groups of problem instances in which the number of files requested by a task was in the range [1–5] or [1–10] or [1–20]. The applications contains $T = 3000$ tasks and $F = 3000$ files where file sizes were random integers ranging from 50 Mbytes to 70 Gbytes.

Similar to [22], to create a heterogeneous system, they use the GridG network topology generator [27]. The generated framework has $P = 32$ processors and up to nine data repositories. The bandwidths of the communication links between repositories and processors were in between 10 Mbit/s and 1 Gbit/s.

Table 5.6 summarizes the results of the experiments conducted to validate the relation between the objective functions proposed for refining task assignments and the turnaround time. That is, these values display the amount of improvements needed in LBTimeEX and EstUBTime to attain one unit of improvement in the turnaround time. As seen in Table 5.6, close to one unit (between 0.97 and 1.29) of improvements are needed in LBTimeEX, whereas the required improvement in EstUBTime is in between 1.28 and 4.42. This shows that the EstUBTime is not a tight upper bound on the turnaround time in the problem instances that

Table 5.6. Effectiveness of the Objective Functions

Heuristic in the first phase	Min		Max		Avg	
	EstUB	LB-EX	EstUB	LB-EX	EstUB	LB-EX
Communication	1.880	1.004	3.151	1.191	2.348	1.071
Computation	1.275	0.973	2.097	1.051	1.682	1.005
Duration	1.699	0.989	2.994	1.273	2.166	1.086
Payoff	1.949	1.031	3.443	1.160	2.484	1.083
Advance	1.951	1.036	4.423	1.287	2.871	1.161

The amount of improvements in LBTimeEX and EstUBTime objective values required to obtain one unit of improvement in the turnaround time, i.e., $\Delta(\text{LBTimeEX})/\Delta(\text{TurnaroundTime})$ and $\Delta(\text{EstUBTime})/\Delta(\text{TurnaroundTime})$, respectively. Here, $\Delta(\text{Obj})$ is the difference between Obj values after the first and the third phases of the proposed heuristic.

Table 5.7. Relative performances of the heuristics for the multiple-repository case

Heuristic	Cost
Iterative-Improvement-Based Heu.	1.005
Sufferage	1.108
MinMin	1.133
MaxMin	1.459

Table shows the averages of the relative performances of every heuristic normalized with respect to the best/fastest heuristic for each scheduling instance.

we consider. It is clear that, different and tighter estimations would increase the efficiency of the proposed heuristic in terms of the turnaround time.

Kaya et al. also perform some experiments to compare the performance of their approach with some greedy constructive scheduling heuristics. They also modify the *MinMin*, *MaxMin* and *Sufferage* to compare their approach with these heuristics. The experimental results show that *Sufferage* is the best among the three heuristics. The proposed iterative-improvement-based heuristic has been reported to be considerably faster than the Sufferage heuristic (approximately 90 times) while obtaining 10% improvement on the average in the turnaround times. Table 5.7 summarizes the results of these experiments.

5.5 Conclusions

This chapter surveys the heuristics for scheduling file-sharing tasks on heterogeneous environments and shows a generic approach to adapt the iterative-improvement-based refinement heuristics to the task scheduling problem.

In this approach, the task scheduling problem is considered as involving two consecutive processes: task assignment which determines the task-to-processor assignments, and execution ordering which determines the order of inter- and intra-processor task executions. This approach enables the use of iterative-improvement heuristics effectively and efficiently in the task assignment process by proposing smooth assignment objective functions that are closely related to the cost of a schedule. This refined task-to-processor assignment is then used to generate a better schedule during execution ordering process.

The presented heuristics are static in the sense that the schedule is determined before the program execution begins. Real-life environments with large and non-dedicated computing platforms may require dynamic scheduling to adapt to the run-time changes such as increases in the workload, processor failures, and link failures. The refinement heuristics seem to be viable to adapt the original schedule to the run-time changes. However, dynamic scheduling methods interact with other system components such as process migration mechanism whose costs should be considered in the refinement heuristics.

The presented heuristics can be used in population-based heuristics. First, the solutions found by the heuristics can be used to generate an initial generation. Note that the randomized part of these heuristics is the creation of the visit orders during the refinement phases. Therefore, we believe that this approach can create initial generations of very small sizes. Second, the refinement heuristics can be used to improve the current individuals before producing the next generation. In other words, the refinement heuristics can be used to let the individuals mature for a while (for a number of refinement phases) before producing the next generation. We have confidence that this approach will deliver promising solutions.

References

1. Ali, S., Siegel, H.J., Maheswaran, M., Hensgen, D., Ali, S.: Task execution time modeling for heterogeneous computing systems. In: Raghavendra, C. (ed.) Proceedings of the 9th Heterogeneous Computing Workshop (HCW 2000), Cancun, Mexico, May 2000, pp. 185–199. IEEE, Los Alamitos (2000)
2. Alpert, C.J., Kahng, A.B.: Recent directions in netlist partitioning: A survey. Integration, The VLSI Journal 19(1-2), 1–81 (1995)
3. Aykanat, C., Pınar, A., Çatalyürek, Ü.V.: Permuting sparse rectangular matrices into block-diagonal form. SIAM Journal on Scientific Computing 25(6), 1860–1879 (2004)
4. Banino, C., Beaumont, O., Carter, L., Ferrante, J., Legrand, A., Robert, Y.: Scheduling strategies for master-slave tasking on heterogeneous processor platforms. IEEE Transactions Parallel and Distributed Systems 15(4), 319–330 (2004)
5. Beaumont, O., Boudet, V., Robert, Y.: A realistic model and an efficient heuristic for scheduling with heterogeneous processors. Technical Report RR-2001-37, LIP, ENS Lyon, France (September 2001)
6. Beaumont, O., Legrand, A., Marchal, L., Robert, Y.: Steady-state scheduling on heterogeneous clusters. International Journal of Foundations of Computer Science 16(2), 163–194 (2005)
7. Beaumont, O., Marchal, L., Robert, Y.: Broadcast trees for heterogeneous platforms. Technical Report RR-2004-46, LIP, ENS Lyon, France (November 2004)
8. Berge, C.: Hypergraphs. North Holland, Amsterdam (1989)
9. Berman, F.: High-performance schedulers. In: Foster, I., Kesselman, C. (eds.) The Grid: Blueprint for a new computing infrastructure, ch. 12, pp. 279–309. Morgan Kaufmann, San Francisco (1999)
10. Berman, F., Wolski, R., Casanova, H., Cirne, W., Dail, H., Faerman, M., Figueira, S.M., Hayes, J., Obertelli, G., Schopf, J.M., Shao, G., Smallen, S., Spring, N.T., Su, A., Zagorodnov, D.: Adaptive computing on the Grid using AppLeS. IEEE Transactions on Parallel and Distributed Systems 14(4), 369–382 (2003)
11. Casanova, H.: Network modeling issues for Grid application scheduling. International Journal of Foundations of Computer Science 16(2), 145–162 (2005)
12. Casanova, H., Legrand, A., Zagorodnov, D., Berman, F.: Heuristics for parameter sweep applications in Grid environments. In: Proc. Ninth Heterogeneous Computing Workshop, pp. 349–363. IEEE Computer Society Press, Los Alamitos (2000)
13. Casanova, H., Obertelli, G., Berman, F., Wolski, R.: The AppLeS parameter sweep template: User-level middleware for the Grid. In: Proceedings of the 2000 ACM/IEEE conference on Supercomputing (CDROM). IEEE Computer Society Press, Los Alamitos (2000)
14. Çatalyürek, Ü.V., Aykanat, C.: A hypergraph model for mapping repeated sparse matrix-vector product computations onto multicomputers. In: Proceedings of The Second International Conference on High Performance Computing, HiPC 1995, Goa, India (1995)

15. Çatalyürek, Ü.V., Aykanat, C.: Hypergraph-partitioning based decomposition for parallel sparse-matrix vector multiplication. IEEE Transactions Parallel and Distributed Systems 10(7), 673–693 (1999)
16. Fidducia, C.M., Mattheyses, R.M.: A linear-time heuristic for improving network partitions. In: 19th ACM/IEEE Design Automation Conference, pp. 175–181 (1982)
17. Giersch, A., Robert, Y., Vivien, F.: Scheduling tasks sharing files on heterogeneous clusters. Technical Report RR-2003-28, LIP, ENS Lyon, France (May 2003)
18. Giersch, A., Robert, Y., Vivien, F.: Scheduling tasks sharing files from distributed repositories. Technical Report RR-2004-04, LIP, ENS Lyon, France (February 2004)
19. Giersch, A., Robert, Y., Vivien, F.: Scheduling tasks sharing files from distributed repositories. In: Danelutto, M., Vanneschi, M., Laforenza, D. (eds.) Euro-Par 2004. LNCS, vol. 3149, pp. 246–253. Springer, Heidelberg (2004)
20. Giersch, A., Robert, Y., Vivien, F.: Scheduling tasks sharing files on heterogeneous master-slave platforms. In: PDP 2004, 12th Euromicro Workshop on Parallel Distributed and Network-based Processing. IEEE Computer Society Press, Los Alamitos (2004)
21. Karypis, G., Kumar, V.: Multilevel k-way partitioning scheme for irregular graphs. Journal of Parallel and Distributed Computing 48(1), 96–129 (1998)
22. Kaya, K., Aykanat, C.: Iterative-improvement-based heuristics for adaptive scheduling of tasks sharing files on heterogeneous master-slave platforms. IEEE Transactions on Parallel and Distributed Systems 17(8), 883–896 (2006)
23. Kaya, K., Uçar, B., Aykanat, C.: Heuristics for scheduling file-sharing tasks on heterogeneous systems with distributed repositories. Journal of Parallel and Distributed Computing 67, 271–285 (2007)
24. Kernighan, B.W., Lin, S.: An efficient heuristic procedure for partitioning graphs. The Bell System Technical Journal 49(2), 291–307 (1970)
25. Khanna, G., Vydyanathan, N., Kurc, T., Çatalyürek, Ü.V., Wyckoff, P., Saltz, J., Sadayappan, P.: A hypergraph partitioning based approach for scheduling of tasks with batch-shared I/O. In: Proceedings of Cluster Computing and Grid (2005)
26. Lengauer, T.: Combinatorial Algorithms for Integrated Circuit Layout. Wiley–Teubner, Chichester (1990)
27. Lu, D., Dinda, P.A.: GridG: Generating realistic computational grids. SIGMETRICS Perform. Eval. Rev. 30(4), 33–40 (2003)
28. Maheswaran, M., Ali, S., Siegel, H.J., Hensgen, D., Freund, R.: Dynamic matching and scheduling of a class of independent tasks onto heterogeneous computing systems. Journal of Parallel and Distributed Computing 59(2), 107–131 (1999)
29. Meuer, H.W., Dongarra, J.J., Strohmaier, E.: TOP500 Supercomputer Sites. In: Proceedings of the IEEE/ACM Supercomputing Conference, SC 2003, 22th edn., Phoenix, USA (2003)
30. Saif, T., Parashar, M.: Understanding the behavior and performance of non-blocking communications in MPI. In: Danelutto, M., Vanneschi, M., Laforenza, D. (eds.) Euro-Par 2004. LNCS, vol. 3149, pp. 173–182. Springer, Heidelberg (2004)
31. Sanchis, L.A.: Multiple-way network partitioning. IEEE Transactions on Computers 38(1), 62–81 (1989)
32. Uçar, B., Aykanat, C.: Encapsulating multiple communication-cost metrics in partitioning sparse rectangular matrices for parallel matrix-vector multiplies. SIAM Journal on Scientific Computing 25(6), 1837–1859 (2004)

6

Advanced Job Scheduler Based on Markov Availability Model and Resource Selection in Desktop Grid Computing Environment

EunJoung Byun[1], SungJin Choi[1], HongSoo Kim[1], ChongSun Hwang[1], and SangKeun Lee[1]

Dept. of Computer Science & Engineering, Korea University, Korea
{vision,lotieye,hera,hwang}@disys.korea.ac.kr, yalphy@korea.ac.kr

Summary. This chapter reviews dynamism in desktop Grid computing and explains the advanced stochastic scheduling scheme with the Markov Job Scheduler based on Availability (MJSA) in the environment.

In recent years, Grid computing [1] has received considerable interest in the field of academics and enterprise. Numerous attempts have been made to organize cost efficient large-scale Grid computing. Desktop Grid computing [13,19,2] is a more flexible paradigm that is used to achieve high performance and high throughput with desktop resources that are less stable and has more inferior performance compared to traditional Grid. It is comprised of a diverse set of desktops interconnected with various network forms ranging from Local Area Network (LAN) to the Internet. Desktop Grid system has played a leading role in the development of large scale aggregated computing power harvested from the edge of the Internet at lower cost. The main goals of the system are to accomplish high throughput and performance by mobilizing the potential colossal computational resources of idle desktops.

However, since a desktop peer is a fluctuating resource that connects to the system, performs computations and disconnects to the network at will, desktop volatility makes the system unstable and unreliable. To develop a reliable desktop Grid computing system, a scheduling scheme must consider the dynamic nature (i.e., volatility) of volunteers and a resource selection scheme should adapt to such a dynamic environment, as the selection is getting complicated due to the uncertain behavior of desktops.

This chapter demonstrates desktop state change modelling and an advanced resource selection scheme, Selection of Credible Resource with Elastic Window (SCREW), to choose reliable resources in dynamic computational desktop Grid environments. Markov modelling of the dynamic state turning provides understanding of the pattern of desktop behavior while SCREW selects qualified desktops that satisfy time requirements to complete given workloads and adapts to the needs of the user and the application on the fly.

Keywords: Desktop Grid computing, Stochastic scheduling, Markov modelling, Hidden Markov Model, Resource selection scheme.

6.1 Introduction

In a traditional Grid system environment, performance becomes both irregular and sporadic due to resource sharing among other applications [7]. The desktop

Grid system is more dynamic and unreliable compared to typical Grid systems since desktop Grid is affected by user activities or owner preferences in use. The diverse network-based architecture helps the system bring together a number of desktops through various networks, whereas dynamic peers threaten system stability and dependability. Particularly, the desktop Grid systems that do not exclude volatile resources have suffered from performance degradation since they cannot prevent unstable and unpredictable system behavior. It is important to recognize that desktop Grid is different from typical Grid in terms of resource stability (i.e. volatility and autonomous donation). The issues related to resources including selection, scheduling (i.e. mapping between job and resource), management, etc. must be dealt with differently from the former Grid. Most desktop Grid computing systems steal widespread available cycles. In a desktop Grid computing environment, voluntary desktops (i.e., resource providers) are free to leave and join independently in the middle of execution. Existing desktop Grid computing systems, however, do not consider volatility in their scheduling procedures. As a result, job execution is often suspended, resulting in delayed completion time and degraded performance and reliability.

Although desktop Grid computing is considered to be a promising solution to compute large-scale problems at a lower cost, it frequently experiences blocked job execution and delayed execution time since variable desktops are engaged or turned off at the whim of individual users. Previous studies do not consider the volatile aspects of desktops. They suffer from the following problems: delayed makespan (i.e., total execution time), degraded performance, unreliability and instability of systems, and unpredictable execution patterns of desktops. The stochastic scheduling scheme based on the reliable resource selection method is demonstrated to face these limitations. The *Markov Job Scheduler based on Availability (MJSA)* consists of two main modules: THE Markov modeling and resource selection part, Selection of Credible Resource with Elastic Window (SCREW). In the modeling, MJSA considers volatility, the state in which a resource is unavailable for utilization caused by participant dynamics, availability, and the responsiveness of participants. The MJSA models the temporal availability of desktops stochastically when measuring the dynamic characteristics of participants as well as determining the probability that a desktop will be activated. The modeled probabilistic resource prediction based on a state transition model considers the flow of time. Since analysis and prediction of the resource state affect the scheduling procedure to select an appropriate desktop, MJSA suggests a refined resource selection scheme, to accurately predict resource state changes in unpredictable and dynamic changing environments. As a result, it can efficiently manage the volatility of a participant. To further improve reliability and stability of the system, MJSA supports three scheduling schemes based on the Markov chain: OPTIMIST, PESSIMIST, and REALIST. These scheduling schemes are based on stochastic modeling of desktop availability. In the OPTIMIST scheme, in which time constraints are relaxed, the MJSA provides reliable resource selection at low cost. In the PESSIMIST scheme, where time constraints are rigid, the MJSA enables stable makespan in strict time. Finally, in the REALIST scheme, where time constraints are only partially relaxed, the MJSA provides enhanced cost efficiency. In conclusion, the MJSA improves performance and reliability by adapting the appropriate scheduling scheme when selecting volunteers according to the needs of applications.

SCREW is designed for MJSA to provide reliable resource management and efficient resource selection. In addition, SCREW provides flexible and tailored selection methods satisfying user and application requirements. The proposed SCREW gives a strong and reliable basis to make a decision whether a particular resource is eligible or suitable for the given condition from the system or workload.

This chapter starts with the description of the Hidden Markov Model. In Section 6.3, the Markov modeling based on availability is demonstrated. Section 6.4 discusses the advanced scheduling schemes based on the Markov model. In Section 6.5, SCREW is explained. Section 6.6 provides performance evaluations through mathematical analysis and experimentation of the MJSA on top of Korea@Home. Finally, Section 6.7 presents the conclusions.

6.2 Hidden Markov Model

Since the MJSA is not based on statistical analysis but on stochastic modeling that presents state transition over time, it gives a well-defined structure to manifest the dynamic facets. Even though statistical modeling has been widely employed to classify patterns at low cost, it is inadequate to present complex and dynamic fluctuating features that differ over time. A more elaborate approach is necessary. The fluctuation of desktop states can be modeled with a parametric random process. In addition, the parameters of stochastic variables can be estimated with higher accuracy. The proposed method supports a reliable method to recognize states and predicts state changes for a volatile resource desktop in the Internet.

The probabilistic model relies upon a hidden Markov model (HMM) [25] in which a Markov process with unknown (i.e. hidden) parameters determines the hidden parameters from observable parameters. The extracted model parameters are used to extract further analysis applications. In a hidden Markov model, the state is not directly visible but variables influenced by the state are visible while each state has a probability matrix among the possible outcomes and the sequence of state products generated by an HMM gives some information about the sequence of desktop states. As shown in Eq. 6.1, the Markov model consists of the following five complete parameters given in Eq. 6.2 through Eq. 6.6, in which N indicates the number of states and M illustrates the number of observable symbols. In addition, A denotes the state transition probabilistic matrix while B presents the observation probability distribution and π means the initial state distribution.

$$\lambda = \{N, M, A, B, \pi\} \tag{6.1}$$

Q consists of the state variable g_t as the state of the system at discrete time t.

$$Q = \{q_1, q_2, q_3, ..., q_N\} \tag{6.2}$$

Observation sequence consists of a series of events. O is known as

$$O = \{o_1, o_2, o_3, ..., o_M\} \tag{6.3}$$

The state transition probability distribution $A = \{a_{ij}\}$ where

$$a_{ij} = P(q_{t+1} = j | q_t = i) \tag{6.4}$$

The observation probability distribution, $B = \{b_j(k)\}$ where

$$b_j(k) = P(O_t = k | q_t = i), \quad i \le k \le M \tag{6.5}$$

The initial state distribution $\pi = \{\pi_i\}$

$$\pi = P[q_1 = i], 1 \le i \le M \tag{6.6}$$

The Markov model designs daily updates while considering each state and calculating the probability of its state transition. The state of each participant is checked and is updated each thirty-minute time unit. The reason why the model updates every thirty-minute is that most subtask units require between twenty and thirty minutes described in [5]. The shorter the window (i.e. less than 30 minutes) the more accurate preclearance of reliability. Conversely, the larger the window (i.e. above 30 minutes) the less accurate the prediction. I close 30 minutes as that is the average fine of assigned tasks.

Three different states are used: Idle (I), Use (U), and Stop (S). The I state indicates a condition that can be used to execute a job without personal activities from the users own. The U state presents a status of being unable to work because of user occupancy or intervention. The S state indicates a disabled state such as machine crash or network disconnection. SCREW focuses on I which is a meaningful state for execution.

$$P(O|Q, \lambda) = \prod_{n=1}^{N} P(o_m | q_n, \lambda), \quad m = \{1, ...M\} \tag{6.7}$$

The forward variable can be derived by

$$\alpha_{t+1}(j) = \sum_{i=1}^{N} (\alpha_t(i) \cdot a_{ij}) \cdot b_j(O_{t+1}) \tag{6.8}$$

The backward variable can be derived by

$$\beta_t(i) = \sum_{i=1}^{N} (a_{ij} \cdot b_j(O_{t+1}) \cdot \beta_{t+1}(j)) \tag{6.9}$$

$$\delta_{t+1}(i) = [\max(\delta_t(i)a_{ij})b_j(o_{t+1})] \tag{6.10}$$

$$P(O|\lambda) = \sum_{r=1}^{k} a_r(j)\beta_r(j) \forall j, 1 \leq j \leq n \tag{6.11}$$

Likelihood L is defined by

$$L = \sum_{i=1}^{n} \sum_{j=1}^{N} (\alpha_t(i) \cdot a_{ij} \cdot b_j(O_{t+1}) \cdot \beta_{t+1}(j)) \tag{6.12}$$

6.3 Markov Job Scheduler Based on Availability

One of the major concerns in desktop Grid systems is reliability. The proposed scheduling scheme resolves system reliability problems resulting from participant volatility through modeling temporal behaviors of each desktop. Although some statistical scheduling approaches [24] are used to predict the available performance of a participant, they do not accurately represent participant execution patterns. The proposed MJSA employs the Markov chain based on availability [18] since the stochastic availability model is used to represent participant experiences in scheduling. The MJSA targets to provide a more reliable execution through the stochastic process. As the Markov chain provides time-travel-like activities of desktop execution [25], the MJSA can accurately determine the intent of all state changes made by a desktop user as a Markov chain is built anew for each time unit. The state of each desktop is checked and updated every thirty minutes. A thirty-minute state update is used because [5] described that most subtask units require between twenty and thirty minutes.

Our model has five parameters in which N is 48 (*2 X 24=48* time units per day while each time unit is thirty-minute), and M is 3 due to state kinds of I, U, and S. A gives the time unit transition probability matrix, B shows the observation probability (i.e. state of I, U, and S), and the initial state is π. The independent observation sequence is represented by Eq. 6.7.

To find the optimal sequence, a Viterbi algorithm [25], which finds most probable sequence with forward and backward algorithm, is introduced as given in Eq. 6.7 through Eq. 6.12.

The state sequence which maximizes $P(O,Q|\lambda)$ to decode with Viterbi is shown in the following equation.

The probability of the entire consecutive sequence is as follows. Eventually, most probable state occurrence in time unit, state credit (SC), presents the most feasible state and derives the probabilities of each state.

Three different states are used: Idle (I), Use (U), and Stop (S). The I state indicates that a desktop can execute a job without user intervention. The U state presents a state of being unable to work owing to user occupancy or intervention while the S state denotes a disabled state caused by machine crash, system termination, network disconnection, etc. The Markov model is updated daily while calculating the probability of each state transition.

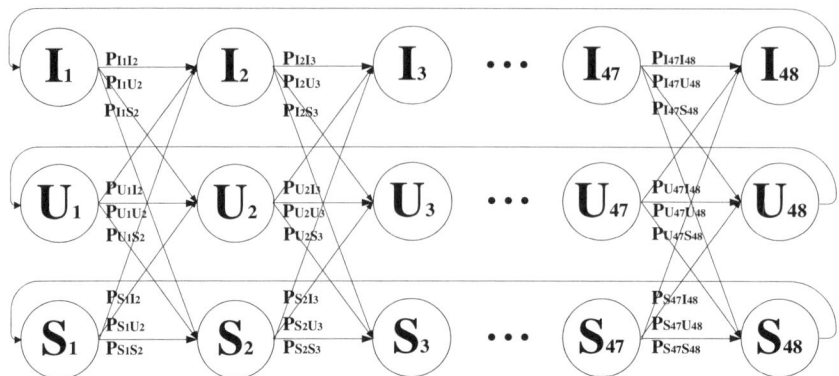

Fig. 6.1. Markov Model for Availability

To reduce server overhead the server evaluates, stores, updates, and maintains state credit values whereas each desktop resource records their state changes. The size of Markov model influences the efficiencies of scheduling in terms of Grid size. As the number of desktop resource increases and the size of desktop Grid system grows, the server in the model could be overloaded. To solve this scalability problem the probability of previous state is simply calculated according to Markov chain property. To reduce server overhead hierarchical architecture supporting sub-manager can be considered.

The Markov model suggests a simple way to process multiple sequences that would otherwise be very difficult to manipulate. Fig. 6.1 presents the Markov chain modeled with forty-eight time units in a twenty-four hour period, and represents each thirty-minute interval. As represented in Fig. 6.1, I_i means I state in ith time unit and $P_{I_i I_j}$ indicates transition probability from I state in ith time unit to I state in jth time unit. Each transition in the Markov chain state diagram has an associated transition probability. The matrix of transition probabilities is shown in Eq. 6.13. The sum of each column is one. The state transition probability matrix, A, a_{ij}, is given by $P_{State_j State_k}$

$$P_{State_j State_k} = \begin{pmatrix} P_{I_j I_k} & P_{I_j U_k} & P_{I_j S_k} \\ P_{U_j I_k} & P_{U_j U_k} & P_{U_j S_k} \\ P_{S_j I_k} & P_{S_j U_k} & P_{S_j S_k} \end{pmatrix} \tag{6.13}$$

The Viterbi algorithm [26] is used to draw an accumulated likelihood score (i. e., probability associated with the state) of the most likely state. To find the most likely transition and update the state probability for each state at the next time unit, the most likely transition probability is found in the given states. Each credit value is calculated based on the Viterbi algorithm applied to the Markov model. The Viterbi algorithm helps to find the most-likely state transition sequence of states that performs sub-optimally. A credit value is introduced, *State*

Credit (SC), which is derived from the probability of a value going through the I state in the Markov chain model. The SC suggests the most-likely probability for each state, indicating a reliability value. The probability of all paths going through the I state is computed by multiplying the *Forward State Credit (FSC)* by the *Backward State Credit (BSC)*. As shown in Eq. 6.2, the FSC is the total sum of the multiplied credit of the previous stage by the transition probability from the previous stage (i.e. *(i-1)th* time unit) to the current stage (i.e. *ith* time unit). In contrast, the BSC, which is a time reversed version of the FSC, is the total sum of the multiplied credit of the next stage by transition probability from the next stage(i.e *(i+1)th* time unit) to the current stage (i.e. *ith* time unit). The state of the desktop calculating probability in each time unit helps to estimate and predict the next state of the desktop. The FSC of I_i and the BSC of I_i are given in Eq. 6.14 and Eq. 6.15.

$$FSC_{I_i} = SC_{I_{i-1}} \times P_{I_{i-1}I_i} + SC_{U_{i-1}} \times P_{U_{i-1}I_i} + SC_{S_{i-1}} \times P_{S_{i-1}I_i} \quad (6.14)$$

$$BSC_{I_i} = SC_{I_{i+1}} \times P_{I_iI_{i+1}} + SC_{U_{i+1}} \times P_{I_iU_{i+1}} + SC_{S_{i+1}} \times P_{I_iS_{i+1}} \quad (6.15)$$

The credit value of all paths going through I_i are calculated as shown in Eq. 6.16 of SC of I_i, SC_{I_i}.

$$SC_{I_i} = FSC_{I_i} \times BSC_{I_i}. \quad (6.16)$$

Initially, the FSC in the first stage without forward transition probabilities in the Markov chain is given by the rate of frequency in generating the Markov chain. That is, the FSC is the ratio of f_{I_1} to sum of f_{I_1}, f_{U_1}, and f_{S_1}, whereas frequency of I_1, U_1, and S_1 are f_{I_1}, f_{U_1}, and f_{S_1}, respectively. In addition, the BSC at the last stage without backward transition probabilities is given by same manner. They are denoted by $FSC_{I_1} = \frac{f_{I_1}}{f_{I_1}+f_{U_1}+f_{S_1}}$ and $BSC_{I_{48}} = \frac{f_{I_{48}}}{f_{I_{48}}+f_{U_{48}}+f_{S_{48}}}$. This predictable availability factor, SC, is measured based on desktop availability. Each of the participants records the state onto profiles every thirty minutes and transmits the current state to the central server whereupon the MJSA in the central server refreshes and updates the Markov model of the desktop for credit values. The MJSA calculates and manages the credit values of the state of the desktops while updating the Markov model. The availability of the participant is assumed to follow hyper-exponential distribution according to [5,6]. The Markov model of the MJSA predicts availability and duration to improve performance, reliability, and execution completion.

6.4 Scheduling Algorithms on the MJSA

As efficient and reliable scheduling schemeing is most formidable in desktop Grid computing since resource management and performance are highly related to the scheme, we propose three scheduling schemes based on the Markov chain previously modeled with desktop availability and duration: OPTIMIST, PESSIMIST, and REALIST. The OPTIMIST scheme uses an advanced availability measurement method as well as resource selection for the desktops. The PESSIMIST aims to provide rapid makespan or to satisfy time requirements by imposing punishment to unpredictable participants enabling more restricted availability estimation and confined resource selection. The REALIST scheme combines elements of the OPTIMIST (ease of use) and the PESSIMIST (accuracy) to obtain benefits from both, providing intermediate load for availability quantification and semi-limited resource selection. The MJSA prevents the problems of unstable resources and frequent job resubmission by selecting an appropriate participant and allocating jobs to them on the basis of these three schemes. Although each scheduling scheme has different features, the basic idea is that the scheduler must reliably consider desktop execution patterns.

6.4.1 OPTIMIST

The OPTIMIST is fundamentally based on the idea that the desktops can complete the workload eventually and therefore workloads can be simply allocated. As a result, there are few severe constraints to resource selection.

The OPTIMIST is used in cases of little or no time restrictions concerning task completion. Applications of this scheme are usually characterized by long, large, independent, and identical workloads. A well-known example is SETI@HOME [11]. However, workload failure probability is not low even though the OPTIMIST outperforms previously existing schemes.

The essence of the OPTIMIST is SC_{I_i}, which is calculated by Eq. 6.16. The algorithm of the OPTIMIST functions are as follows. First, the scheduler sets a starting point to the present time unit in the Markov chain of a desktop. Second, it forms a window from the starting point. Finally, all SC_{I_i} in the window are summed up and compared with Required Time Factor (RTF). As given in Eq. 6.17, where n is window size and RTF depicts the time required to finish the workload, when the sum of SC_{I_i} is larger than RTF, the scheduler allocates the workload to the desktop. If not, the window is enlarged in a forward direction until its value reaches the given Due Time Factor (DTF) describing the allowed time period. Whereupon the sum of SC reaches the RTF, the scheduler allocates the workload to the desktop. If not, the MJSA repeats this process to ascertain eligible desktops.

$$\sum_{i=1}^{n} SC_{I_i} > RTF \qquad (6.17)$$

6.4.2 PESSIMIST

In contrast to the OPTIMIST, the PESSIMIST is based on the assumption that the desktops can fail to deliver owing to their inherent volatility. In the PESSIMIST, the workload must be allocated to reliable desktops. To do so, the PESSIMIST is based on assigning a penalty for unpredictable or irregular execution using the entropy of availability in order to more accurately predict desktop execution patterns. The PESSIMIST is applied in case of strict time constraints, for example large amounts of business or government data (such as census processing) that must adhere to deadlines or realtime problem solving and mission-critical projects. In these cases, completion time is more important. However, this scheme has some overhead to estimate availability. The PESSIMIST is useful in cases where desktops outnumber workloads. The PESSIMIST uses a severe factor, Representative Credit (RC) of I state in ith time unit, RC_{I_i}, representing availability with high probability, since SC_{I_i} is not enough to determine whether the desktop is available or not, or how long it will remain available. RC_{I_i} considers executable probability to credit values as given by Eq. 6.18. RC provides prediction of ability and duration of desktop execution by adapting mining data about desktop state transition. We introduce the concept of entropy, information entropy measuring the amount of information from an unknown quantity, to consider dynamism. The entropy suggests how predictably a desktop participates and how efficiently desktop Grid computing systems can handle scheduling. In particular, the penalty is computed through execution patterns by introducing desktop execution entropy as represented in Eq. 6.18. We modeled a probability system for RC_{I_i} in each time unit in the Markov chain to derive entropy. The PESSIMIST requires stern penalties in terms of execution predictability and accuracy for prediction. In case of strict time constraints, a reasonable penalty for expected reliability deviation is imposed. RC improves estimation accuracy by considering the probability of I state once again to SC. The manifest credit value, RC, is given by Eq. 6.18.

$$RC_{I_i} = (FSC_{I_i} \times P_{I_{i-1}I_i}) \times (P_{I_iI_{i+1}} \times BSC_{I_i}) \qquad (6.18)$$

We propose the condition of inspecting whether the desktop is suitable to execute reliably or not, according to a given workload. The eligibility condition of the desktops to execute a job is given by Eq. 6.19, where n is the size of windows.

$$(\sum_{i=1}^{n} RC_{I_i} + \theta) \cdot (1 - \varphi) > RTF \qquad (6.19)$$

Entropy of RC_{I_i}, $\varepsilon_{RC_{I_i}}$, we obtain:

$$\theta = \frac{\kappa}{\varepsilon_{RC_{I_i}}} \qquad (6.20)$$

$$\varepsilon_{RC_{I_i}} = -\sum_{i=1}^{48} P_{RC_{I_i}} \log P_{RC_{I_i}} \qquad (6.21)$$

In Eq. 6.20, θ is a penalty value imposed to irregular, uncertain or otherwise volatile execution. θ deals with unpredictability of the desktop appropriately by using the entropy with specific intensity. κ is a constant value assigned for each thirty-minute interval, which is changeable according to system policy for penalty intensity. The entropy is derived from the probability system of RC_{I_i} modeled with a rate of each RC_{I_i} to the sum of RC_{I_i}. φ suggests accuracy of prediction implicitly where φ stands for fault rate and $1 - \varphi$ stands for success rate. Fault rate φ is computed by $\varphi = \frac{\#of failure}{\#of failure + \#of success}$.

6.4.3 REALIST

The REALIST is introduced in cases of partial time constraints or performance enhancement. This scheme solves the problems of time limitation and delays at lower cost. As described in [23], uncompleted workloads must be redistributed efficiently since total completion time (i.e. makespan or turnaround time) is delayed by them. In this chapter, we emphasize the reallocation of the unfinished workload to stable and reliable desktops through the hybrid scheme.

The REALIST uses elements of the two above-mentioned schemes by varying degrees. Loose and strict policies in time limit are applied to the REALIST according to workload requirements. The REALIST is a hybrid scheduling scheme of the OPTIMIST and the PESSIMIST, utilizing the OPTIMIST in usual situations and incorporating the PESSIMIST in urgent cases resulting from execution failure or time constraints. Actually this scheme was motivated by [23], which reported that job completion time is delayed due to workload failure near the end of the execution. Therefore, we propose the hybrid REALIST scheme to prevent highly volatile and unpredictable desktop resources with low cost. In the REALIST, the scheduler selects a desktop by using the OPTIMIST according to the time required to finish the work load at the time of job allocation. After a specific timeout, the scheduler chooses another desktop by using the PESSIMIST according to strict time limitations for more reliable job completion. The REALIST not only enhances performance but also guarantees job completion within time constraints.

6.5 Selection of Credible Resource with Elastic Window

This section describes Selection of Credible Resource with Elastic Window (SCREW) and depicts how it works in detail. Since a desktop is an autonomous computing unit, the desktop is free to take part in computation in desktop Grids. Desktops freely share their resources and, in the same breath, can terminate their participation. Some desktops may even disappear without producing any results. If a desktop fails to finish a workload, the system must reschedule the failed workload to another desktop, thus delaying total execution time (i.e. completion time or makespan). According to [19], uncompleted workloads and repeated reallocation cause deferred makespan. The systems which are negligent to distinguish between volatile resources suffer from unstable and unpredictable system behavior. To reduce such deterioration in system performance in terms of time and computational capacity due to the volatile nature of desktop participation, SCREW provides advanced resource selection sorting of reliable desktops as well as flexibility in meeting application requirements. As a result, SCREW can choose more dedicated desktop peers, which can complete the allocated workloads within the time requirement. Through this chapter, we will illustrate how SCREW enables the system to select a reliable, sound, and stable desktop by using an elastic window. SCREW is a novel resource selection scheme to overcome the restrictions caused by resource volatility while taking into account dynamic state changes. SCREW concentrates on solving critical system stability and performance degradation problems caused by desktop volatility. On the other hand, it supports a self-adjusting formula for workload requirements and selects eligible desktops according to the user and the application requests. We reach a solution for problems caused by volatility with a stochastic approach, considering the dynamic features and their behavioral patterns as time varies, unlike the existing approaches which use a simple statistical approach. Furthermore, we add flexibility, meeting the requirements of the workloads and adapting the needs from the user or the application dynamically.

The fundamental characteristics of a desktop platform are autonomy, volatility, malice, etc. To provide a reliable and stable system environment, the system must consider these unstable features. SCREW confronts volatile resources to select reliable resources using a flexible window based on stochastic state modeling in desktop Grid environments. In addition, it forms the basis of desktop availability and duration. Elastic Window (EW) is a self-adjusting window that supports curtailing and expanding the window and calculating a consecutive SC while SC represents the probability of state occurrence (i.e. especially I state). The EW is based on the probabilistic observation prediction model solving the time-unit-oriented architecture and transition relationships with representative probabilities. The EW is used to select a desktop more reliably as well as to adapt to workload requirements flexibly while giving the system what it needs, no more and no less, based on a probabilistic model. The window size is related to time constraint and is operated by time-related value needed by workloads or desired by users or the application. SCREW computes the desktops' usable window to consider how much execution desktops can process over time. This elastic window moves to the right as SCREW calculates SC_{Ii} and compares it with SC_{need} in which SC_{Ii} represents a desktop's SC of I state in i_{th} time unit and SC_{need} denotes needed SC to complete the workload. It has a flexible structure supporting extension and contraction operations on window size. To control the size of the window, SCREW supports forward and backward operation in which the starting point is the left most edge representing the first part of the checking spot and the ending point is the right most edge representing the last part of the checking spot. There are two cases to lengthen the window: increase of SC_{Ii} due to stable execution of the desktop, and the relief of the required time amount coming from workload changes or user requirement. The enlarged window, through extension operations, includes a much wider time scope (i.e., the number of states). On the other hand, there are two cases to shorten the window: the decrease of SC_{Ii} and the increase of the required time amount in the same manner. The contraction operation curtails window size in order to impose penalties to prevent volatile activities. The reason why the window is

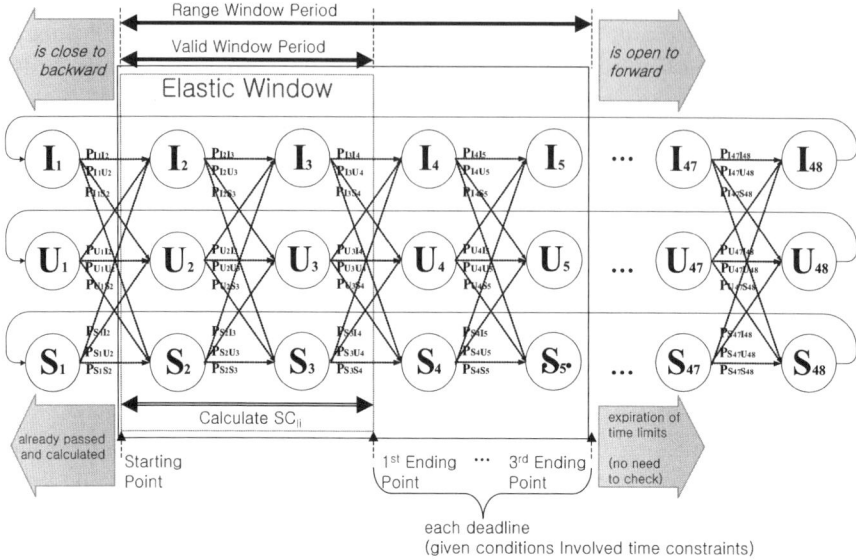

Fig. 6.2. Selection of Credible Resource with Elastic Window

called EW is that it can extend or contract according to workload requirements and conditions. The left edge of window is fixed (i.e. closed), leaving the right side extended. Since the left edge is set as the starting point, the left edge of the window cannot move backward. On the other hand, the right edge of the window is open to advance to the right, allowing for condition checks. Also the right edge contracts to the left side by applying penalties. The ending point of the Window Period presents the deadline in which the ending point of Basic Window Period should finish by the deadline. SCREW shrinks the window by moving the right edge to the left. If the left edge reaches the right edge, it is called a point window. In the point window, since SCREW just checks the SC_{Ii} of the pointed time unit, the point window is used to check the present state of the desktop. We illustrate the dynamics of the elastic window operation and activity and show how it calculates the SC_{Ii} within given time constraints and checks conditions to decide eligibility.

SCREW has to keep checking the state from the starting time unit to the deadline until the measurement meets requirements related to workload execution completion time. SCREW must adjust the window to the number of time units available under workload conditions. The window expands only in a forward direction because we focus on predicting desktop state (i.e. activity) in subsequent time units. The EW is interested in the flow of time in the model based on time-travel structure. Fig. 6.2 shows the position changes of the window in the probabilistic model according to contraction and expansion. If the requirements of the workload vary as time goes by, the size of the window expands. SCREW extends its window and calculates SC_{Ii}. SCREW repeats this process until it satisfies the given condition about the expected value of SC_{Ii}. This process is related to the workload characteristics, resulting in the expansion of the window size. SCREW controls the measurement and selection quantifying the amount of workload that a peer desktop can provide before allocation. In an extreme case, a desktop cannot process the workload allocated since the desktop frequently suspends execution or declines to donate the necessary time or physical capacity necessary to complete the workload. We define a window that is imposed on the availability model of the desktop. The system allocates a workload amount is measured by EW. To accomplish dependable resource selection, SCREW uses EW. EW spans a portion of the desktop capability containing the time that the desktop can execute before workload allocation. The amount of time capability, SC_{Ii} being calculated in the probabilistic model, is measured and adjusted by EW.

6.6 Performance Evaluation

6.6.1 Mathematical Analysis of Execution Completion Probability

In mathematical analysis, the execution completion probability is chosen to determine how much the MJSA is leveraged for system performance. In this chapter, we assume that the availability of desktop follows a hyper-exponential distribution according to [5] indicating that desktop availability in enterprise and the Internet fits the hyper-exponential or Weibull distribution. The probability of execution success measures the probability of continuous execution. The success probability of each scheduling scheme is compared with the proposed MJSA as shown in equations Eq. 6.22 through Eq. 6.26, where MJSA-P, MJSA-O, MJSA-R represents PESSIMIST, OPTIMIST, and REALIST, respectively.

$$P\{S_{exe}^{FCFS}\} = \sum_{i=1}^{n} p_i \lambda_i e^{-\lambda_i(\Delta + (1-(\alpha+\beta))\Delta)} \tag{6.22}$$

$$P\{S_{exe}^{Eager}\} = \sum_{i=1}^{n} p_i \lambda_i e^{-\lambda_i(\Delta + (1-\alpha)\Delta)} \tag{6.23}$$

$$P\{S_{exe}^{MJSA-O}\} = \sum_{i=1}^{n} p_i \lambda_i e^{-\lambda_i \Delta} \tag{6.24}$$

$$P\{S_{exe}^{MJSA-P}\} = \sum_{i=1}^{n} p_i \lambda_i e^{-\lambda_i \Delta}, \lambda_i = min\{\lambda_1, \lambda_2, ...\} \tag{6.25}$$

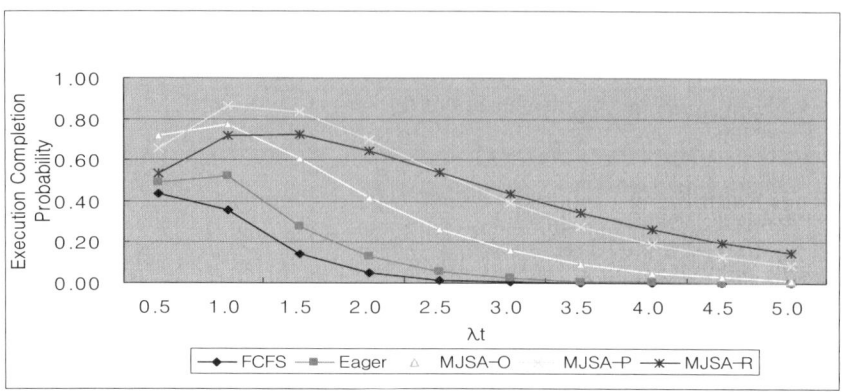

Fig. 6.3. Comparisons of Execution Completion Probability

$$P\{S_{exe}^{MJSA-R}\} = \sum_{i=1}^{n} p_i \lambda_i e^{-\lambda_i \Delta}, \lambda_i = mid\{\lambda_i, min\{\lambda_1, \lambda_2, ...\}\} \quad (6.26)$$

λ means execution stop rate and Δ means the time required to execute a workload. Execution time deteriorates when the workload is repeatedly allocated.

The FCFS scheme requires time for reallocation and delays; α and β, respectively, in Eq. 6.22. Eq. 6.23 shows that eager scheduling exceeds execution time because it makes Δ increase by requiring additional reallocation time, denoted by α. In these schemes, execution time worsens when the workload is reallocated repeatedly. The results indicate steadily lower execution success probability. In Eq. 6.24, MJSA-O maintains the initial execution time, promising a high execution success probability since it considers the quality of execution, (i.e. duration of availability) in advance. In addition, MJSA-P obtains good success probability by declining to $min\{\lambda_1, \lambda_2, ...\}$ as shown in Eq. 6.25 while MJSA-R attains λ_i with median of λ_i and $min\{\lambda_1, \lambda_2, ...\}$ in Eq. 6.26. Fig. 6.3 indicates that the proposed schemes result in greater execution success as the stop rate increases.

6.6.2 Problem Solving with SCREW

This section shows how SCREW solves the given problems from user needs or meets application requirements with mathematical approaches. The probability estimation for each sequence of states in the model, the most probable sequence, as well as the probability of the visits meaningful states and remains the sequence of the states. In addition, we illustrate how SCREW meets user and application requirements. The probability of computation and resource selection are done by SCREW, given diverse conditions through probability in terms of state sequence, selection range, accuracy rate, etc.

Solicitation of Certain Consecutive States Problem

Problem: Users require the desktop to have the probability that the desktop state has ten successive state sequences such as "$I\ I\ I\ S\ S\ I\ U\ I\ I\ S$" (i.e. successive states of "*idle-idle-idle-stop-stop-idle-used-idle-idle-stop*") corresponding to the model. What is the probability of consecutive state conditions?

Solution: The observation sequence (OS) is given as "$I\ I\ I\ S\ S\ I\ U\ I\ I\ S$" desktop state conditions over a ten-time unit period. In this problem, we summarized the states and the observations. To calculate the probability of observation sequence O given in the model, $P(q_1, q_2, ..., q_{48}|O, \lambda)$, we can calculate

$$\begin{aligned} P(O|\lambda) &= P[I, I, I, I, S, S, I, U, I, I, S|\lambda] \\ &= P[I]P[I|I]^2 P[S|I]P[S|S]P[I|S]P[U|I]P[I|U]P[S|I]^2 \quad (6.27) \\ &= \pi_1 \cdot (b_{II})^2 b_{IS} b_{SS} b_{SI} b_{SU} b_{UI} b_{II} b_{IS} \end{aligned}$$

Satisfaction of Probability of Durability Problem

Problem: How can SCREW compute the probability of remaining in a certain state in which the system is in a known state?

Solution: The duration is an expected number of state observations. The probability of durability is calculated from the expected number of observations in a state by taking the expectation (i.e. mean) of the quantity. The probability of the duration is evaluated with the probability of the observation given the model using the Bayesian Rule as follows.

$$P(O|\lambda, q_1 = i) = \frac{P(O, q_1 = i|\lambda)}{P(q_1 = i)} =$$
$$= \frac{\pi_i \cdot (b_{II})^{d-1} \cdot (1 - b_{II})}{\pi_i} = (b_{II})^{d-1} \cdot (1 - b_{II}) = P_i(d)$$

Since the measure $P_i(d)$ is the probability distribution function (pdf) of duration d in state I, we expect the number of successive desktop states as

$$\bar{d}_i = \sum_{d=1}^{\infty} dP_i(d) = \sum_{d=1}^{\infty} d(b_{II})^{d-1}(1-b_{II}) = \frac{1}{1-b_{II}} \quad (6.28)$$

K-step transition probabilities Problem

Problem: What is the probability that the model makes remain certain states after k-step (i.e. length of time unit)?

Solution: The k-step transition probabilities of the model are developed from the previous state at step k. The k-step transition matrix is given as

$$P_{ij}^{(k)} = Pr\{X_{n+k} = j | X_n = i\} \forall k \geq 1 \quad (6.29)$$

$$P^{(1)} = [P_{ij}{}^{(k)}]_{i,j \in S} = [P_{ij}]_{i,j \in S} = P, where P_{ij}{}^{(1)} = P_{ij} \quad (6.30)$$

$$P^{(0)} = [P_{ij}{}^{(0)}]_{i,j \in S} = I \quad (6.31)$$

$$P^{(k)} = P^k \ \forall k \geq 0 \quad (6.32)$$

After all, the k-step transition matrix is the power k of the one step transition. This fact helps prevent radical increase of the amount of the calculation as the length of the step (i.e. the number of path state).

$$P_{ij}{}^k = \begin{cases} P_{ij}, k = 1 \\ P_{ij}{}^k, k > 1 \end{cases} \quad (6.33)$$

$$P_{ij}{}^k == \begin{cases} \pi^{(0)} P, \pi^{(0)} = initial probability \\ \pi^{(0)} P^n, \pi^{(0)} = initial probability \end{cases} \quad (6.34)$$

Satisfaction of Revisit Probabilities Problem

Problem: How can SCREW compute the probability that the model makes revisit in a certain state?

Solution: The total number of visits state is important to measure quality of resource donation of the desktop. This means that the state stays at j state in the first n steps, starting from state i. The expected number of visits is given as

$$\varepsilon[\nu_{ij}{}^{(n)}] = \sum_{k=0}^{n-1} P_{ij}{}^k = \sum_{k=0}^{n-1} P^{(k)} = \sum_{k=0}^{n-1} P^k, where \varepsilon[I_{ij}{}^{(n)}] = P\{I_{ij}^{(k)} = 1\} = P_{ij}{}^k \quad (6.35)$$

6.6.3 Experimental Evaluation of Makespan

To evaluate the performance of these scheduling schemes, the experiment was tested on Korea@Home [17]. Korea@Home attempts to harness massive computing power using great numbers of personal desktops distributed over the Internet based on P2P technology. The MJSA was tested as an independent module on top of Korea@Home to improve reliability. We tested the scheduling schemes using the MJSA with New Drug Candidate Discovery running on fifty different desktops at Korea University and at the Korea Institute of Science and Technology Information (KISTI), with ten-unit increases from ten to fifty. We measured the availability of each volunteer every thirty minutes. We measured and updated the Markov chain over one month. Fig. 6.4 compares the makespan according to each scheduling scheme. Usually, the makespan is parameterized to measure performance reflecting computational power.

As shown in Fig. 6.4, the MJSA-P outperforms in makespan, especially when the number of desktops increases. The MJSA-R performs quite well, which possesses merits in both performance and cost efficiency. The MJSA-O performs better than either FCFS or eager scheduling in terms of reliability and stability. The MJSA-P adds ingenuity to the system by providing a more precise basis for reliable desktop selection, thus enhancing performance as well as stability.

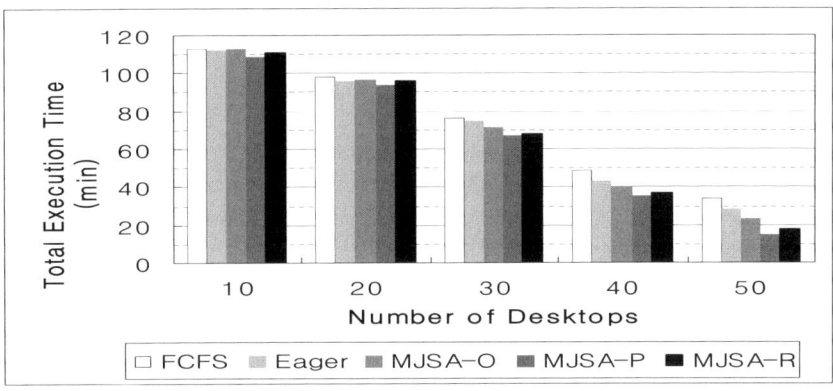

Fig. 6.4. Comparisons of Makespan

6.7 Conclusions

To solve the problems arising from volatility, this chapter demonstrates advanced desktop state modeling with the Markov chain, MJSA and a sophisticated resource selection scheme based on the stochastic model. A Markov-chain-based scheduler for desktop Grids consisting of machines that have high volatility, is presented. The Markov chain (three types of states for each forty-eight time units with thirty minutes intervals making up a full day) is represented with a credit assignment for the idle states on which the scheduler bases its decisions. In addition, the credit values, FSC, BSC, and SC, to measure the degree of desktop state reliability is presented. By using these accurate metrics, the MJSA improves system performance in terms of makespan and execution success when compared to earlier studies. In practice, the MJSA provides three advanced scheduling schemes: OPTIMIST, PESSIMIST, and REALIST in order to overcome limitations such as increasing makespan, decreasing performance, and unpromising job completion.

SCREW aims at giving a stable execution environment and suggests dependable resource selection. SCREW helps desktop systems to ensure requirements for reliable execution by measuring execution capability of each desktop through the elastic window (EW) based on the Markov desktop state model. EW is a method to derive a typical sequence of states for pattern recognition of state transition and compares it with similarities. The proposed SCREW adjusts a window size according to volatility as well as time constraint in order to adapt to an unpredictable and dynamic changing environment when predicting a resource's state in a selection procedure. SCREW advances system performance and improves system readability while satisfying needs from users and applications in dynamically changing circumstances. The approach provides good quality recognition performance on desktop behavior for a variety of desktop activities.

Enhanced performance has been achieved from mathematical analysis as well as experimental results. Consequently, the proposed MJSA helps to improve not only system performance but also stability.

References

1. Berman, F., Fox, G.C., Hey, J.G.: Grid Computing: Making the Global Infrastructure a Reality. Wiley, Chichester (2003)
2. Entropia, http://www.entropia.com
3. Bhagwan, R., Savage, S., Voelker, G.M.: Understanding availability. In: Kaashoek, M.F., Stoica, I. (eds.) IPTPS 2003. LNCS, vol. 2735, pp. 256–267. Springer, Heidelberg (2003)
4. Chu, J., Labonte, K., Levine, B.N.: Availability and Popularity Measurements of Peer-to-Peer File systems, Technical report 04-36 (2004)
5. Nurmi, D., Brevik, J., Wolski, R.: Modeling Machine Availability in Enterprise and Wide-Area Distributed Computing Environments. In: Cunha, J.C., Medeiros, P.D. (eds.) Euro-Par 2005. LNCS, vol. 3648, pp. 432–441. Springer, Heidelberg (2005)
6. Mutka, M.W., Livny, M.: The Available Capacity of a Privately Owned Workstation Environmont. Performance Evaluation 12(4), 269–284 (1991)
7. Yang, L., Schopf, J.M., Foster, I.: Conservative scheduling: Using predicted variance to improve scheduling decisions in dynamic environment. In: Proceedings of the SC 2003, pp. 31–46 (2003)
8. Sarmenta, L.F.G., Hirano, S.: Bayanihan: Building and Studying Web-Based Volunteer Computing Systems Using Java. Future Generation Computer Systems 15(5-6), 675–686 (1999)
9. Neary, M.O., Christiansen, B.O., Cappello, P., Schauser, K.E.: Javelin: Parallel computing on the Internet. Future Generation Computer Systems 15(5-6), 659–674 (1999)
10. Neary, M.O., Cappello, P.: Advanced Eager Scheduling for Java Based Adaptively Parallel Computing. In: Proceedings of the JGI 2002, pp. 56–65 (2002)
11. Anderson, D.P., Cobb, J., Korpela, E., Lebofsky, M., Werthimer, D.: SETI@home: an experiment in public-resource computing. Communications of the ACM 45(11) (November 2002)
12. Lau, L.F., Ananda, A.L., Tan, G., Wong, W.F.: Gucha: Internet-based parallel computing using Java. In: Proceedings of the ICA3PP, pp. 397–408 (2000)
13. Fedak, G., Germain, C., Neri, V., Cappello, F.: XtremWeb: A Generic Global Computing System. In: Workshop on Global Computing on Personal Devices, Proceedings of the CCGRID 2001, pp. 582–587 (2001)
14. Baratloo, A., Karaul, M., Kedem, Z.M., Wijckoff, P.: Charlotte: Metacomputing on the Web. Future Generation Computer Systems 15(5-6), 559–570 (1999)
15. Nisan, N., London, S., Regev, O., Camiel, N.: Globally Distributed Computation over Internet? The POPCORN Project. In: Proceedings of the ICDCS 1998, pp. 592–601 (1998)
16. Morrison, J.P., Kennedy, J.J., Power, D.A.: WebCom: A Web Based Volunteer Computer. The Journal of Supercomputing 18(1), 47–61 (2001)
17. Korea@Home, http://www.koreaathome.org
18. Byun, E.J., Choi, S.J., Baik, M.S., Park, C.Y., Jung, S.Y., Hwang, C.S.: Scheduling Scheme based on Dedication Rate in Volunteer Computing Environment. In: Proceedings of the ISPDC 2005, pp. 234–241 (2005)

19. Kondo, D., Fedak, G., Cappello, F., Chien, A., Casanova, H.: Characterizing resource availability in enterprise desktop grids. Future Generation Computing Systems 23(7), 888–903 (2007)
20. Casanova, H., Legrand, A., Zagorodnov, D., Berman, F.: Heuristics for Scheduling Parameter Sweep Applications in Grid Environments. In: Proceedings of the 9th Heterogeneous Computing Workshop, pp. 349–363 (2000)
21. Wolski, R., Spring, N., Hayes, J.: The Network Weather Service: A Distributed Resource Performance Forecasting Service for Metacomputing. Future Generation Computing Systems 15(5-6), 757–768 (1999)
22. Litzkow, M., Livny, M., Mutka, M.: Condor - A Hunter of Idle Workstations. In: Proceedings of the ICDCS 1998, pp. 104–111 (1988)
23. Kondo, D., Chien, A., Casanova, H.: Resource Management for Rapid Application Turnaround on Enterprise Desktop Grids. In: Proceedings of the SC 2004, pp. 17–29 (2004)
24. Schopt, J., Berman, F.: Stochastic Scheduling. In: Proceedings of the SC 1999, p. 48 (1999)
25. Rainer, L.R.: A Tutorial on Hidden Markov Models and Selected Application in Speech Recognition. Proceedings of the IEEE 77(2), 257–286 (1989)
26. Forney Jr., G.D.: The Viterbi Algorithm. Proceedings of the IEEE 61(3), 268–278 (1973)

7

Workflow Scheduling Algorithms for Grid Computing

Jia Yu, Rajkumar Buyya, and Kotagiri Ramamohanarao

Grid Computing and Distributed Systems (GRIDS) Laboratory
Department of Computer Science and Software Engineering
The University of Melbourne, VIC 3010 Austraila
{jiayu,raj,rao}@csse.unimelb.edu.au

Summary. Workflow scheduling is one of the key issues in the management of workflow execution. Scheduling is a process that maps and manages execution of inter-dependent tasks on distributed resources. It introduces allocating suitable resources to workflow tasks so that the execution can be completed to satisfy objective functions specified by users. Proper scheduling can have significant impact on the performance of the system. In this chapter, we investigate existing workflow scheduling algorithms developed and deployed by various Grid projects.

Keywords: Workflow scheduling, Inter-dependent tasks, Distributed resources, Heuristics.

7.1 Introduction

Grids [22] have emerged as a global cyber-infrastructure for the next-generation of e-Science and e-business applications, by integrating large-scale, distributed and heterogeneous resources. A number of Grid middleware and management tools such as Globus [21], UNICORE [1], Legion [27] and Gridbus [13] have been developed, in order to provide infrastructure that enables users to access remote resources transparently over a secure, shared scalable world-wide network. More recently, Grid computing has progressed towards a service-oriented paradigm [7,24] which defines a new way of service provisioning based on utility computing models. Within utility Grids, each resource is represented as a service to which consumers can negotiate their usage and Quality of Service.

Scientific communities in areas such as high-energy physics, gravitational-wave physics, geophysics, astronomy and bioinformatics, are utilizing Grids to share, manage and process large data sets. In order to support complex scientific experiments, distributed resources such as computational devices, data, applications, and scientific instruments need to be orchestrated while managing the application workflow operations within Grid environments [36]. Workflow is concerned with the automation of procedures, whereby files and other data are passed between participants according to a defined set of rules in order to achieve an overall goal [30]. A workflow management system defines, manages and executes workflows on computing resources.

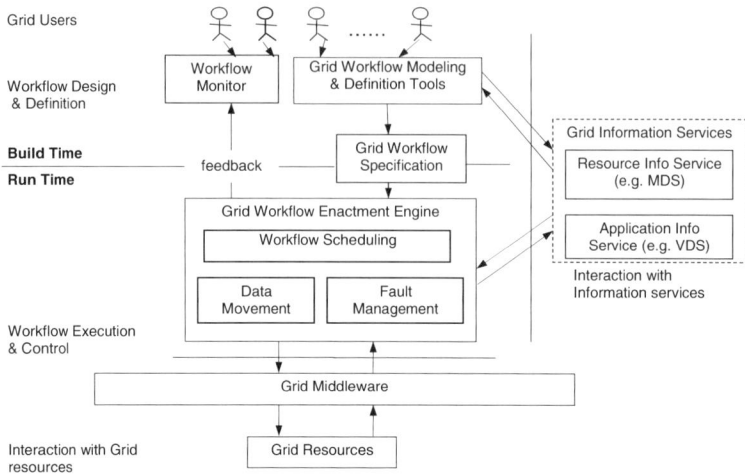

Fig. 7.1. Grid Workflow Management System

Fig. 7.1 shows an architecture of workflow management systems for Grid computing. In general, a workflow specification is created by a user using workflow modeling tools, or generated automatically with the aid of Grid information services such as MDS(Monitoring and Discovery Services) [20] and VDS (Virtual Data System) [23] prior to the run time. A workflow specification defines workflow activities (tasks) and their control and data dependencies. At run time, a workflow enactment engine manages the execution of the workflow by utilizing Grid middleware. There are three major components in a workflow enactment engine: the workflow scheduling, data movement and fault management. Workflow scheduling discovers resources and allocates tasks on suitable resources to meet users' requirements, while data movement manages data transfer between selected resources and fault management provides mechanisms for failure handling during execution. In addition, the enactment engine provides feedback to a monitor so that users can view the workflow process status through a Grid workflow monitor. Workflow scheduling is one of the key issues in the workflow management [59].

A scheduling is a process that maps and manages the execution of interdependent tasks on the distributed resources. It allocates suitable resources to workflow tasks so that the execution can be completed to satisfy objective functions imposed by users. Proper scheduling can have significant impact on the performance of the system. In general, the problem of mapping tasks on distributed services belongs to a class of problems known as NP-hard problems [53]. For such problems, no known algorithms are able to generate the optimal solution within polynomial time. Solutions based on exhaustive search are impractical as the overhead of generating schedules is very high. In Grid environments, scheduling decisions must be made in the shortest time possible, because there are many users competing for resources, and time slots desired by one user could be taken up by another user at any moment.

Many heuristics and meta-heuristics based algorithms have been proposed to schedule workflow applications in heterogeneous distributed system environments. In this chapter, we discuss several existing workflow scheduling algorithms developed and deployed in various Grid environments.

7.2 Workflow Scheduling Algorithms for Grid Computing

Many heuristics [33] have been developed to schedule inter-dependent tasks in homogenous and dedicated cluster environments. However, there are new challenges for scheduling workflow applications in a Grid environment, such as:

- Resources are shared on Grids and many users compete for resources.
- Resources are not under the control of the scheduler.
- Resources are heterogeneous and may not all perform identically for any given task.
- Many workflow applications are data-intensive and large data sets are required to be transferred between multiple sites.

Therefore, Grid workflow scheduling is required to consider non-dedicated and heterogeneous execution environments. It also needs to address the issue of large data transmission across various data communication links.

The input of workflow scheduling algorithms is normally an *abstract workflow model* which defines workflow tasks without specifying the physical location of resources on which the tasks are executed. There are two types of abstract workflow model, *deterministic* and *non-deterministic*. In a deterministic model, the dependencies of tasks and I/O data are known in advance, whereas in a non-deterministic model, they are only known at run time.

The workflow scheduling algorithms presented in the following sections are based on the deterministic type of the abstract workflow model and are represented as a Directed Acyclic Graph (DAG). Let Γ be the finite set of tasks $T_i (1 \leq i \leq n)$. Let Λ be the set of directed edges. Each edge is denoted by (T_i, T_j), corresponding to the data communication between task T_i and T_j, where T_i is called an immediate parent task of T_j, and T_j the immediate child task of T_i. We assume that a child task cannot be executed until all of its parent tasks are completed. Then, the workflow application can be described as a tuple $\Omega(\Gamma, \Lambda)$.

In a workflow graph, a task which does not have any parent task is called an *entry task*, denoted as T_{entry} and a task which does not have any child task is called an *exit task*, denoted as T_{exit}. If a workflow scheduling algorithm requires a single entry task or a single exit task, and a given workflow contains more than one entry task or exit task in the workflow graph, we can produce a new workflow by connecting entry points to a zero-cost pseudo entry and exiting nodes to an exit task, without without affecting the schedule [45].

To date, there are two major types of workflow scheduling (see Fig. 7.2), *best-effort based* and *QoS constraint based* scheduling. The best-effort based scheduling attempts to minimize the execution time ignoring other factors such as the

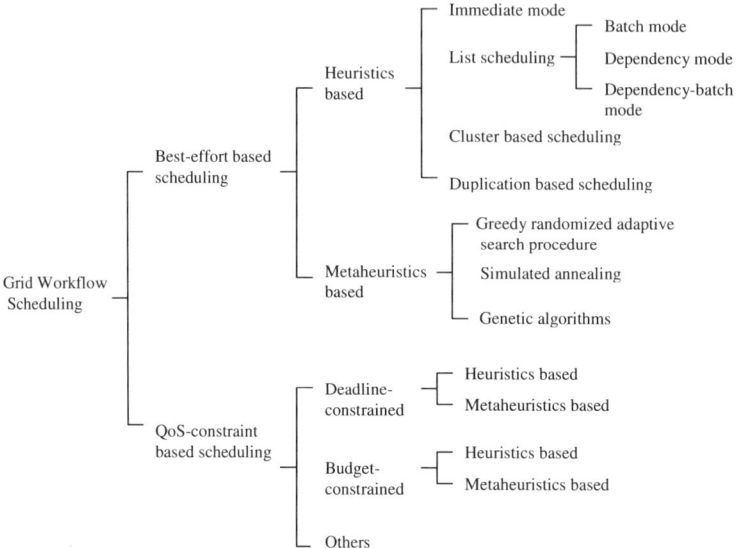

Fig. 7.2. A taxonomy of Grid workflow scheduling algorithms

monetary cost of accessing resources and various users' QoS satisfaction levels. On the other hand, QoS constraint based scheduling attempts to minimize performance under most important QoS constraints, for example time minimization under budget constraints or cost minimization under deadline constraints.

7.3 Best-Effort Based Workflow Scheduling

Best-effort based workflow scheduling algorithms are targeted towards Grids in which resources are shared by different organizations, based on a community model (known as *community Grid*). In the community model based resource allocation, monetary cost is not considered during resource access. Best-effort based workflow scheduling algorithms attempt to complete execution at the earliest time, or to minimize the makespan of the workflow application. The makespan of an application is the time taken from the start of the application, up until all outputs are available to the user [14].

In general, best-effort based scheduling algorithms are derived from either *heuristics based* or *meta-heuristics based* approach. The heuristic based approach is to develop a scheduling algorithm which fit only a particular type of problem, while the meta-heuristic based approach is to develop an algorithm based on a meta-heuristic method which provides a general solution method for developing a specific heuristic to fit a particular kind of problem [29].Table 7.1 7.2 show the overview of best-effort based scheduling algorithms.

Table 7.1. Overview of Best-effort Workflow Scheduling Algorithms (Heuristics)

Scheduling Method		Algorithm	Project	Organization	Application
Individual task scheduling		Myopic	Condor DAG Man	University of Wisconsin-Madison, USA.	N/A
List scheduling	Batch mode	Min-Min	vGrADS	Rice University, USA.	EMAN bio-imaging
			Pegasus	University of Southern California, USA.	Montage astronomy
		Max-min	vGrADS	Rice University, USA.	EMAN bio-imaging
		Sufferage	vGrADS	Rice University, USA.	EMAN bio-imaging
	Dependency mode	HEFT	ASKALON	University of Innsbruck, Austria.	WIEN2K quantum chemistry & Invmod hydrological
	Dependency-batch mode	Hybrid	Sakellarious & Zhao	University of Manchester, UK.	Randomly generated task graphs
Cluster based scheduling Duplication based scheduling		THAN	Ranaweera & Agrawal	University of Cincinnati, USA	Randomly generated task graphs

Table 7.2. Overview of Best-effort Workflow Scheduling Algorithms (Meta-heuristics)

Scheduling Method	Project	Organization	Application
Greedy randomized adaptive search procedure (GRASP)	Pegasus	University of Southern California, USA.	Montage astronomy
Genetic algorithms (GA)	ASKALON	University of Innsbruck, Austria.	WIEN2K quantum chemistry
Simulated annealing (SA)	ICENI	London e-Science Centre, UK.	Randomly generated task graphs

7.3.1 Heuristics

In general, there are four classes of scheduling heuristics for workflow applications, namely *individual task scheduling*, *list scheduling*, and *cluster and duplication based scheduling*.

Individual task scheduling

The individual task scheduling is the simplest scheduling method for scheduling workflow applications and it makes schedule decision based only on one individual task. The *Myopic* algorithm [55] has been implemented in some Grid systems such as Condor DAGMan [49]. The detail of the algorithm is shown in Algorithm 7.1. The algorithm schedules an unmapped ready task to the resource that is expected to complete the task earliest, until all tasks have been scheduled.

Algorithm 7.1. Myopic scheduling algorithm
1: **while** $\exists t \in \Gamma$ is not completed **do**
2:     $task \leftarrow$ get a ready task whose parent tasks have been completed
3:     $r \leftarrow$ for $t \in task$, get a resource which can complete t at the earliest time
4:     schedule t on r
5: **end while**

List scheduling

A list scheduling heuristic prioritizes workflow tasks and scheldules the tasks based on their priorities. There are two major phases in a list scheduling heuristic, the *task prioritizing* phase and the *resource selection* phase [33]. The task prioritizing phase sets the priority of each task with a rank value and generates a scheduling list by sorting the tasks according to their rank values. The resource selection phase selects tasks in the order of their priorities and map each selected task on its optimal resource.

Different list scheduling heuristics use different attributes and strategies to decide the task priorities and the optimal resource for each task. We categorize workflow-based list scheduling algorithms as either *batch*, *dependency* or *dependency-batch* mode.

The batch mode scheduling group workflow tasks into several independent tasks and consider tasks only in the current group. The dependency mode ranks workflow tasks based on its weight value and the rank value of its inter-dependent tasks, while the dependency-batch mode further use a batch mode algorithm to re-ranks the independent tasks with similar rank values.

Batch mode

Batch mode scheduling algorithms are initially designed for scheduling parallel independent tasks, such as bag of tasks and parameter tasks, on a pool of resources. Since the number of resources is much less than the number of tasks, the tasks need to be scheduled on the resources in a certain order. A batch mode algorithm intends to provide a strategy to order and map these parallel tasks on the resources, in order to complete the execution of these parallel tasks at earliest time. Even though batch mode scheduling algorithms aim at the scheduling problem of independent tasks; they can also be applied to optimize the execution time of a workflow application which consists of a lot of independent parallel tasks with a limited number of resources.

Batch Mode Algorithms. *Min-Min, Max-Min, Sufferage* proposed by Maheswaran et al. [39] are three major heuristics which have been employed for scheduling workflow tasks in vGrADS [4] vGrADS [4] and pegasus [11]. The heuristics is based on the performance estimation for task execution and I/O data transmission. The definition of each performance metric is given in Table 7.3.

Table 7.3. Performance Matrices

Symbol	Definition
$EET(t,r)$	**Estimated Execution Time:** the amount of time the resource r will take to execute the task t, from the time the task starts to execute on the resource.
$EAT(t,r)$	**Estimated Availability Time:** the time at which the resource r is available to execute task t.
$FAT(t,r)$	**File Available Time:** the earliest time by which all the files required by the task t will be available at the resource r.
$ECT(t,r)$	**Estimated Completion Time:** the estimated time by which task t will complete execution at resource r: $ECT(t,r) = EET(t,r) + \max(EAT(t,r), FAT(t,r))$
$MCT(t)$	**Minimum Estimated Completion Time:** minimum ECT for task t over all available resources.

Algorithm 7.2. Min-Min and Max-Min task scheduling algorithms

1: **while** $\exists t \in \Gamma$ is not scheduled **do**
2:    $availTasks \leftarrow$ get a set of unscheduled ready tasks whose parent tasks have been completed
3:    schedule($availTasks$)
4: **end while**
5: **PROCEDURE:** schedule($availTasks$)
6: **while** $\exists t \in availTasks$ is not scheduled **do**
7:    **for all** $t \in availTasks$ **do**
8:       $availResources \leftarrow$ get available resources for t
9:       **for all** $r \in availResources$ **do**
10:          compute $ECT(t,r)$
11:       **end for**
12:       // get $MCT(t,r)$ for each resource
          $R_t \leftarrow \min_{r \in availResources} ECT(t,r)$
13:    **end for**
14:    // Min-Min: get a task with minimum $ECT(t,r)$ over tasks
      $T \leftarrow \arg\min_{t \in availTasks} ECT(t, R_t)$
      // Max-Min: get a task with maximum $ECT(t,r)$ over tasks
      $T \leftarrow \arg\max_{t \in availTasks} ECT(t, R_t)$
15:    schedule T on R_T
16:    remove T from $availTasks$
17:    update $EAT(R_T)$
18: **end while**

Algorithm 7.3. Sufferage task scheduling algorithm.

1: **while** $\exists t \in \Gamma$ is not completed **do**
2:    $availTasks \leftarrow$ get a set of unscheduled ready tasks whose parent tasks have been completed
3:    schedule($availTasks$)
4: **end while**
5: **PROCEDURE:** schedule($availTasks$)
6: **while** $\exists t \in availTasks$ is not scheduled **do**
7:    **for all** $t \in availTasks$ **do**
8:       $availResources \leftarrow$ get available resources for t
9:       **for all** $r \in availResources$ **do**
10:          compute $ECT(t,r)$
11:       **end for**
12:       // compute earliest ECT
         $R_t^1 \leftarrow arg \min_{r \in availResources} ECT(t,r)$
13:       // compute second earliest ECT
         $R_t^2 \leftarrow arg \min_{r \in availResources \& r \neq R_t^1} ECT(t,r)$
14:       // compute sufferage value for task t
         $suf_t \leftarrow ECT(t, R_t^2) - ECT(t, R_t^1)$
15:    **end for**
16:    $T \leftarrow arg \max_{t \in availTasks} suf_t$
17:    schedule T on R_T^1
18:    remove T from $availTasks$
19:    update $EAT(R_T)$
20: **end while**

The Min-Min heuristic schedules sets of independent tasks iteratively (Algorithm 7.2: 1-4). For each iterative step, it computes ECTs(Early Completion Time) of each task on its every available resource and obtains the MCT(Minimum Estimated Completion Time) for each task (Algorithm 7.2: 7-12). A task having minimum MCT value over all tasks is chosen to be scheduled first at this iteration. It assigns the task on the resource which is expected to complete it at earliest time.

The Max-Min heuristic is similar to the Min-Min heuristic. The only difference is the Max-Min heuristic sets the priority to the task that requires longest execution time rather than shortest execution time. After obtaining MCT values for each task (Algorithm 7.2: 7-13), a task having maximum MCT is chosen to be scheduled on the resource which is expected to complete the task at earliest time. Instead of using minimum MCT and maximum MCT, the Sufferage heuristic sets priority to tasks based on their sufferage value. The sufferage value of a task is the difference between its earliest completion time and its second earliest completion time (Algorithm 7.3: 12-14).

Comparison of batch mode algorithms. The overview of three batch mode algorithms is shown in Table 7.4. The Min-Min heuristic schedules tasks having shortest execution time first so that it results in the higher percentage of

Table 7.4. Overview of batch mode algorithms

Algorithm	Features
$Min-Min$	It sets high scheduling priority to tasks which have the shortest execution time.
$Max-Min$	It sets high scheduling priority to tasks which have long execution time.
$Sufferage$	It sets high scheduling priority to tasks whose completion time by the second best resource is far from that of the best resource which can complete the task at earliest time.

tasks assigned to their best choice (which can complete the tasks at earlist time) than Max-Min heuristics [12]. Experimental results conducted by Maheswaran et al. [39] and Casanova et al. [14] have proved that Min-Min heuristic outperform Max-Min heuristic. However, since Max-min schedule tasks with longest execution time first, a long execution execution task may have more chance of being executed in parallel with shorter tasks. Therefore, it might be expected that the Max-Min heuristic perform better than the Min-Min heuristic in the cases where there are many more short tasks than long tasks [12, 39].

On the other hand, since the Sufferage heuristic considers the adverse effect in the completion time of a task if it is not scheduled on the resource having with minimum completion time [39], it is expected to perform better in the cases where large performance difference between resources. The experimental results conducted by Maheswaran et al. shows that the Sufferage heuristic produced the shortest makespan in the high heterogeneity environment among three heuristics discussion in this this section. However, Casanova et al. [14] argue that the Sufferage heuristic could perform worst in the case of data-intensive applications in multiple cluster environments.

Extended batch mode algorithms. XSufferage is an extension of the Suffereage heuristic. It computes the sufferage value on a cluster level with the hope that the files presented in a cluster can be maximally reused. A modified Min-Min heuristic, *QoS guided Min-Min*, is also proposed in [28]. In addition to comparing the minimum completion time over tasks, it takes into account different levels of quality of service (QoS) required by the tasks and provided by Grid resources such as desirable bandwidth, memory and CPU speed. In general, a task requiring low levels of QoS can be executed either on resources with low QoS or resources with high QoS, whereas the task requiring high levels of QoS can be processed only on resources with high QoS. Scheduling tasks without considering QoS requirements of tasks may lead to poor performance, since low QoS tasks may have higher priority on high QoS resources than high QoS tasks, while resources with low QoS remain idle [28]. The QoS guided Min-Min heuristic starts to map low QoS tasks until all high QoS tasks have been mapped. The priorities of tasks with the same QoS level are set in the same way of the Min-Min heuristic.

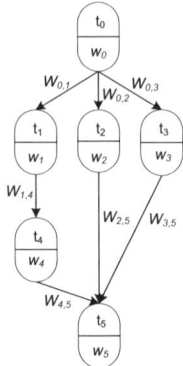

Fig. 7.3. A weighted task graph example

Dependency Mode

Dependency mode scheduling algorithms are derived from the algorithms for scheduling a task graph with interdependent tasks on distributed computing environments. It intends to provide a strategy to map workflow tasks on heterogeneous resources based on analyzing the dependencies of the entire task graph, in order to complete these interdependent tasks at earliest time. Unlike batch mode algorithms, it ranks the priorities of all tasks in a workflow application at one time.

Many dependency mode heuristics rank tasks based on the weights of task nodes and edges in a task graph. As illustrated in Fig. 7.3, a weight w_i is assigned to a task T_i and a weight $w_{i,j}$ is assigned to an edge (T_i, T_j). Many list scheduling schemes [33] developed for scheduling task graphs on homogenous systems set the weight of each task and edge to be equal to its estimation execution time and communication time, since in a homogenous environment, the execution times of a task and data transmission time on all available resources are identical. However, in a Grid environment, resources are heterogeneous. The computation time varies from resource to resource and the communication time varies from data link to data link between resources. Therefore, it needs to consider processing speeds of different resources and different transmission speeds of different data links and an approximation approach to weight tasks and edges for computing the rank value.

Zhao and Sakellariou [62] proposed six possible approximation options, *mean value*, *median value*, *worst value*, *best value*, *simple worst value*, and *simple best value*. These approximation approaches assign a weight to each task node and edge as either the average, median, maximum, or minimum computation time and communication time of processing the task over all possible resources. Instead of using approximation values of execution time and transmission time, Shi and Dongarra [46] assign a higher weight task with less capable resources. Their motivation is quite similar to the QoS guided min-min scheduling, i.e., it may cause longer delay if tasks with scarce capable resources are not scheduled first, because there are less choices of resources to process these tasks.

Dependency Mode Algorithm. The *Heterogeneous-Earliest-Finish-Time* (HEFT) algorithm proposed by Topcuoglu et al. [51] has been applied by the ASKALON project [18,55] to provide scheduling for a quantum chemistry application, WIEN2K [10], and a a quantum chemistry application, and a hydrological application, Invmod [43], on the Austrian Grid [2].

As shown in Algorithm 7.4, the algorithm first calculates average execution time for each task and average communication time between resources of two successive tasks. Let $time(T_i, r)$ be the execution time of task T_i on resource r and let R_i be the set of all available resources for processing T_i. The average execution time of a task T_i is defined as

$$\varpi_i = \frac{\sum_{r \in R_i} time(T_i, r)}{|R_i|} \qquad (7.1)$$

Let $time(e_{ij}, r_i, r_j)$ be the data transfer time between resources r_i and r_j which process the task Ti and task T_j respectively. Let R_i and R_j be the set of all available resources for processing Ti and T_j respectively. The average transmission time from T_i to T_j is defined by:

$$\overline{c_{ij}} = \frac{\sum_{r_i \in R_i, r_j \in R_j} time(e_{ij}, r_i, r_j)}{|R_i||R_j|} \qquad (7.2)$$

Then tasks in the workflow are ordered in HEFT based on a rank fuction. For a exit task T_i, the rank value is:

$$rank(T_i) = \varpi_i \qquad (7.3)$$

The rank values of other tasks are computed recursively based on Eqs. 7.1, 7.2, 7.3 as shown in 7.4.

Algorithm 7.4. Heterogeneous-Earliest-Finish-Time (HEFT) algorithm

---
1: compute the average execution time for each task $t \in \Gamma$ according to equation 7.1
2: compute the average data transfer time between tasks and their successors according to equation 7.2
3: compute rank value for each task according to equations 7.3 and 7.4
4: sort the tasks in a scheduling list Q by decreasing order of task rank value
5: **while** Q is not empty **do**
6:     $t \leftarrow$ remove the first task from Q
7:     $r \leftarrow$ find a resource which can complete t at earliest time
8:     schedule t to r
9: **end while**

$$rank(T_i) = \varpi_i + \max_{T_j \epsilon succ(T_i)} (\overline{c_{ij}} + rank(T_j)) \qquad (7.4)$$

where $succ(T_i)$ is the set of immediate successors of task T_i. The algorithm then sorts the tasks by decreasing order of their rank values. The task with higher rank value is given higher priority. In the resource selection phase, tasks are scheduled in the order of their priorities and each task is assigned to the resource that can complete the task at the earliest time.

Even though original HEFT proposed by Topcuoglu et al. [51] computes the rank value for each task using the mean value of the task execution time and communication time over all resources, Zhao and Sakellariou [62] investigated and compared the performances of the HEFT algorithm produced by other different approximation methods on different cases. The results of the expeiments showed that the mean value method is not the most effiecient choice, and the performance could differ significantly from one application to another [62].

Dependency-Batch Mode

Sakellariou and Zhao [45] proposed a hybrid heuristic for scheduling DAG on heterogeneous systems. The heuristic combines dependency mode and batch mode. As described in Algorithm 7.5, the heuristic first compute rank values of each task and ranks all tasks in the decreasing order of their rank values (Algorithm 7.5: line 1-3). And then it creates groups of independent tasks (Algorithm 7.5:line 4-11). In the grouping phase, it processes tasks in the order of their rank values and add tasks into the current group. Once it finds a task which has a dependency with any task within the group, it creates another new group. As a result, a number of groups of independent tasks are generated. And the group number is assigned

---

Algorithm 7.5. Hybrid heuristic

1: compute the weight of each task node and edge according to equations 7.1 and 7.2

2: compute the rank value of each task according to equations 7.3 and 7.4
3: sort the tasks in a scheduling list Q by decreasing order of task rank value
4: create a new group G_i and $i = 0$
5: **while** Q is not empty **do**
6:    $t \leftarrow$ remove the first task from Q
7:    **if** t has a dependence with a task in G_i **then**
8:       $i++$; create a new group G_i
9:    **end if**
10:   add t to G_i
11: **end while**
12: $j = 0$
13: **while** $j <= i$ **do**
14:    scheduling tasks in G_i by using a batch mode algorithm
15:    $j++$
16: **end while**

---

based on the order of rank values of their tasks, i.e., if $m > n$, the ranking value of tasks in group m is higher than that of the tasks in group n. Then it schedules tasks group by group and uses a batch mode algorithm to reprioritize the tasks in the group.

Cluster based and Duplication based scheduling

Both cluster based scheduling and duplication based scheduling are designed to avoid the transmission time of results between data interdependent tasks, such that it is able to reduce the overall execution time. The cluster based scheduling clusters tasks and assign tasks in the same cluster into the same resource, while the duplication based scheduling use the idling time of a resource to duplicate some parent tasks, which are also being scheduled on other resources.

Bajai and Agrawal [3] proposed a *task duplication based scheduling algorithm for network of heterogeneous systems*(TANH) . The algorithm combines cluster based scheduling and duplication based scheduling and the overview of the algorithm is shown in Algorithm 7.6. It first traverses the task graph to compute parameters of each node including earliest start and completion time, latest start and completion time, critical immediate parent task, best resource and the level of the task. After that it clusters tasks based on these parameters. The tasks in a same cluster are supposed to be scheduled on a same resource. If the number of the cluster is greater than the number of resources, it scales down the number of clusters to the number of resources by merging some clusters. Otherwise, it utilizes the idle times of resources to duplicate tasks and rearrange tasks in order to decrease the overall execution time.

Algorithm 7.6. TANH algorithm

1: compute parameters for each task node
2: cluster workflow tasks
3: **if** the number of clusters greater than the number of available resources **then**
4:     reducing the number of clusters to the number of available resources
5: **else**
6:     perform duplication of tasks
7: **end if**

7.3.2 Meta-heuristics

Meta-heuristics provide both a general structure and strategy guidelines for devoping a heuristic for solving computational problems. They are generally applied to a large and complicated problem. They provide an efficient way of moving quickly toward a very good solution. Many metahuristics have been applied for solving workflow scheduling problmes, including GRASP, Genetic Algorithms and Simulated Annealing. The details of these algorithms are presented in the sub-sections that follow.

Greedy Randomized Adaptive Search Procedure (GRASP)

A *Greedy Randomized Adaptive Search Procedure* (GRASP) is an iterative randomized search technique. Feo and Resende [19] proposed guidelines for developing heuristics to solve combinatorial optimization problems based on the GRASP concept. Binato et al. [8] have shown that the GRASP can solve job-shop scheduling problems effectively. Recently, the GRASP has been investigated by Blythe et al. [11] for workflow scheduling on Grids by comparing with the Min-Min heuristic on both computational- and data-intensive applicaitons.

Algorithm 7.7. GRASP algorithm

1: **while** stopping criterion not satisfied **do**
2:     $schedule \leftarrow$ createSchedule($workflow$)
3:     **if** $schedule$ is better than $bestSchedule$ **then**
4:         $bestSchedule \leftarrow schedule$
5:     **end if**
6: **end while**
7: **PROCEDURE:** createSchedule($workflow$)
8: $solution \leftarrow$ constructSolution($workflow$)
9: $nSolution \leftarrow$ localSearch($solution$)
10: **if** $nSolution$ is better than solution **then**
11:     **return** $nSolution$
12: **end if**
13: **return** $solution$
14: **END** createSchedule
15: **PROCEDURE:** constructSolution($workflow$)
16: **while** schedule is not completed **do**
17:     $T \leftarrow$ get all unmapped ready tasks
18:     make a RCL for each $t \in T$
19:     $subSolution \leftarrow$ select a resource randomly for each $t \in T$ from its RCL
20:     $solution \leftarrow solution \bigcup subSolution$
21:     update information for further RCL making
22: **end while**
23: **return** solution
24: **END** constructSolution
25: **PROCEDURE:** localSearch($solution$)
26: $nSolution \leftarrow$ find a optimal local solution
27: **return** nSolution
28: **END** localSearch

Algorithm 7.7 describes a GRASP. In a GRASP, a number of iterations are conducted to search a possible optimal solution for scheduling tasks on resources. A solution is generated at each iterative step and the best solution is kept as the final schedule (Algorithm 7.7:line 1-6). A GRASP is terminated when the specified termination criterion is satisfied, for example, after completing a certain number of interations. In general, there are two phases in each interation: *construction phase* and *local search phase*.

The construction phase (Algorithm 7.7:line 8 and line 15-24) generates a feasible solution. A feasible solution for the workflow scheduling problem is required to meet the following conditions: a task must be started after all its predecessors have been completed; every task appears once and only once in the schedule. In the construction phase, a *restricted candidate list* (RCL) is used to record the best candidates, but not necessarily the top candidate of the resources for processing each task. There are two major mechanisms that can be used to generate the RCL, *cardinality-based RCL* and *value-based RCL*.

Algorithm 7.8. Construction phase procedure for workflow scheduling

1: **PROCEDURE:** constructSolution(Ω)
2: **while** schedule is not completed **do**
3:     $availTasks \leftarrow$ get unmapped ready tasks
4:     $subSolution \leftarrow$ schedule($availTasks$)
5:     $solution \leftarrow solution \bigcup subSolution$
6: **end while**
7: **return** solution
8: **END** constructSolution
9: **PROCEDURE:** schedule($tasks$)
10: $availTasks \leftarrow tasks$
11: $pairs \leftarrow$
12: **while** $\exists t \in tasks$ not scheduled **do**
13:     **for all** $t \in availTasks$ **do**
14:       $availResources \leftarrow$ get available resources for t
15:       **for all** $r \in availResources$ **do**
16:          compute $increaseMakespan(t,r)$
17:          $pairs \leftarrow pairs \bigcup (t,r)$
18:       **end for**
19:     **end for**
20:     $minI \leftarrow$ minimum makespan increase over $availPairs$
21:     $maxI \leftarrow$ maximum makespan increase over $availPairs$
22:     $availPairs \leftarrow$ select pairs whose makespan increase is less than $minI + \alpha(maxI - minI)$
23:     $(t',r') \leftarrow$ select a pair at random from $availPairs$
24:     remove t' from $availTasks$
25:     $solution \leftarrow solution \bigcup (t',r')$
26: **end while**
27: **return** solution
28: **END** schedule

The cardinality-based RCL records the k best rated solution components, while the value-based RCL records all solution components whose performance evaluated values are better than a better than a given threshold [31]. In the GRASP, resource allocated to each task is randomly selected from its RCL (Algorithm 7.7: line 19). After allocating a resource to a task, the resource information is updated and the scheduler continues to process other unmapped tasks.

Algorithm 7.8 shows the detailed implementation of the construction phase for workflow scheduling presented by Blythe et al. [11] which uses a value-based RCL method. The scheduler estimates the makespan increase for each unmapped ready task (Algorithm 7.8: line 3-4 and line 13-19) on each resource that is able to process the task. A *makespan increase* of a task t on a resource r is the increase of the execution length to the current completion length (makespan) if r is allocated to t. Let $minI$ and $maxI$ be the lowest and highest makespan increase found respectively. The scheduler selects a task assignment randomly from the task and resource pair whose makespan increase is less than $minI + \alpha(maxI - minI)$, where is a parameter to determine how much variation is allowed for creating RCL for each task and $0 \leq \alpha \leq 1$.

Once a feasible solution is constructed, a local search is applied into the solution to improve it. The local search process searches local optima in the neighborhood of the current solution and generates a new solution. The new solution will replace the current constructed solution if its overall performance is better (i.e. its makespan is shorter than that of the solution generated) in the construction phase. Binato et al. [8] implementation of the local search phase for job-shop scheduling. It identifies the critical path in the disjunctive graph of the solution generated in the construction phase and swaps two consecutive operations in the critical path on the same machine. If the exchange improves the performance, it is accepted.

Genetic Algorithms (GAs)

Genetic Algorithms (GAs) [25] provide robust search techniques that allow a high-quality solution to be derived from a large search space in polynomial time by applying the principle of evolution. Using genetic algorithms to schedule task graphs in homogeneous and dedicated multiprocessor systems have been proposed in [31, 56, 64]. Wang et al. [54] have developed a genetic-algorithm-based scheduling to map and schedule task graphs on heterogeneous enviorments. Prodan and Fahringer [42] have employed GAs to schedule WIEN2k workflow [10] on Grids. Spooner et al. [47] have employed GAs to schedule sub-workflows in a local Grid site.

A genetic algorithm combines exploitation of best solutions from past searches with the exploration of new regions of the solution space. Any solution in the search space of the problem is represented by an individual (chromosome). A genetic algorithm maintains a population of individuals that evolves over generations. The quality of an individual in the population is determined by a *fitness function* . The fitness value indicates how good the individual is compared to others in the population.

A typical genetic algorithm is illustrated in Fig. 7.4. It first creates an initial population consisting of randomly generated solutions. After applying genetic operators, namely selection, crossover and mutation, one after the other, new offspring are generated. Then the evaluation of the fitness of each individual in the population is conducted. The fittest individuals are selected to be carried over next generation. The above steps are repeated until the termination condition

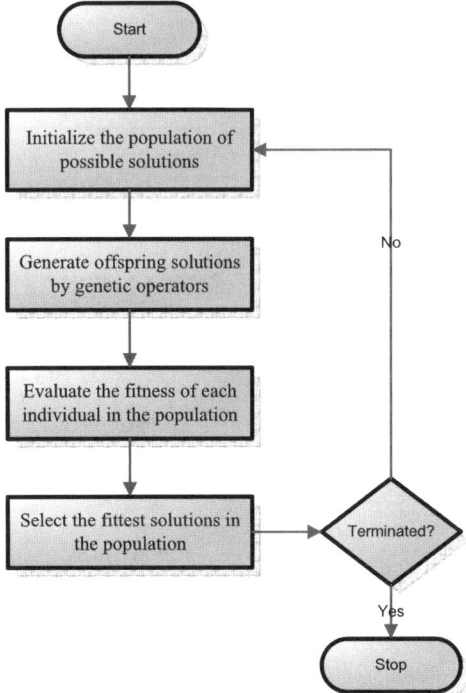

Fig. 7.4. Genetic Algorithms

Table 7.5. Fitness Values and Slots for Roulette Wheel Selection

Individual	Fitness value	Slot Size	Slot
1	0.45	0.25	0.25
2	0.30	0.17	0.42
3	0.25	0.14	0.56
4	0.78	0.44	1
Total	1.78	1	

is satisfied. Typically, a GA is terminated after a certain number of iterations, or if a certain level of fitness value has been reached [64].

The construction of a genetic algorithm for the scheduling problem can be divided into four parts [32]: the choice of representation of individual in the population; the determination of the fitness function; the design of genetic operators; the determination of probabilities controlling the genetic operators.

As genetic algorithms manipulate the code of the parameter set rather than the parameters themselves, an encoding mechanism is required to represent individuals in the population. Wang et al. [54] encoded each chromosome with two separated parts: the *matching string* and the *scheduling string*. Matching string

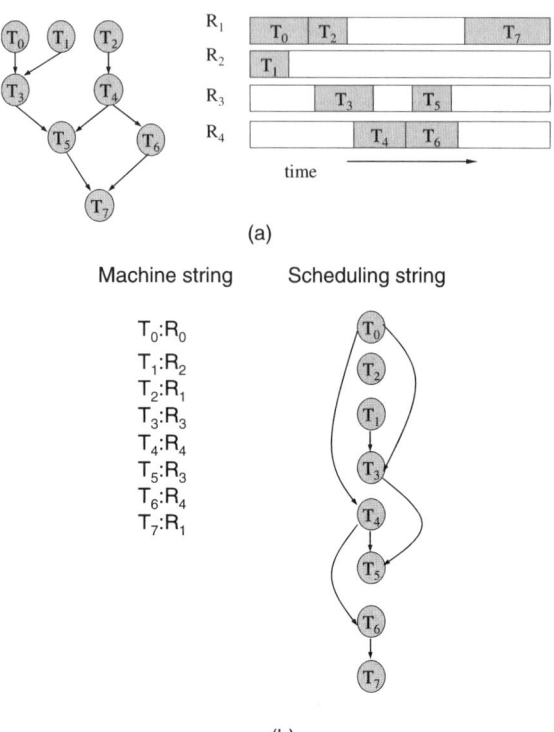

Fig. 7.5. (a) Workflow application and schedule. (b) seperated machine string and scheduling string. (c) two-dimensional string.

Table 7.6. Fitness Values and Slots for Rank Selection

Individual	Fitness value	Rank	Slot Size	Slot
1	0.45	3	0.3	0.3
2	0.30	2	0.2	0.5
3	0.25	1	0.1	0.6
4	0.78	4	0.4	1

represents the assignment of tasks on machines while scheduling string represents the execution order of the tasks (Fig. 7.5a.). However, a more intuitive scheme, *two-dimensional coding scheme* is employed by many [32, 56, 64] for scheduling tasks in distributed systems. As illustrated in Fig. 7.5c, each schedule is simplified by representing it as a 2D string. One dimension represents the numbers of resources while the other dimension shows the order of tasks on each resource.

A fitness function is used to measure the quality of the individuals in the population. The fitness function should encourage the formation of the solution to achieve the objective function. For example, the fitness function developed in [32] is $C_{max} - FT(I)$, where C_{max} is the maximum completion time observed so far and $FT(I)$ is the completion time of the individual I. As the objective function is to minimize the execution time, an individual with a large value of fitness is fitter than the one with a small value of fitness.

After the fitness evaluation process, the new individuals are compared with the previous generation. The selection process is then conducted to retain the fittest individuals in the population, as successive generations evolve. Many methods for selecting the fittest individuals have been used for solving task scheduling problems such as *roulette wheel selection, rank selection* and *elitism*.

The roulette wheel selection assigns each individual to a slot of a roulette wheel and the slot size occupied by each individual is determined by its fitness value. For example, there are four individuals (see Table 7.5) and their fitness values are 0.45, 0.30, 0.25 and 0.78, respectively. The slot size of an individual is calculated by dividing its fitness value by the sum of all individual fitness in the population. As illustrated in Fig. 7.6, *individual 1* is placed in the slot ranging from $0 - 0.25$ while *individual 2* is in the slot ranging from $0.26 - 0.42$. After that, a random number is generated between 0 and 1, which is used to determine which individuals will be preserved to the next generation. The individuals with a higher fitness value are more likely to be selected since they occupy a larger slot range.

The roulette wheel selection will have problems when there are large differences between the fitness values of individuals in the population [41]. For example, if the best fitness value is 95% of all slots of the roulette wheel, other individuals will have very few chances to be selected. Unlike the roulette wheel selection in which the slot size of an individual is proportional to its fitness value, a rank selection process firstly sorts all individuals from best to worst according to their fitness values and then assigns slots based on their rank. For example, the size of slots for each individual implemented by DOĞAN and Özgüner [16] is proportional to their rank value. As shown in Table 7.6, the size of the slot for individual I is defined as $PI = \frac{R(I)}{\sum_{i=1}^{n} R(i)}$, where $R(I)$ is the rank value of I and n is the number of all individuals. Both the roulette wheel selection and the rank selection select individuals according to their fitness value. The higher the fitness value, the higher the chance it will be selected into the next generation. However, this does not guarantee that the individual with the highest value goes to the next generation for reproduction. Elitism can be incorporated into these two selection methods, by first copying the fittest individual into the next generation and then using the rank selection or roulette wheel selection to construct the rest of the population. Hou et al. [32] showed that the elitism method can improve the performance of the genetic algorithm.

In addition to selection, crossover and mutation are two other major genetic operators. Crossovers are used to create new individuals in the current population by combining and rearranging parts of the existing individuals. The idea behind the crossover is that it may result in an even better individual by combining two fittest individuals [32]. Mutations occasionally occur in order to allow a

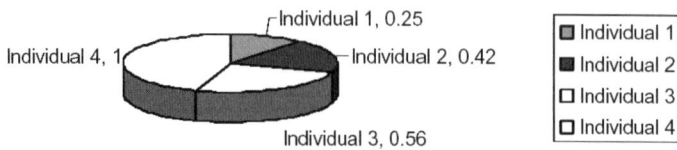

Fig. 7.6. Roulette Wheel Selection Example

certain child to obtain features that are not possessed by either parent. It helps a genetic algorithm to explore new and potentially better genetic material than was previously considered. The frequency of mutation operation occurrence is controlled by the mutation rate whose value is determined experimentally [32].

Simulated Annealing (SA)

Simulated Annealing (SA) [38] derives from the Monte Carlo method for statistically searching the global optimum that distinguishes between different local optima. The concept is originally from the way in which crystalline structures can be formed into a more ordered state by use of the annealing process, which repeats the heating and slowly cooling a structure. SA has been used by YarKhan and Dongarra [57] to select a suitable size of a set of machines for scheduling a ScaLAPACK applicaton [9] in a Grid environment. Young et al. [58] have investigated performances of SA algorithms for scheduling workflow applications in a Grid envrionment.

A typical SA algorithm is illustrated in Fig. 7.7. The input of the algorithm is an initial solution which is constructed by assigning a resource to each task at random. There are several steps that the simulated annealing algorithm needs to go through while the temperature is decreased by a specified rate. The annealing process runs through a number of iterations at each temperature to sample the search space. At each cycle, it generates a new solution by applying random change on the current solution. Young et al. [58] implemented this randomization by moving one task onto a different resource. Whether or not the new solution is accepted as a current solution is determined by the Metropolis algorithm [38, 58] shown in Algorithm 7.9. In the Metropolis algorithm, the new solution and the current solution are compared and the new solution is unconditionally accepted if it is better than the current one. In the case of the minimization problem of workflow scheduling, the better solution is one which has a lower execution time and the improved value is denoted as $d\beta$. In other cases, the new solution is accepted with the Boltzmann probability $e^{\frac{-d\beta}{T}}$ [38] where T is the current temperature. Once a specified number of cycles have been completed, the temperature is decreased. The process is repeated until the lowest allowed temperature has been reached. During this process, the algorithm keeps the best solution so far, and returns this solution at termination as the final optimal solution.

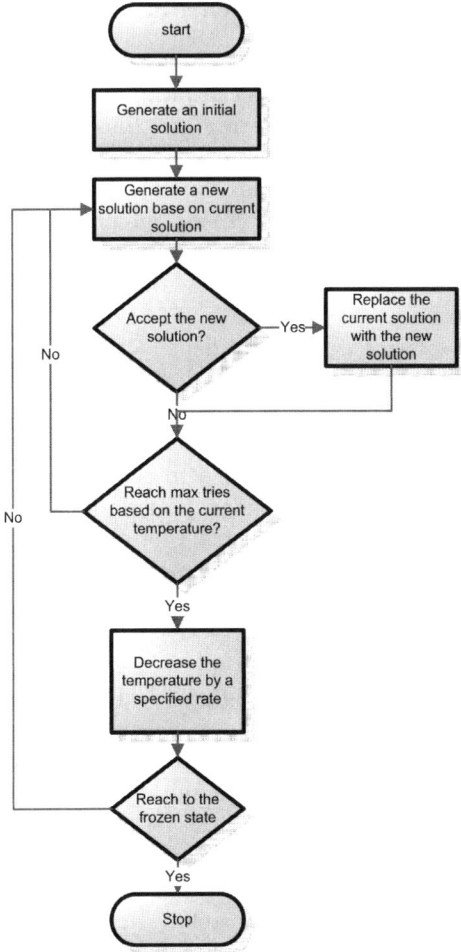

Fig. 7.7. Simulated Annealing

Algorithm 7.9. Metropolis algorithm

1: **if** $d\beta$ **then**
2:     **return** $true$
3: **else if** a random number less than $e^{\frac{-d\beta}{T}}$ **then**
4:     **return** $true$
5: **else**
6:     **return** $false$
7: **end if**

7.3.3 Comparison of Best-Effort Scheduling Algorithms

The overview of the best effort scheduling is presented in Table 7.7 and 7.8. In general, the heuristic based algorithms can produce a reasonable good solution in a polynomial time. Among the heuristic algorithms, individual task scheduling is simplest and only suitable for simple workflow structures such as a pipeline in which several tasks are required to be executed in sequential. Unlike individual task scheduling, list scheduling algorithms set the priorities of tasks in order to make an efficient schedule in the situation of many tasks compete for limited number of resources. The priority of the tasks determines their execution order. The batch mode approach orders the tasks required to be executed in parallel based on their execution time whereas the dependency mode approach orders the tasks based on the length of their critical path. The advantage of the dependency mode approach is that it intent to complete tasks earlier whose interdependent tasks required longer time in order to reduce the overall execution time. However, its complexity is higher since it is required to compute the critical path of all tasks. Another drawback of the dependency mode approach is that it cannot efficiently solve resource competition problem for a workflow consisting of many parallel tasks having the same length of their critical path. The dependency-batch mode approach can take advantage of both approaches, and Sakellariou and Zhao [45] shows that it outperforms the dependency mode approach in most cases. However, computing task priorities based on both batch mode and dependency mode approach results in higher scheduling time.

Even though data transmission time has been considered in the list scheduling approach, it still may not provide an efficient schedule for data intensive workflow applications, in which the majority of computing time is used for transferring data of results between the inter-dependent tasks. The main focus of the list scheduling is to find an efficient execution order of a set of parallel tasks and the determination of the best execution resource for each task is based only on the information of current task. Therefore, it may not assign data interdependent tasks on resources among which an optimized data transmission path is provided. Both cluster based and duplication based scheduling approach focus on reducing communication delay among interdependent tasks. The clustering based approach minimizes the data transmission time by grouping heavily communicating tasks to a same task cluster and assigns all tasks in the cluster to one resource, in order to minimize the data transmission time, while duplication based approach duplicates data-interdependent tasks to avoid data transmission. However, the restriction of the algorithms based on these two approaches up to date may not be suitable for all Grid workflow applications, since it assumes that heavily communicating tasks can be executed on a same resource. Tasks in Grid workflow applications can be highly heterogeneous and require different type of resources.

The meta-heuristics based workflow scheduling use guided random search techniques and exploit the feasible solution space iteratively. The GRASP generates a randomized schedule at each iteration and keeps the best solution as the final solution. The SA and GAs share the same fundamental assumption that

Table 7.7. Comparison of Best-effort Workflow Scheduling Algorithms (Heuristics)

Scheduling Method		Algorithm	Complexity*	Features
Individual task scheduling		Myopic	$O(vm)$	Decision is based on one task.
List scheduling	Batch mode	Min-min	$O(vgm)$	Decision based on a set of parallel independent tasks.
	Dependency mode	HEFT	$O(v^2m)$	Decision based on the critical path of the task.
	Dependency-batch mode	Hybrid	$O(v^2m + vgm)$	Ranking tasks based on their critical path and re-ranking adjacent independent tasks by using a batch mode algorithm.
Cluster based scheduling Duplication based scheduling		THAN	$O(v^2)$	Replicating tasks to more than one resources in order to reduce transmission time.

*where v is the number of tasks in the workflow, m is the number of resources and g is the number of tasks in a group of tasks for the batch mode scheduling.

Table 7.8. Comparison of Best-effort Workflow Scheduling Algorithms (Meta-heuristics)

Scheduling Method	Features
Greedy randomized adaptive search procedure (GRASP)	Global solution obtained by comparing differences between randomized schedules over a number of iteration.
Genetic algorithms (GA)	Global solution obtained by combining current best solutions and exploiting new search region over generations.
Simulated annealing (SA)	Global solution obtained by comparing differences between schedules which are generated based on current accepted solutions over a number of iterations, while the acceptance rate is decreased.

an even better solution is more probably derived from good solutions. Instead of creating a new solution by randomized search, SA and GAs generate new solutions by randomly modifying current already know good solutions. The SA uses a point-to-point method, where only one solution is modified in each iteration, whereas GAs manipulate a population of solutions in parallel which reduce the probability of trapping into a local optimum [65]. Another benefit of producing a collection of solutions at each iteration is the search time can be significantly decreased by using some parallelism techniques.

Compared with the heuristics based scheduling approaches, the advantage of the meta-heuristics based approaches is that it produces an optimized scheduling solution based on the performance of entire workflow, rather than the partial of the workflow as considered by heuristics based approach. Thus, unlike heuristics based approach designed for a specified type of workflow application, it can produce good quality solutions for different types of workflow applications (e.g. different workflow structure, data- and computational-intensive workflows, etc). However, the scheduling time used for producing a good quality solution required by meta-heuristics based algorithms is significantly higher. Therefore, the heuristics based scheduling algorithms are well suited for a workflow with a simple structure, while the meta-heuristics based approaches have a lot of potential for solving large and complex structure workflows. It is also common to incorporate these two types of scheduling approaches by using a solution generated by a heuristic based algorithm as a start search point for the meta-heuristics based algorithms to generate a satisfactory solution in shorter time.

7.3.4 Dynamic Scheduling Techniques

The heuristics presented in last sections assume that the estimation of the performance of task execution and data communication is accurate. However, it is difficult to predict accurately execution performance in community Grid environments due to its dynamic nature. In a community Grid, the utilization and availability of resources varies over time and a better resource can join at any time. Constructing a schedule for entire workflow before the execution may result in a poor schedule. If a resource is allocated to each task at the beginning of workflow execution, the execution environment may be very different at the time of task execution. A 'best' resource may become a 'worst' resource. Therefore, the workflow scheduler must be able to adapt the resource dynamics and update the schedule using up-to-date system information. Several approaches have been proposed to address these problems. In this section, we focus on the approaches which can apply the algorithms into dynamic environments.

For individual task and batch mode based scheduling, it is easy for the scheduler to use the most up-to-date information, since it takes into account only the current task or a group of independent tasks. The scheduler could map tasks only after their parent tasks become to be executed.

For dependency mode and metahueristics based scheduling, the scheduling decision is based on the entire workflow. In other words, scheduling current tasks require information about its successive tasks. However, it is very difficult to estimate execution performance accurately, since the execution environment may change a lot for the tasks which are late executed. The problems appear more significant for a long lasting workflow. In general, two approaches, task partitioning and iterative re-computing, have been proposed to allow these scheduling approaches to allocate resources more efficiently in a dynamic environment.

Task partitioning is proposed by Deelman et al. [17]. It partitions a workflow into multiple sub-workflows which are executed sequentially. Rather than mapping the entire workflow on Grids, allocates resources to tasks in one sub-workflow at a time. A new sub-workflow mapping is started only after the last mapped sub-workflow has begun to be executed. For each sub-workflow, the scheduler applies a workflow scheduling algorithm to generate an optimized schedule based on more up-to-date information.

Iterative re-computing keeps applying the scheduling algorithm on the remaining unexecuted partial workflow during the workflow execution. It does not use the initial assignment to schedule all workflow tasks but reschedule unexecuted tasks when the environment changes. A low-cost rescheduling policy has been developed by developed by Sakellariou and Zhao [44]. It reduces the overhead produced by rescheduling by conducting rescheduling only when the delay of a task execution impacts on the entire workflow execution.

In addition to mapping tasks before execution using up-to-date information, task migration [4, 42] has been widely employed to reschedule a task to another resource after it has been executed. The task will be migrated when the task execution is timed out or a better resource is found to improve the performance.

7.4 QoS-Constraint Based Workflow Scheduling

Many workflow applications require some assurances of quality of services (QoS) . For example, a workflow application for maxillo-facial surgery planning [16] needs results to be delivered before a certain time. For thus applications, workflow scheduling is required to be able to analyze users' QoS requirements and map workflows on suitable resources such that the workflow execution can be completed to satisfy users' QoS constraints.

However, whether the execution can be completed within a required QoS not only depend on the global scheduling decision of the workflow scheduler but also depend on the local resource allocation model of each execution site. If the execution of every single task in the workflow cannot be completed as what the scheduler expects, it is impossible to guarantee the entire workflow execution. Instead of scheduling tasks on community Grids, QoS-constraint based schedulers should be able to interact with service-oriented Grid services to ensure resource availability and QoS levels. It is required that the scheduler can negotiate with service providers to establish a service level agreement (SLA) which is a contract specifying the minimum expectations and obligations between service providers and consumers. Users normally would like to specify a QoS constraint for entire workflow. The scheduler needs to determine a QoS constraint for each task in the workflow, such that the QoS of entire workflow is satisfied.

In general, service-oriented Grid services are based on utility computing models. Users need to pay for resource access and service pricing is based on the QoS level and current market supply and demand. Therefore, unlike the scheduling strategy deployed in community Grids, QoS constraint based scheduling may not always need to complete the execution at earliest time. They sometimes may prefer to use cheaper services with a lower QoS that is sufficient to meet their requirements.

To date, supporting QoS in scheduling of workflow applications is at a very preliminary stage. Most QoS constraint based workflow scheduling heuristics are based on either *time* or *cost* constraints. Time is the total execution time of the

Table 7.9. Overview of deadline constrained workflow scheduling algorithms

Algorithm	Project	Organization	Application
Back-tracking	Menascé & Casalicchio	George Mason University, USA Univ. Roma "Tor Vergata", Italy	N/A
Deadline distribution	Gridbus	University of Melbourne, Australia	Randomly generated task graphs
Genetic algorithms	Gridbus	University of Melbourne, Australia	Randomly generated task graphs

Table 7.10. Overview of budget constrained workflow scheduling algorithms

Algorithm	Project	Organization	Application
LOSS and GAIN	CoreGrid	University of Cyprus, Cyprus University of Manchester, UK	Randomly generated task graphs
Genetic algorithms	Gridbus	University of Melbourne, Australia	Randomly generated task graphs
Genetic algorithms	Gridbus	University of Melbourne, Australia	Randomly generated task graphs

workflow (known as *deadline*). Cost is the total expense for executing workflow execution including the usage charges by accessing remote resources and data transfer cost (known as *budget*). In this section, we present scheduling algorithms based on these two constraints, called *Deadline constrained* scheduling and *Budget constrained* scheduling. Table 7.9 and 7.10 presents the overview of QoS constrained workflow scheduling algorithms.

7.4.1 Deadline Constrained Scheduling

Some workflow applications are time critical and require the execution can be completed within a certain timeframe. Deadline constrained scheduling is designed for these applications to deliver results before the deadline. The distinction between the deadline constrained scheduling and the best-effort scheduling is that the deadline constrained scheduling also need to consider monetary cost when it schedules tasks. In general, users need to pay for service assess. The price is based on their usages and QoS levels. For example, services which can process faster may charges higher price. Scheduling the tasks based on the best-effort based scheduling algorithms presented in the previous sections, attempting to minimize the execution time will results in high and unnecessary cost. Therefore, a deadline constrained scheduling algorithm intends to minimize the execution cost while meeting the specified deadline constraint.

Two heuristics have been developed to minimize the cost while meeting a specified time constraint. One is proposed by Menasc and Casalicchio [37] denoted as *Back-tracking*, and the other is proposed by Yu et al. [60] denoted as *Deadline Distribution*.

Back-tracking

The heuristic developed by Menascè and Casalicchio assigns available tasks to least expensive computing resources. An available task is an unmapped task whose parent tasks have been scheduled. If there is more than one available task, the algorithm assigns the task with the largest computational demand to the fastest resources in its available resource list. The heuristic repeats the procedure until all tasks have been mapped. After each iterative step, the execution time of current assignment is computed. If the execution time exceeds the time constraint, the heuristic back-tracks the previous step and remove the least expensive resource from its resource list and reassigns tasks with the reduced resource set. If the resource list is empty the heuristic keep back-tracking to the previous step, reduces corresponding resource list and reassign the tasks.

Deadline/Time Distribution (TD)

Instead of back-tracking and repairing the initial schedule, the TD heuristic partitions a workflow and distributes overall deadline into each task based on their workload and dependencies. After deadline distribution, the entire workflow scheduling problem has been divided into several sub-task scheduling problems.

As shown in Fig. 7.8, in workflow task partitioning, workflow tasks are categorized as either *synchronization tasks* or *simple tasks*. A synchronization task is defined as a task which has more than one parent or child task. For example, T_1, T_{10} and T_{14} are synchronization tasks. Other tasks which have only one parent task and child task are simple tasks. For example, $T_2 - T_9$ and $T_{11} - T_{13}$ are simple tasks. Simple tasks are then clustered into a *branch*. A branch is a set of interdependent simple tasks that are executed sequentially between two synchronization tasks. For example, the branches in the example are $\{T_2, T_3, T_4\}$ and $\{T_5, T_6\}$, $\{T_7\}$, $\{T_8, T_9\}$, $\{T_{11}\}$ and $\{T_{12}, T_{13}\}$.

After task partitioning, workflow tasks $\varGamma$ are then clustered into partitions and the overall deadline is distributed over each partition. The deadline assignment strategy considers the following facts:

- The cumulative expected execution time of a simple path between two synchronization tasks is same.
- The cumulative expected execution time of any path from an entry task to an exit task is equal to the overall deadline.

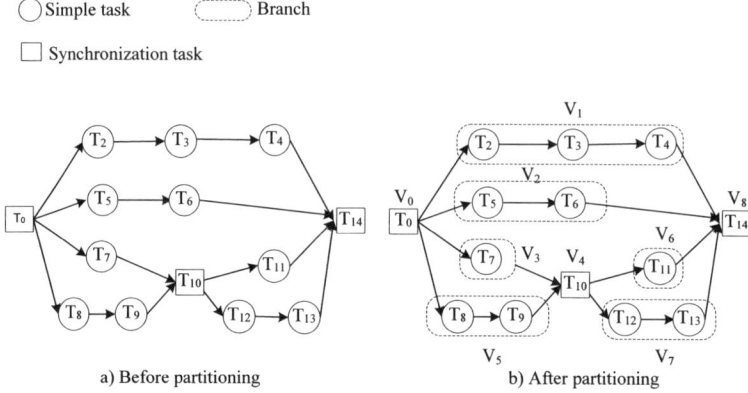

Fig. 7.8. Workflow Task Partition

- The overall deadline is divided over task partitions in proportion to their minimum processing time.

After distributing overall deadline into task partitions, each task partition is assigned a deadline. There are three attributes associated with a task partition V_i : deadline($dl[V_i]$), ready time ($rt[V_i]$), and expected execution time($eet[V_i]$). The ready time of V_i is the earliest time when its first task can be executed. It can be computed according to its parent partitions and defined by:

$$rt[V_i] = \begin{cases} 0 & , T_{entry} \in V_i \\ \max_{V_j \in PV_i} dl[V_j] & , otherwise \end{cases} \quad (7.5)$$

where PV_i is the set of parent task partitions of V_i. The relation between three attributes of a task partition V_i follows that:

$$eet[V_i] = dl[V_i] - rt[V_i] \quad (7.6)$$

A sub-deadline can be also assigned to each task based on the deadline of its task partition. If the task is a synchronization task, its sub-deadline is equal to the deadline of its task partition. However, if a task is a simple task of a branch, its sub-deadline is assigned by dividing the deadline of its partition based on its processing time. Let P_i be the set of parent tasks of T_i and S_i is the set of resources that are capable to execute T_i. t_i^j is the sum of input data transmission time and execution time of executing T_i on S_i. The sub-deadline of task in partition is defined by:

$$dl[T_i] = eet[T_i] + rt[V] \quad (7.7)$$

where

$$eet[T_i] = \frac{\min\limits_{1\leq j\leq |S_i|} t_i^j}{\sum\limits_{T_k \in V} \min\limits_{1\leq l\leq |S_k|} t_k^l} eet[V]$$

$$rt[T_i] = \begin{cases} 0, & T_i = T_{entry} \\ \max\limits_{T_j \in P_i} dl[T_j], & otherwise \end{cases}$$

Once each task has its own sub-deadline, a local optimal schedule can be generated for each task. If each local schedule guarantees that their task execution can be completed within their sub-deadline, the whole workflow execution will be completed within the overall deadline. Similarly, the result of the cost minimization solution for each task leads to an optimized cost solution for the entire workflow. Therefore, an optimized workflow schedule can be constructed from all local optimal schedules. The schedule allocates every workflow task to a selected service such that they can meet its assigned sub-deadline at low execution cost.

7.4.2 Budget Constrained Scheduling

As the QoS guaranteed resources charges access cost, users would like to execute workflows based on the budget they available. Budget constrained scheduling intends to minimize workflow execution time while meeting users' specified budgets. Tsiakkouri et al. [52] present budget constrained scheduling called *LOSS* and *GAIN*.

LOSS and GAIN

LOSS and GAIN scheduling approach adjusts a schedule which is generated by a time optimized heuristic and a cost optimized heuristic to meet users' budget constraints, respectively. A time optimized heuristic attempts to minimize execution time while a cost optimization attempts to minimize execution cost.

If the total execution cost generated by time optimized schedule is not greater than the budget, the schedule can be used as the final assignment; otherwise, the LOSS approach is applied. The idea behinds LOSS is to gain a minimum loss in execution time for the maximum money savings, while amending the schedule to satisfy the budget. The algorithm repeats to re-assign the tasks with smallest values of the *LossWeight* until the budget constraint is satisfied. The LossWeight value for each task to each available resource is computed and it is defined by:

$$LossWeight(i, r) = \frac{T_{new} - T_{old}}{C_{old} - C_{new}} \qquad (7.8)$$

where T_{old} and C_{old} are the execution time and corresponding cost of task T_i on the original resource assigned by the time optimized scheduling, T_{new} and C_{new} are the execution time of task T_i on resource r respectively. If C_{old} is not greater than C_{new}, the value of LossWeight is set to zero.

If the total execution cost generated by a cost optimized scheduler is less than the budget, the GAIN approach is applied to use surplus to decrease the execution time. The idea behinds GAIN is to gain the maximum benefit in execution time for the minimum monetary cost, while amending the schedule. The algorithm repeats to re-assign the tasks with biggest value of the GainWeight until the cost exceeds the budget. The GainWeight value for each task to each available resource is computed and it is defined by:

$$GainWeight(i, r) = \frac{T_{old} - T_{new}}{C_{new} - C_{old}} \quad (7.9)$$

where T_{new}, T_{old}, C_{new} and C_{old} have the same meaning as in the LOSS approach. If T_{new} is greater than T_{old} or C_{new} is equal to C_{old}, the value of GainWeight is set to zero.

7.4.3 Meta-heuristic Based Constrained Workflow Scheduling

A genetic algorithm [61] is also developed to solve the deadline and budget constrained scheduling problem. It defines a fitness function which consists of two components, *cost-fitness* and *time-fitness*. For the budget constrained scheduling, the cost-fitness component encourages the formation of the solutions that satisfy the budget constraint. For the deadline constrained scheduling, it encourages the genetic algorithm to choose individuals with less cost. The cost fitness function of an individual I is defined by:

$$F_{cost}(I) = \frac{c(I)}{B^{\alpha}(maxCost^{(1-\alpha)})}, \alpha = \{0, 1\} \quad (7.10)$$

where $c(I)$ is the sum of the task execution cost and data transmission cost of I, $maxCost$ is the most expensive solution of the current population and B is the budget constraint. α is a binary variable and $\alpha = 1$ if users specify the budget constraint, otherwise $\alpha = 0$.

For the budget constrained scheduling, the time-fitness component is designed to encourage the genetic algorithm to choose individuals with earliest completion time from the current population. For the deadline constrained scheduling, it encourages the formation of individuals that satisfy the deadline constraint. The time fitness function of an individual I is defined by:

$$F_{time}(I) = \frac{t(I)}{D^{\beta}(maxTime^{(1-\beta)})}, \beta = \{0, 1\} \quad (7.11)$$

where $t(I)$ is the completion time of I, $maxTime$ is the largest completion time of the current population and D is the deadline constraint. β is a binary variable and $\beta = 1$ if users specify the deadline constraint, otherwise $\beta = 0$.

For the deadline constrained scheduling problem, the final fitness function combines two parts and it is expressed as:

$$F(I) = \begin{cases} F_{time}(I), & if F_{time}(I) > 1 \\ F_{cost}(I), & otherwise \end{cases} \quad (7.12)$$

For the budget constrained scheduling problem, the final fitness function combines two parts and it is expressed as:

$$F(I) = \begin{cases} F_{cost}(I), & if F_{cost}(I) > 1 \\ F_{time}(I), & otherwise \end{cases} \quad (7.13)$$

In order to applying mutation operators in Grid environment, it developed two types of mutation operations, *swapping mutation* and *replacing mutation*. Swapping mutation aims to change the execution order of tasks in an individual that compete for a same time slot. It randomly selects a resource and swaps the positions of two randomly selected tasks on the resource. Replacing mutation reallocates an alternative resource to a task in an individual. It randomly selects a task and replaces its current resource assignment with a resource randomly selected in the resources which are able to execute the task.

7.4.4 Comparison of QoS Constrained Scheduling Algorithms

The overview of QoS constrained scheduling is presented in Table 7.11 7.12. Comparing two heuristics for the deadline constrained problem, the back-tracking approach is more nave. It is like a constrained based myopic algorithm since it makes a greedy decision for each ready task without planning in the view of entire workflow. It is required to track back to the assigned tasks once it finds the deadline constraint cannot be satisfied by the current assignments. It is restricted to many situations such as data flow and the distribution of execution time and cost of workflow tasks. It may be required to go through many iterations to modify the assigned schedule in order to satisfy the deadline constraint. In contrast, the deadline distribution makes a scheduling decision for each task based on a planned sub-deadline according to the workflow dependencies and overall deadline. Therefore, it has a better plan while scheduling current tasks and does not require tracing back the assigned schedule. However, different deadline distribution strategies may affect the performance of the schedule produced from one workflow structure to another.

To date, the LOSS and GAIN approach is the only heuristic that addresses the budget constrained scheduling problem for Grid workflow applications. It takes advantage of heuristics designed for a single criteria optimization problem such as time optimization and cost optimization scheduling problem to solve a multi-criteria optimization problem. It amends the schedule optimized for one factor to satisfy the other factor in the way that it can gain maximum benefit or minimum loss. Even though the original heuristics are targeted at the budget-constrained scheduling problem, such concept is easy to apply to other constrained scheduling. However, there exist some limitations. It relies on the results generated by an optimization heuristics for a single objective. Even though time optimization based heuristics have been developed over two decades, there is a lack of workflow optimization heuristics for other factors such as monitory cost based on different workflow application scenarios. In addition, large scheduling computation time could occur for data-intensive applications due to the weight re-computation for each pair of task and resource after amending a task assignment.

Table 7.11. Comparison of deadline constrained workflow scheduling algorithms

Algorithm	Features
Back-tracking	It assigns ready tasks whose parent tasks have been mapped to the least expensive computing resources and back-tracks to previous assignment if the current aggregative execution time exceeds the deadline.
Deadline distribution	It distributes the deadline over task partitions in workflows and optimizes execution cost for each task partition while meeting their sub-deadlines.
Genetic algorithms	It uses genetic algorithms to search a solution which has minimum execution cost within the deadline.

Table 7.12. Comparison of budget constrained workflow scheduling algorithms

Algorithm	Features
LOSS and GAIN	It iteratively adjusts a schedule which is generated by a time optimized heuristic or a cost optimized heuristic based on its corresponding LOSS or GAIN weight rate of each task-resource pair, until the total execution cost meets users' budget constraint.
Genetic algorithms	It uses genetic algorithms to search a solution which has minimum execution time within the budget.

Unlike best-effort scheduling in which only one single objective (either optimizing time or system utilization) is considered, QoS constrained scheduling needs to consider more factors such as monetary cost and reliability. It needs to optimize multiple objectives among which some objectives are conflicting. However, with the increase of the number of factors and objectives required to be considered, it becomes infeasible to develop a heuristic to solve QoS constrained scheduling optimization problems. For this reason, we can believe that metahueristics based scheduling approach such as genetic algorithms will play more important role for the multi-objective and multi-constraint based workflow scheduling.

7.5 Simulation Results

In this section, we show an example of experimental comparisons for workflow scheduling algorithms. Basically, we compares deadline constrained scheduling heuristics which are presented in previous section.

7.5.1 Workflow Applications

Given that different workflow applications may have a different impact on the performance of the scheduling algorithms, a task graph generator is developed to automatically generate a workflow based on the specified workflow structure, and the range of task workload and the I/O data. Since the execution requirements for tasks in scientific workflows are heterogeneous, the service type attribute is used to represent different types of services. The range of service types in the workflow can be specified. The width and depth of the workflow can also be adjusted in order to generate workflow graphs of different sizes.

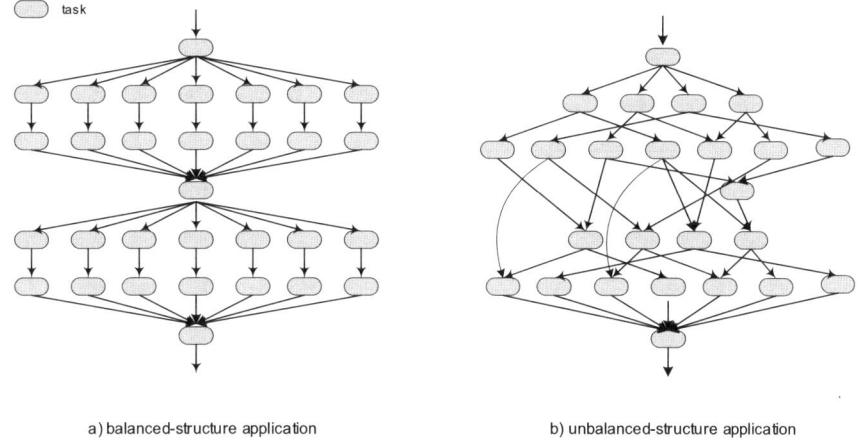

a) balanced-structure application   b) unbalanced-structure application

Fig. 7.9. Small portion of workflow applications

According to many Grid workflow projects [11, 35, 55], workflow application structures can be categorized as either *balanced structure* or *unbalanced structure*. Examples of balanced structure include Neuro-Science application workflows [63] and EMAN refinement workflows [35], while the examples of unbalanced structure include protein annotation workflows [40] and Montage workflows [11]. Fig. 7.9 shows two workflow structures, a *balanced-structure application* and an *unbalanced-structure application*, used in our experiments. As shown in Fig. 7.9a, the balanced-structure application consists of several parallel pipelines, which require the same types of services but process different data sets. In Fig. 7.9b, the structure of the unbalanced-structure application is more complex. Unlike the balanced-structure application, many parallel tasks in the unbalanced structure require different types of services, and their workload and I/O data varies significantly.

7.5.2 Experiment Setting

GridSim [48] is used to simulate a Grid environment for experiments. Fig. 7.10 shows the simulation environment, in which simulated services are discovered by

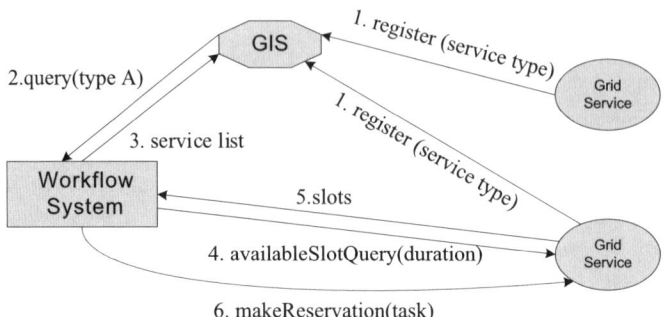

Fig. 7.10. Simulation environment

Table 7.13. Service speed and corresponding price for executing a task

Service ID	Processing Time(sec)	Cost($/sec)
1	1200	300
2	600	600
3	400	900
4	300	1200

Table 7.14. Transmission bandwidth and corresponding price

Bandwidth(Mbps)	Cost ($/sec)
100	1
200	2
512	5.12
1024	10.24

querying the GridSim Index Service (GIS). Every service is able to provide free slot query, and handle reservation request and reservation commitment.

There are 15 types of services with various price rates in the simulated Grid testbed, each of which was supported by 10 service providers with various processing capability. The topology of the system is such that all services are connected to one another, and the available network bandwidths between services are 100Mbps, 200Mbps, 512Mbps and 1024Mbps.

For the experiments, the cost that a user needs to pay for a workflow execution comprises of two parts: processing cost and data transmission cost. Table 7.13 shows an example of processing cost, while Table 7.14 shows an example of data transmission cost. It can be seen that the processing cost and transmission cost are inversely proportional to the processing time and transmission time respectively.

In order to evaluate algorithms on a reasonable deadline constraint we also implemented a time optimization algorithm, HEFT, and a cost optimization algorithm, *Greedy Cost*(GC). The HEFT algorithm is a list scheduling algorithm which attempts to schedule DAG tasks at minimum execution time on a heterogeneous environment. The GC approach is to minimize workflow execution cost by assigning tasks to services of lowest cost. The deadline used for the experiments are based on the results of these two algorithms. Let T_{max} and T_{min} be the total execution time produced by GC and HEFT respectively. Deadline D is defined by:

$$D = T_{min} + k(T_{max} - T_{min}) \tag{7.14}$$

The value of k varies between 0 and 10 to evaluate the algorithm performance from tight constraint to relaxed constraint. As k increases, the constraint is more relaxed.

7.5.3 Backtracing(BT) vs. Deadline/Time Distribution (TD)

In this section, TD is compared with BackTracking denoted as BT on the two workflow applications, balanced and unbalanced. In order to show the results more clearly, we normalize the execution time and cost. Let C_{value} and T_{value} be the execution time and the monetary cost generated by the algorithms in the experiments respectively. The execution time is normalized by using T_{value}/D, and the execution cost by using C_{value}/C_{min}, where C_{min} is the minimum cost achieved Greedy Cost. The normalized values of the execution time should be no greater than one, if the algorithms meet their deadline constraints.

A comparison of the execution time and cost results of the two deadline constrained scheduling methods for the balanced-structure application and

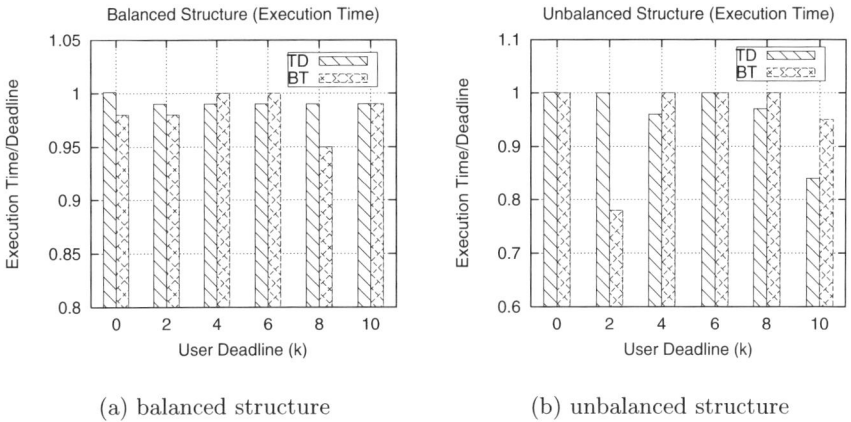

(a) balanced structure (b) unbalanced structure

Fig. 7.11. Execution time for scheduling balanced- and unbalanced-structure applications

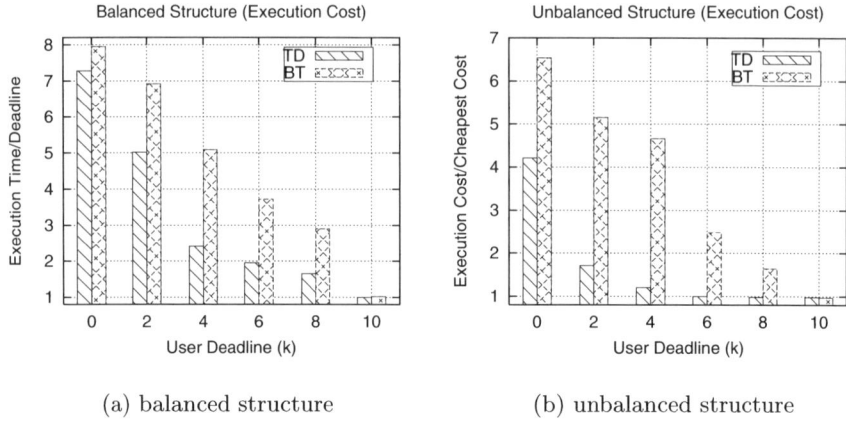

(a) balanced structure (b) unbalanced structure

Fig. 7.12. Execution cost for scheduling balanced- and unbalanced-structure applications

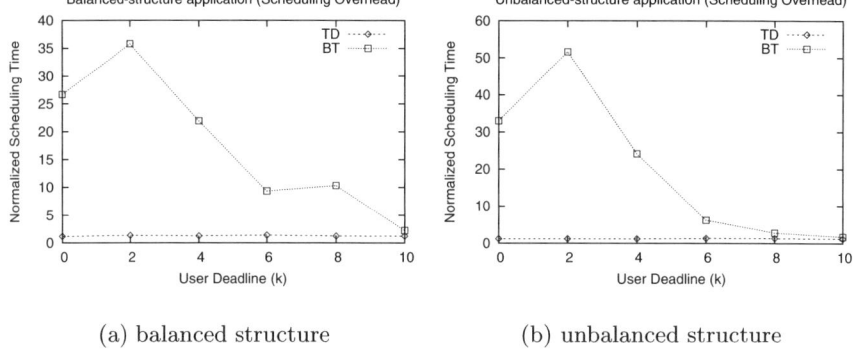

(a) balanced structure (b) unbalanced structure

Fig. 7.13. Scheduling overhead for deadline constrained scheduling

unbalanced-structure application is shown in Fig. 7.11 and Fig. 7.12 respectively. From Fig. 7.11, we can see that TD slightly exceeds deadline at $k = 0$, while BT can satisfy deadlines each time. For execution cost required by the two approaches shown in Fig. 7.12, TD significantly outperforms BT. TD saves almost 50% execution cost when deadlines are relatively low. However, the two approaches produce similar results when deadline is greatly relaxed.

Fig. 7.13 shows the comparison of scheduling running time for two approaches. The scheduling time required by TD is much lower than BT. As the deadline varies, BT requires more running time when deadlines are relatively tight. For example, scheduling times at $k = 0, 2, 4$ are much longer than at $k = 6, 8, 10$. This is because it needs to back-track for more iterations to adjust previous task assignments in order to meet tight deadlines.

7.5.4 TD vs. Genetic Algorithms

In this section, the deadline constrained genetic algorithm is compared with the non-GA heuristics (i.e. TD) on the two workflow structures, balanced and unbalanced workflows.

The genetic algorithm is investigated by starting with two different initial populations. One initial population consists of randomly generated solutions, while the other initial population consists of a solution produced by TD together with other randomly generated solutions. In the result presentation, the results generated by GA with a completely random initial population is denoted by GA, while the results generated by GA which include an initial individual produced by the TD heuristic are denoted as GA+TD. The parameter settings used as the default configuration for the proposed genetic algorithm are listed in Table 7.15.

Fig. 7.14 and Fig. 7.15 compare the execution time and cost of using three scheduling approaches for scheduling the balanced-structure application and unbalanced-structure application with various deadlines respectively.

We can see that it is hard for both GA and TD to successfully meet the low deadline individually. As shown in Fig. 7.14a and 7.15a, the normalized execution times produced by TD and GA exceed 1 at tight deadline ($k = 0$), and

Table 7.15. Default settings

Parameter	Value/Type
Population size	10
Maximum generation	100
Crossover probability	0.9
Reordering mutation probability	0.5
Replacing mutation probability	0.5
Selection scheme	elitism-rank selection
Initial individuals	randomly generated

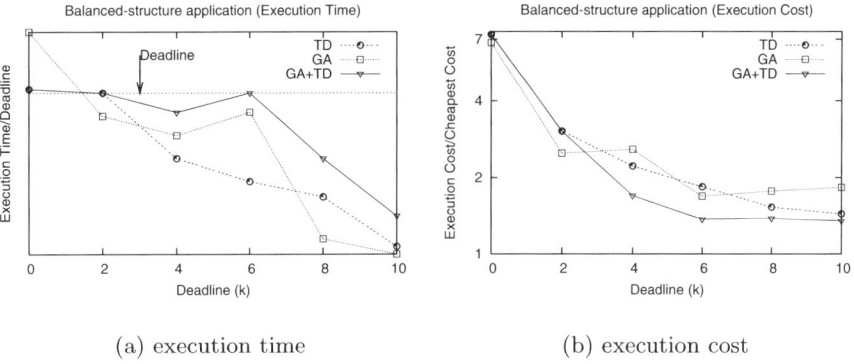

(a) execution time   (b) execution cost

Fig. 7.14. Normalized Execution Time and Cost for Scheduling Balanced-structure Application

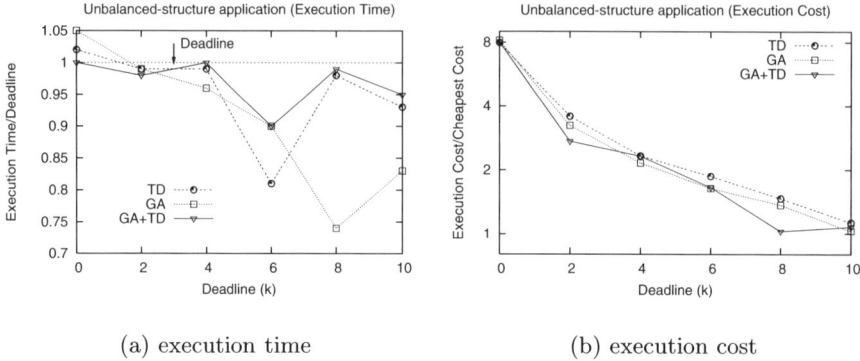

(a) execution time (b) execution cost

Fig. 7.15. Normalized Execution Time and Cost for Scheduling Unbalanced-structure Application

GA performs worse than TD since its values is higher than TD, especially for balanced-structure application. However, the results are improved when incorporating GA and TD together by putting the solution produced by TD into the initial population of GA. As shown in Fig. 7.15a, the value of GA+TD is much lower than that of GA and TD at the tight deadline.

As the deadline increases, both GA and TD can meet the deadline (see Fig. 7.14a and 7.15a) and GA can outperform TD. For example, execution time (see Fig. 7.14a) and cost (see Fig. 7.14b) generated by GA at $k = 2$ are lower than that of TD. However, as shown in Fig. 7.14b) the performance of GA is reduced and TD can perform better, when the deadline becomes very large ($k = 8$ and 10). In general, GA+TD performs best. This shows that the genetic algorithm can improve the results returned by other simple heuristics by employing these heuristic results as individuals in its initial population.

7.6 Conclusions

In this chapter, we have presented a survey of workflow scheduling algorithms for Grid computing. We have categorized current existing Grid workflow scheduling algorithms as either best-effort based scheduling or QoS constraint based scheduling.

Best-effort scheduling algorithms target on community Grids in which resource providers provide free access. Several heuristics and metahueristics based algorithms which intend to optimize workflow execution times on community Grids have been presented. The comparison of these algorithms in terms of computing time, applications and resources scenarios has also been examined in detail. Since service provisioning model of the community Grids is based on best effort, the quality of service and service availability cannot be guaranteed. Therefore, we have also discussed several techniques on how to employ the scheduling algorithms in dynamic Grid environments.

QoS constraint based scheduling algorithms target on utility Grids in which service level agreements are established between service providers and consumers. In general, users are charged for service access based on the usage and QoS levels. The objective functions of QoS constraint based scheduling algorithms are determined by QoS requirements of workflow applications. In this chapter, we have focused on examining scheduling algorithms which intend to solve performance optimization problems based on two typical QoS constraints, deadline and budget.

Acknowledgment

We would like to thank Hussein Gibbins and Chee Shin Yeo for their comments on this paper. This work is partially supported through Australian Research Council (ARC) Discovery Project grant.

References

1. Almond, J., Snelling, D.: UNICORE: Uniform Access to Supercomputing as an Element of Electronic Commerce. Future Generation Computer Systems 15, 539–548 (1999)
2. The Austrian Grid Consortium, http://www.austrangrid.at
3. Bajaj, R., Agrawal, D.P.: Improving Scheduling of Tasks in a Heterogeneous Environment. IEEE Transactions on Parallel and Distributed Systems 15, 107–118 (2004)
4. Berman, F., et al.: New Grid Scheduling and Rescheduling Methods in the GrADS Project. International Journal of Parallel Programming (IJPP) 33(2-3), 209–229 (2005)
5. Berriman, G.B., et al.: Montage: a Grid Enabled Image Mosaic Service for the National Virtual Observatory. In: ADASS XIII, ASP Conference Series (2003)
6. Berti, G., et al.: Medical Simulation Services via the Grid. In: HealthGRID 2003 conference, Lyon, France, January 16-17 (2003)
7. Benkner, S., et al.: VGE - A Service-Oriented Grid Environment for On-Demand Supercomputing. In: The 5th IEEE/ACM International Workshop on Grid Computing (Grid 2004), Pittsburgh, PA, USA (November 2004)
8. Binato, S., et al.: A GRASP for job shop scheduling. In: Essays and surveys on meta-heuristics, pp. 59–79. Kluwer Academic Publishers, Dordrecht (2001)
9. Blackford, L.S., et al.: ScaLAPACK: a linear algebra library for message-passing computers. In: The Eighth SLAM Conference on Parallel Processing for Scientific Computing (Minneapolis, MN, 1997), Philadelphia, PA, USA, p. 15 (1997)
10. Blaha, P., et al.: WIEN2k: An Augmented Plane Wave plus Local Orbitals Program for Calculating Crystal Properties. Institute of Physical and Theoretical Chemistry, Vienna University of Technology (2001)
11. Blythe, J., et al.: Task Scheduling Strategies for Workflow-based Applications in Grids. In: IEEE International Symposium on Cluster Computing and the Grid (CCGrid 2005) (2005)
12. Braun, T.D., Siegel, H.J., Beck, N.: A Comparison of Eleven static Heuristics for Mapping a Class of Independent Tasks onto Heterogeneous Distributed Computing Systems. Journal of Parallel and Distributed Computing 61, 801–837 (2001)

13. Buyya, R., Venugopal, S.: The Gridbus Toolkit for Service Oriented Grid and Utility Computing: An overview and Status Report. In: The 1st IEEE International Workshop on Grid Economics and Business Models, GECON 2004, Seoul, Korea, April 23 (2004)
14. Casanova, H., et al.: Heuristics for Scheduling Parameter Sweep Applications in Grid Environments. In: The 9th Heterogeneous Computing Workshop (HCW 2000) (April 2000)
15. Cooper, K., et al.: New Grid Scheduling and Rescheduling Methods in the GrADS Project. In: NSF Next Generation Software Workshop, International Parallel and Distributed Processing Symposium, Santa Fe (April 2004)
16. Doğan, A., Özgüner, F.: Genetic Algorithm Based Scheduling of Meta-Tasks with Stochastic Execution Times in Heterogeneous Computing Systems. Cluster Computing 7, 177–190 (2004)
17. Deelman, E., et al.: Pegasus: Mapping scientific workflows onto the grid. In: European Across Grids Conference, pp. 11–20 (2004)
18. Fahringer, T., et al.: ASKALON: a tool set for cluster and Grid computing. Concurrency and Computation: Practice and Experience 17, 143–169 (2005)
19. Feo, T.A., Resende, M.G.C.: Greedy Randomized Adaptive Search Procedures. Journal of Global Optimization 6, 109–133 (1995)
20. Fitzgerald, S., et al.: A Directory Service for Configuring High-Performance Distributed Computations. In: The 6th IEEE Symposium on High-Performance Distributed Computing, Portland State University, Portland, Oregon, August 5-8 (1997)
21. Foster, I., Kesselman, C.: Globus: A Metacomputing Infrastructure Toolkit. International Journal of Supercomputer Applications 11(2), 115–128 (1997)
22. Foster, I., Kesselman, C. (eds.): The Grid: Blueprint for a Future Computing Infrastructure. Morgan Kaufmann Publishers, USA (1999)
23. Foster, I., et al.: Chimera: A Virtual Data System for Representing, Querying and Automating Data Derivation. In: The 14th Conference on Scientific and Statistical Database Management, Edinburgh, Scotland (July 2002)
24. Foster, I., et al.: The Physiology of the Grid, Open Grid Service Infrastructure WG. In: Global Grid Forum (2002)
25. Goldberg, D.E.: Genetic Algorithms in Search, Optimization, and Machine Learning. Addison-Wesley, Reading (1989)
26. Goldberg, D.E., Deb, K.: A comparative analysis of selection schemes used in genetic algorithms. Foundations of Genetic Algorithms, 69–93 (1991)
27. Grimshaw, A., Wulf, W.: The Legion vision of a worldwide virtual computer. Communications of the ACM 40(1), 39–45 (1997)
28. He, X., Sun, X., von Laszewski, G.: QoS Guided Min-Min Heuristic for Grid Task Scheduling. Journal of Computer Science and Technology 18(4), 442–451 (2003)
29. Hillier, F.S., Lieberman, G.J.: Introduction to Operations Research. McGraw-Hill Science, New York (2005)
30. Hollinsworth, D.: The Workflow Reference Model, Workflow Management Coalition, TC00-1003 (1994)
31. Hoos, H.H., Stützle, T.: Stochastic Local Search: Foundation and Applications. Elsevier Science and Technology (2004)
32. Hou, E.S.H., Ansari, N., Ren, H.: A Genetic Algorithm for Multiprocessor Scheduling. IEEE Transactions on Parallel and Distributed Systems 5(2), 113–120 (1994)
33. Kwok, Y.K., Ahmad, I.: Static Scheduling Algorithms for Allocating Directed Task Graphs to Multiprocessors. ACM Computing Surveys 31(4), 406–471 (1999)

34. Ludtke, S., Baldwin, P., Chiu, W.: EMAN: Semiautomated software for high-resolution single-particle reconstructions. Journal of Structural Biology 128, 82–97 (1999)
35. Mandal, A., et al.: Scheduling Strategies for Mapping Application Workflows onto the Grid. In: IEEE International Symposium on High Performance Distributed Computing (HPDC 2005) (2005)
36. Mayer, A., et al.: Workflow Expression: Comparison of Spatial and Temporal Approaches. In: Workflow in Grid Systems Workshop, GGF-10, Berlin, March 9 (2004)
37. Menascè, D.A., Casalicchio, E.: A Framework for Resource Allocation in Grid Computing. In: The 12th Annual International Symposium on Modeling, Analysis, and Simulation of Computer and Telecommunications Systems (MASCOTS 2004), Volendam, The Netherlands, October 5-7 (2004)
38. Metropolis, N., et al.: Equations of state calculations by fast computing machines. Joural of Chemistry and Physics 21, 1087–1091 (1953)
39. Maheswaran, M., et al.: Dynamic Matching and Scheduling of a Class of Independent Tasks onto Heterogeneous Computing Systems. In: The 8th Heterogeneous Computing Workshop (HCW 1999), San Juan, Puerto Rico, April 12 (1999)
40. O'Brien, A., Newhouse, S., Darlington, J.: Mapping of Scientific Workflow within the e-Protein project to Distributed Resources, UK e-Science All Hands Meeting, Nottingham, UK (2004)
41. Obitko, M.: Introduction to Genetic Algorithms (March 2006), http://cs.felk.cvut.cz/~xobitko/ga/
42. Prodan, R., Fahringer, T.: Dynamic Scheduling of Scientific Workflow Applications on the Grid using a Modular Optimisation Tool: A Case Study. In: The 20th Symposium of Applied Computing (SAC 2005), Santa Fe, New Mexico, USA, March 2005. ACM Press, New York (2005)
43. Rutschmann, P., Theiner, D.: An inverse modelling approach for the estimation of hydrological model parameters. Journal of Hydroinformatics (2005)
44. Sakellariou, R., Zhao, H.: A Low-Cost Rescheduling Policy for Efficient Mapping of Workflows on Grid Systems. Scientific Programming 12(4), 253–262 (2004)
45. Sakellariou, R., Zhao, H.: A Hybrid Heuristic for DAG Scheduling on Heterogeneous Systems. In: The 13th Heterogeneous Computing Workshop (HCW 2004), Santa Fe, New, Mexico, USA, April 26 (2004)
46. Shi, Z., Dongarra, J.J.: Scheduling workflow applications on processors with different capabilities. Future Generation Computer Systems 22, 665–675 (2006)
47. Spooner, D.P., et al.: Performance-aware Workflow Management for Grid Computing. The Computer Journal (2004)
48. Sulistio, A., Buyya, R.: A Grid Simulation Infrastructure Supporting Advance Reservation. In: The 16th International Conference on Parallel and Distributed Computing and Systems (PDCS 2004), November 9-11. MIT, Cambridge (2004)
49. Tannenbaum, T., et al.: Condor - A Distributed Job Scheduler. In: Computing with Linux. MIT Press, Cambridge (2002)
50. Thickins, G.: Utility Computing: The Next New IT Model. Darwin Magazine (April 2003)
51. Topcuoglu, H., Hariri, S., Wu, M.Y.: Performance-Effective and Low-Complexity Task Scheduling for Heterogeneous Computing. IEEE Transactions on Parallel and Distributed Systems 13(3), 260–274 (2002)
52. Tsiakkouri, E., et al.: Scheduling Workflows with Budget Constraints. In: Gorlatch, S., Danelutto, M. (eds.) The CoreGRID Workshop on Integrated research in Grid Computing, Technical Report TR-05-22, University of Pisa, Dipartimento Di Informatica, Pisa, Italy, November 28-30, pp. 347–357 (2005)

53. Ullman, J.D.: NP-complete Scheduling Problems. Journal of Computer and System Sciences 10, 384–393 (1975)
54. Wang, L., et al.: Task Mapping and Scheduling in Heterogeneous Computing Environments Using a Genetic-Algorithm-Based Approach. Journal of Parallel and Distributed Computing 47, 8–22 (1997)
55. Wieczorek, M., Prodan, R., Fahringer, T.: Scheduling of Scientific Workflows in the ASKALON Grid Enviornment. ACM SIGMOD Record 34(3), 56–62 (2005)
56. Wu, A.S., et al.: An Incremental Genetic Algorithm Approach to Multiprocessor Scheduling. IEEE Transactions on Parallel and Distributed Systems 15(9), 824–834 (2004)
57. YarKhan, A., Dongarra, J.J.: Experiments with Scheduling Using Simulated Annealing in a Grid Environment. In: Parashar, M. (ed.) GRID 2002. LNCS, vol. 2536. Springer, Heidelberg (2002)
58. Young, L., et al.: Scheduling Architecture and Algorithms within the ICENI Grid Middleware. In: UK e-Science All Hands Meeting, pp. 5–12. IOP Publishing Ltd., Bristol, UK, Nottingham, UK (2003)
59. Yu, J., Buyya, R.: A Taxonomy of Workflow Management Systems for Grid Computing. Journal of Grid Computing 3(3-4), 171–200 (2005)
60. Yu, J., Buyya, R., Tham, C.K.: A Cost-based Scheduling of Scientific Workflow Applications on Utility Grids. In: The first IEEE International Conference on e-Science and Grid Computing, Melbourne, Australia, December 5-8 (2005)
61. Yu, J., Buyya, R.: Scheduling Scientific Workflow Applications with Deadline and Budget Constraints using Genetic Algorithms. Scientific Programming 14(3-4), 217–230 (2006)
62. Zhao, H., Sakellariou, R.: An experimental investigation into the rank function of the heterogeneous earliest finish time shceulding algorithm. In: Kosch, H., Böszörményi, L., Hellwagner, H. (eds.) Euro-Par 2003. LNCS, vol. 2790, pp. 189–194. Springer, Heidelberg (2003)
63. Zhao, Y., et al.: Grid Middleware Services for Virtual Data Discovery, Composition, and Integration. In: The Second Workshop on Middleware for Grid Computing, Toronto, Ontario, Canada (2004)
64. Zomaya, A.Y., Ward, C., Macey, B.: Genetic Scheduling for Parallel Processor Systems: Comparative Studies and Performance Issues. IEEE Transactions on Parallel and Distributed Systems 10(8), 795–812 (1999)
65. Zomaya, A.Y., Teh, Y.H.: Observations on Using Genetic Algorithms for Dynamic Load-Balancing. IEEE Transactions on Parallel and Distributed Systems 12(9), 899–911 (2001)

8
Decentralized Grid Scheduling Using Genetic Algorithms

George Iordache[1], Marcela Boboila[1], Florin Pop[2], Corina Stratan[2], and Valentin Cristea[2]

[1] Stony Brook University, USA
georgei@cs.sunysb.edu, mboboila@cs.sunysb.edu
[2] University *"Politehnica"* of Bucharest, Romania
florinpop@cs.pub.ro, corina@cs.pub.ro, valentin@cs.pub.ro

Summary. The chapter describes a solution to the key problem of ensuring high performance behavior of the Grid, namely the scheduling of tasks. It presents a distributed, fault-tolerant, scalable and efficient solution for optimizing task assignment. The scheduler uses a combination of genetic algorithms and lookup services for obtaining a scalable and highly reliable optimization tool. The experiments have been carried out on the Mon*ALISA* monitoring environment and its extensions. The results demonstrate very good behavior in comparison with other scheduling approaches.

Keywords: Decentralized Grid Scheduling, Genetic Algorithms, Task assignment, Lookup services.

8.1 Introduction

The increased interest in scheduling in heterogeneous computing systems, is due partly to the fact that a single parallel architecture may not be adequate for exploiting needs for parallelism especially when dealing with a computational power Grid for wide-area parallel and distributed computing. In some cases, heterogeneous systems have been shown to produce higher performance for lower costs than a single large computing machine. Grid computing developed in recent years in response to challenges raised by complex problems solving and resource sharing in collaborative, dynamic environments. Grid computing concerns large-scale interconnected systems and has the main purpose to aggregate and to efficiently exploit the power of widely distributed resources. This means, among other things, a proper assignment of tasks to the available resources. In grid computing, load-balancing plays an essential role, in cases where one is concerned with optimized use of resources. A well-balanced task distribution contributes to reducing execution time for jobs and to using resources, such as processors, efficiently, in the system. On the other hand, the problem of scheduling heterogeneous tasks onto heterogeneous resources is intractable, thus making room for good heuristic solutions. We denote heterogeneous tasks as tasks that have different execution times, memory and storage requirements. Concerning the platforms, heterogeneity refers to hardware, software, communication characteristics and protocols, network irregularities, etc.

Scheduling in Grid computing must take into account additional issues such as resource consumer and owner requirements, the need to continuously adapt to changes in the availability of resources, etc. Based on this Grid characteristic, a number of challenging issues need to be addressed: maximization of system throughput and user satisfaction, the sites' autonomy (the Grid is composed of resources owned by different users, which retain control over them), scalability, and fault-tolerance.

Various strategies for scheduling have been developed, in order to achieve optimized task planning in distributed systems. Researchers have directed their studies toward static schedulers [1, 24, 28], in which the assignment of tasks to processors and the time at which tasks start execution are determined a priori. In the static model, every task is assigned only once to a resource. A realistic prediction of the cost of the computation can be made before to the actual execution. The static model adopts a "global view" of tasks and computational costs. One of the major benefits is the easy of implementation. On the other hand, static strategies cannot be applied in a scenario where tasks appear a-periodically, and the environment undergoes various state changes. Cost estimate does not adapt to situations in which one of the nodes selected to perform a computation fails, becomes isolated from the system due to network failures, is so heavily loaded with jobs that its response time becomes longer than expected, or a new computing node enters the system. These changes are possible in Grids.

In dynamic scheduling techniques, which have been widely explored in literature [2, 3, 5, 26, 32], tasks are allocated dynamically at their arrival. Dynamic scheduling is usually applied when it is difficult to estimate the cost of applications, or jobs are coming online dynamically (in this case, it is also called online scheduling). Dynamic task scheduling has two major components: one for system state estimation (other than cost estimation in static scheduling) and one for decision making. System state estimation involves collecting state information through Grid monitoring and constructing an estimate. On this basis, decisions are made to assign tasks to selected resources. Since the cost for an assignment is not always available, a natural way to keep the whole system healthy is by balancing the loads of all resources.

To-date research on the subject has been focused on both centralized and decentralized scheduling approaches. In centralized scheduling algorithms [1, 2, 3, 5], a single processor collects the load information in the system and determines the optimal allocation. Due to the overall control, this organization has various advantages, including speed, easy management, simple deployment, and the ability to co-allocate resources. Unfortunately, because of the Grid organizational model, the centralized approach lacks the scalability, robustness, and fault-tolerance.

Decentralized algorithms [7, 12, 15, 21] come with reliable solutions for robust systems, at the expense of high communication costs. The centralized control is substituted in distributed approaches by an increased level of decision-making authority for the nodes involved in running the scheduling algorithm. It

8 Decentralized Grid Scheduling Using Genetic Algorithms 217

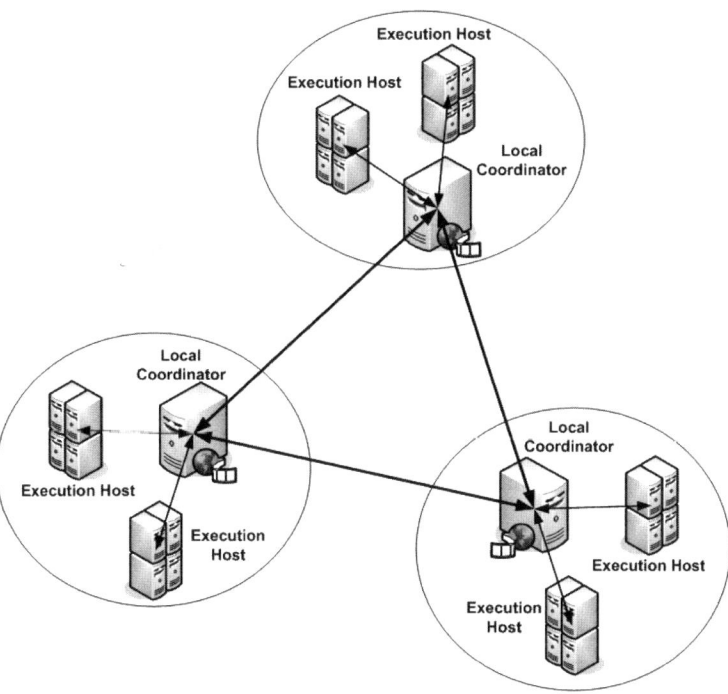

Fig. 8.1. Decentralized Scheduler Architecture. *There are many collaborating sites, each of them having a local coordinator.*

naturally addresses the issues of fault-tolerance, scalability, site-autonomy, and multi-policy scheduling. However, decentralized organizations introduce several problems of management, usage tracking, co-allocation. The coordination between controllers running the scheduling algorithm introduces an overhead but, at the same time, increases the efficiency of the resulting schedules.

A decentralized scheme (see Fig. 8.1) distributes the responsibility of scheduling to every site [7, 12, 15]. Each site in the Grid acts as both a scheduler and a computational resource. User applications are submitted to the local Grid scheduler where the applications originate. The local scheduler is responsible for scheduling its local applications, thus it possibly maintains a local queue to hold its own pending applications. Meanwhile, it should be able to respond to other schedulers requests by acknowledging or denying it. Since the responsibility of scheduling is distributed, the failure of a single scheduler does not affect others working. So the decentralized scheme delivers better fault-tolerance and reliability than the centralized scheme. But the lack of a global scheduler, which knows the information of all applications and resources, usually results in low efficiency. Nevertheless, different scheduling policies on the local sites are possible. Therefore, site-autonomy can be achieved easily as the local schedulers can be specialized for the site owners needs.

Since optimal schedules are difficult to compute, current research aims to find algorithms for suboptimal solutions. The approximate algorithms use formal computational models, but instead of searching the entire solution space for an optimal solution, they stop when a solution that is sufficiently "good" is found. The applicability of this approach depends on the availability of a function to evaluate a solution, the time required to evaluate a solution, and the availability of a mechanism for intelligently pruning the solution space. The heuristic solutions are based on experiments in the real world or on simulation. Not restricted by formal assumptions, heuristic algorithms are more adaptive to the Grid scenarios where both resources and applications are highly diverse and dynamic. Heuristic approaches include algorithms such as: Opportunistic Load Balancing, Minimum Execution Time, Minimum Completion Time, Min-min, Max-min, genetic algorithms etc.

Genetic Algorithms are used for searching large solution spaces. Multiple possible mappings are computed, which are considered chromosomes in the population. Each chromosome has a fitness value, which is the result of an objective function designed in accordance with the performance criteria of the problem (for example maxpan). At each iteration, all of the chromosomes in the population are evaluated based on their fitness value, and only the best of them survive in the next population, where new allocations are generated based on crossover and mutation operators. The algorithm usually stops after a predefined number of steps, or when no noticeable improvements are foreseen.

Genetic algorithms have been largely used for the task allocation problem [1, 2, 3, 5]. The successful results obtained by means of GAs have proved their robustness and efficiency in the field. Research has been done recently, particularly in the area of hybrid algorithms, which use problem-specific knowledge to speed up the search or lead to a better solution [2]. In a novel approach, Wu et al. [1] focus on a thorough exploration of the search space by means of an incremental fitness function and a flexible representation of chromosomes. Moreover, the optimization of scheduling via GAs using the load balancing performance metric has been a key concern for genetic research [1, 3, 5].

Genetic algorithms (GAs) have been widely used to solve difficult NP complete problems like scheduling problem. Hao Yin at al. [19] present an improved genetic algorithm for scheduling independent tasks in Grid environment, which can increase search efficiency with a limited number of iterations by improving the evolutionary process while meeting a feasible result. A fault tolerance-genetic algorithm for Grid task scheduling using check point was proposed by Baghavathi Priya et al. in [29]. Aggarwal et al. developed a genetic algorithm based scheduler for computational grids that minimize make-span, idle time of the available computational resources, turn-around time and the specified deadlines provided by users. The proposed architecture is hierarchical and the scheduler is usable at either the lowest or the higher tiers. It can also be used in both the intra-grid of a large organization and in a research Grid consisting of large clusters, connected through a high bandwidth dedicated network [6].

This paper presents DIOGENES ("DIstributed near-Optimal GENEtic algorithm for Grid applications Scheduling"), a decentralized solution for task scheduling in heterogeneous environments [16]. The chapter is structured as follows: Section 8.2 is a general presentation of the DIOGENES features. Section 8.3 describes the structure and functionality of the proposed system. Section 8.4 introduces the main implementation issues and decribe the genetic algortm. We describe and comment on the experimental results in the 5th section. Section 8.6 contains conclusions and directions for future research.

8.2 General Presentation of the DIOGENES Features

In this section we describe the scheduling requirements approached in our solution and present its main characteristics. Our approach deals with three types of possible requirements: timing constraints (deadlines), resource requirements, and priorities.

Timing Constrains

In Grid systems, the time at which the results of real tasks are delivered are as important as the logical soundness of the results [13]. Depending on the consequences of missing a deadline, real-time tasks are typically classified as hard real-time systems, in which catastrophic consequences may result from missing the deadlines (e.g. space stations, patient monitoring systems, nuclear plant control, and avionics control systems), firm real-time systems, in which the results produced by a task are not useful after the expiration of its deadline (e.g. online transactions processing systems, such as airline reservation and banking), and soft real-time systems, in which usefulness of results produced by a task decreases over time after the deadline expires without causing any damage to the controlled environment (e.g. telephone switching systems and image processing applications).

Resource Requirements

Resource Requirements are also important and must be compared with available resources. CPU-intensive or data-intensive applications require computation resources with a specific processing power or available memory. We have focused on the resource characteristics described by the tuple: <CPU Power, Free Memory, Swap Memory>.

Priority

Priority is a measure of the precedence in a group of tasks. If two tasks are allocated to the same computation resource, the task with the higher precedence is executed first. Therefore, the priorities determine the order of execution of tasks assigned to a specific resource. If p_i is the priority of task i and p_j is the priority of task j, we have:

$$p_i > p_j \Rightarrow t_{spi} < t_{spj} \qquad (8.1)$$

In this forumla we considered that t_{spi} is the time at which task i starts execution on processor p and t_{spj} is the time at which task j starts execution on processor p.

Some of the Grid tasks are periodic in nature, and need to be cyclically executed at constant rates. Other real-time tasks are a-periodic, and they are activated only upon the occurrence of particular events [14]. Hence, periodic tasks consist of an infinite sequence of identical tasks that are regularly activated at a constant rate. Each particular task in the sequence represents an instance of the same task.

Performance Metrics

Different scheduling performance metrics can be taken into account in the design of a feasible scheduling algorithm. They can also represent optimization criteria and are based on various constraints such as deadline, guaranteed completion time, average service time, start and end time, etc. Some of the metrics that can be used to measure the performances of a Grid scheduling algorithm are summarized in Table 8.1.

The description of a scheduling request specifies the tasks' requirements together with other information of interest such as the task ID, path to the

Table 8.1. Performance Metrics Description

Metric	Description
global job success rate	the percentage of submitted jobs that were finished successfully before their deadline
local job kill rate	percentage of local jobs that have been killed
total load	average percentage of busy processors over the entire system
global load	percentage of the total computing power that is used for computing the global jobs. It represents the effective computing power that the scheduler has been able to get from the grid
processor wasted time	percentage of the total computing power (MIPS) that is wasted because of claiming processors before the actual deadlines of jobs
makespan	total execution time of tasks in the system, and is practically equal to the largest processing time over all processors
average processor utilization	average times of processors' utilization relative to the maximum execution time
load-balance	the uniformity of the tasks disposal on the processors, with the purpose to obtain similar execution times on processors, and reduce idle times and overloading

```xml
<task>
    <taskId>24</taskId>
    <path>/home/student/pi/mpi999999999.sh</path>
    <arrivingDate>2007/05/04</arrivingDate>
    <arrivingTime>01:45:05</arrivingTime>
    <arguments>999999999</arguments>
    <input></input>
    <output>mpi.out</output>
    <error>mpi.err</error>
    <requirements>
        <memory>2.95MB</memory>
        <swapSpace>2.95MB</swapSpace>
        <cpuPower>2682.41MHZ</cpuPower>
        <processingTime>39</processingTime>
        <deadlineTime>2006/06/09 00:00:01</deadlineTime>
        <schedulePriority>10</schedulePriority>
    </requirements>
    <nrexec>1</nrexec>
</task>
```

Fig. 8.2. The XML description of a task

executable, the arguments, the input data file, the output and error files, and the arriving time. The task description adopted in our system is presented in Fig. 8.2.

The requirements specified for each task include:

- resource requirements (CPU Power, Free Memory, Free Swap)
- restrictions (deadlines), and
- priorities.

Some functional and descriptive information about the task are indicated in the XML description, such as:

- the *path* to the executable
- the *arguments* received by the executable, in case they are needed
- the *input* file received by the program
- files for redirection of: *output*, and *error*
- arriving date (*arrivingDate*) and time (*arrivingTime*) of the task, for a possible insertion in the task queue ordered by arriving time.

Requirements are specified for each task, as follows:

- *memory*, disk space (*swapSpace*) and CPU power (*cpuPower*) requirements
- processing time (*processingTime*), which is an initial value of the time necessary for the task to be processed in the conditions specified by the memory and CPU power requirements
- deadline restrictions (*deadlineTime*), representing the date and time by which the task must finish execution

- schedule priority (*schedulePriority*), representing a priority associated with the task, and denoting its precedence over other tasks considered for allocation, which have a smaller priority.

Moreover, the task description specifies the number of executions (*nrexec*) of the specific task that may occur.

A user may ask for the scheduling of more than one task at a time. The assignment of a task to a given computing node is conditioned by meeting the resource requirements. We have focused our study on classes of independent tasks, as described in [23], which avoids communication costs due to dependencies. We have built a model based on a real scenario in which groups of tasks are submitted by independent users, to be executed on a group of nodes.

In our scheduling scheme, tasks may arrive simultaneously and resources may dynamically join or leave the system. Aspects of heterogeneity of tasks and processors are also considered in [3], which reports results of simulated experiments. In our work, we present a simulation study, and supplement it with experimental results obtained in existing monitoring and job execution platforms. For experiments, we used the Mon*ALISA* monitoring platform and its extensions [18, 20]. Another major accomplishment of this research is the migration towards a decentralized scheduler by means of lookup services. Using this feature, we overcome one of the main drawbacks of centralized schedulers, which is the lack of robustness in realistic scenarios. Decentralized scheduling approaches have focused on partitioning the task sets or the computation resources into subparts and on running the algorithm on each of them [7, 12, 15, 21]. The results generally indicate high overloads and low balancing, which lead to scarce performance. We also directed our research towards speeding up the convergence of genetic algorithms by using multiple agents (see Section 8.3.2) and different populations to schedule sets of tasks. The experimental results show that the number of generations necessary for the algorithm to converge is significantly reduced. The use of multiple initial search points in the problem space favors a high probability to converge towards a global optimum. Combined with the lookup services, this approach offers a solution to high scalability and reliability.

8.3 Architecture

Designing the proposed system started from the requirements for an efficient Grid scheduling solution. The DIOGENES architecture takes into account two main issues that are intrinsic to a de-centralized Grid scheduler: how to monitor resources and how to discover available services.

With DIOGENES, the scheduler receives job execution requests from users and maps the application to resources and services according to some optimization criteria. In order to address the first issue, related to resource monitoring, we observe that the Grid scheduler doesn't control the clusters/resources directly. Rather, it gathers information about the available resources by calling available Grid monitoring services, and submits the schedules for execution by calling the local schedulers of the target systems.

8 Decentralized Grid Scheduling Using Genetic Algorithms   223

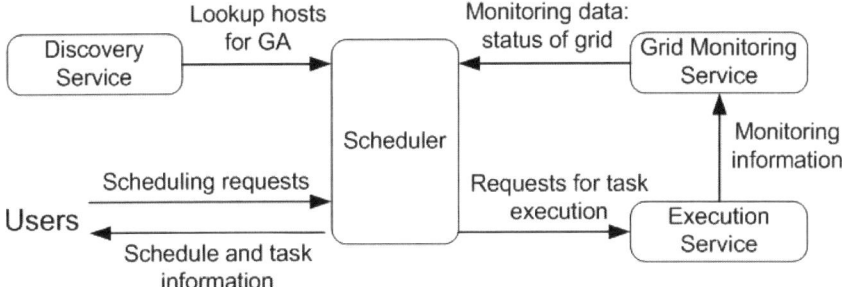

Fig. 8.3. Anatomy of the DIOGENES system

The second issue, concerning the discovery of available resources, is closely related to the de-centralized nature of the DIOGENES scheduler. Several scheduler components must run on different nodes, which have the common characteristics of Grid resources: dynamicity, volatility, etc. Since the usual solution is to use a discovery service, it was adopted also for our system.

8.3.1 System Anatomy

A schematic view of the DIOGENES system is presented in Fig. 8.3. Users submit Scheduling requests. A near-optimal schedule is computed by the Scheduler based on the Scheduling requests and the Monitoring data provided by the Grid Monitoring Service (Mon*ALISA*). The schedule is then sent as a Request for task execution to the Execution Service. The user receives feedback related to the solution determined by the scheduler, as well as to the status of the executed jobs in the form of the Schedule and task information. Furthermore, the system can easily integrate new hosts in the scheduling process, or overcome failure situations by means of the Discovery Service.

The characteristics and functionalities of the services that interact with the scheduler - Grid Monitoring Service, Execution Service and Discovery Service - are further detailed.

Grid Monitoring Service

The Grid Monitoring Service gathers real-time information in a heterogeneous and dynamic environment such as the Grid. It plays an essential role in the scheduling system, since it collects data about the shared resources provided in the distributed environment. The monitoring information is used by the internal DIOGENES scheduling algorithm to generate automated decisions that maintain and optimize the assignation of jobs on the resources of the computational Grid.

We are using the Mon*ALISA* [18] distributed service system in conjunction with ApMon [20]. ApMon is a library that can be used to send status information in the form of UDP datagrams to Mon*ALISA* services. Mon*ALISA* provides system information for computer nodes and clusters, network information for WAN and LAN, monitoring information about the performance of applications, jobs or services. It proved its reliability and scalability in several large scale distributed systems [18]. We have extended the existing implementation of the Mon*ALISA* Web Service Client with new features, in order to connect to the monitoring service via proxy servers and obtain data for the genetic algorithm. A daemon application performs task monitoring is using ApMon. This daemon provides information regarding task status parameters on each node (amount of memory, disk and CPU time used by the tasks). The up-to-date information offered by the Grid Monitoring Service leads to realistic execution times for assigned tasks, as shown by experimental results.

Execution Service

Given its capability to dynamically load modules that interface existing job monitoring with batch queuing applications and tools (e.g. Condor [9], PBS [27], SGE [31], LSF [30]), the Execution Service can send execution requests to an already installed batch queuing system on the computing node to which a particular group of tasks was assigned. Sets of tasks are dynamically sent to computing nodes in the form of a specific command.

The time ordering policy established by the genetic algorithm for tasks assigned on the same processor is preserved at execution time. Discovery Service. Lookup processes are triggered by the Discovery Service and determine the possibility of achieving a decentralized schedule by increasing the number of hosts involved in the genetic scheduling. The apparition or dysfunction of agents in the system can easily be intercepted, resulting in a scalable and highly reliable optimization tool. If one agent ceases to function, the system as a whole is not prejudiced, but the probability of reaching a less optimal solution for the same number of generations increases.

8.3.2 Functional Aspects

In our approach, the grid nodes are part of a group, according to their specific function, as described below:

- **Scheduling Group.** The computers in this group run the scheduling algorithm. They receive requests for tasks to be scheduled, and return a near-optimal schedule according to the genetic algorithm.
- **Execution Group.** The computers in this group execute the tasks that have been previously scheduled and assigned to them. The use of real-time monitoring information about the Execution Group by means of Mon*ALISA* implies that computers in the Grid may well take part in both scheduling and execution, provided that they have enough resources to support the load produced by the scheduling algorithm. In the case of less stringent deadline requirements, the solution of using the computers for both scheduling and execution is valid, and employs a reduced number of resources.

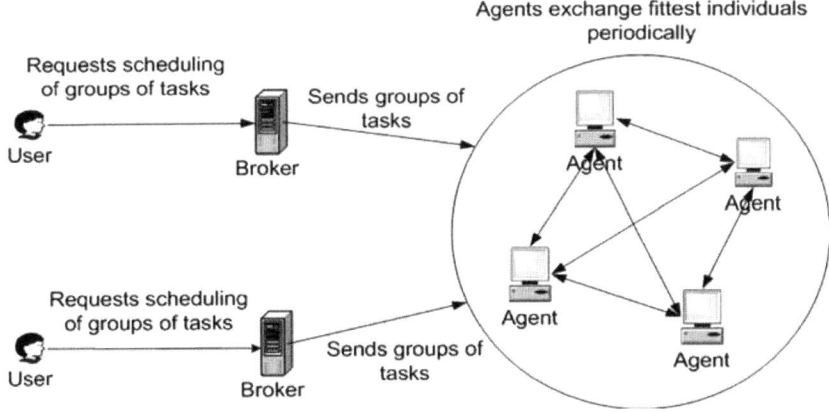

Fig. 8.4. Communication model

The computers in the Scheduling Group are either brokers or agents. The brokers are designated to receive user requests, and the agents run the genetic algorithm in order to find a near-optimal solution.

The Broker develops communication in two directions (see Fig. 8.4): with the user on one side and with all the other nodes in the Grid on the other side. The following features are implemented by the Broker:

- A listener to the input, which perceives user quests on that node. The quest input is namely a file and contains a XML description of the tasks to be scheduled. In this way, a user may request the scheduling of more than one task at a time;
- A parser of the input file that results in creating an object of type "batch of tasks", containing all the tasks to be scheduled.

Each Agent has a local task queue in which the tasks are inserted sorted according to one of the following criteria:

- the arriving time,
- the scheduling priority assigned by the user.

The genetic algorithm starts when the task queue is not empty and either a predefined waiting period has elapsed or there are enough tasks in the queue to complete a chromosome. At the same time the agents communicate with each other, exchanging best individuals. In this way, the fittest chromosomes determined by each Agent may be selected and subsequently used at the following step and implicitly at the final step. The migration of the best individuals leads to a better optimal result in a shorter time.

We have experimented three types of communication among the agents running the Genetic Algorithm:

1. synchronous, with best scheduling chromosome exchanged after a given period of time has passed (TIMEOUT)
2. asynchronous, with each of the agents sending their current best scheduling chromosome after a number of generations. Depending on the running speed of each agent, they may not get to the same number of generations in an equal amount of time
3. without communication; the working nodes are sending the Best Scheduling Chromosome at the end of the algorithm

One major objective of our research is to fit in real scenarios of Grid scheduling. In this context, DIOGENES can well be mapped on two scenarios that may appear. In the first scenario, the Scheduling requests are sent from remote sites by means of a portal to the computers in the Scheduling Group. In this case, the brokers provide the input for our scheduling algorithm, but do not actually run it. In the second scenario, all the computers in the Scheduling Group run the scheduling algorithm and are open to Scheduling requests at the same time. The execution of the genetic algorithm is not limited to only a part of the computers in the Scheduling Group, and that translates into an efficiency increase. The first approach is preferable when users are somewhere remote and would rather not overload their computer with running the scheduling algorithm. In the second case, the resources available are not preferentially treated and therefore the number of resources used is reduced. Depending on the situation, the intended objectives or cost limits, one of the two scenarios may be adopted.

8.4 Genetic Algorithm

Genetic algorithms (GAs) are a well known optimization heuristic, especially when dealing with combinatorial problems, and particularly with NP-Complete ones [17], which are not solvable in a polynomial time. The GA was first introduced by Holland (1975) and represent a machine learning optimization method based on a metaphor of the evolution process observed in nature. These search procedures are based on the evolution of a population of individuals (a vector of possible solutions). GAs are part of evolutionary supervised learning methods, as they require an evaluation function provided by the one who creates of the algorithm (instead of reinforcement learning agents for example).

As described by Goldberg [17], in general terms, a genetic algorithm consists of four parts, each of them with a well defined role (see Fig. 8.5):

1. Generate an initial population
2. Select pair of individuals based on the fitness function
3. Produce next generation from the selected pairs by performing random changes on the selected parents (by applying pre-selected genetic operators).
4. Test for stopping criterion:
   - return the solutions/individuals if satisfied, or
   - go to step 2. if not.

8 Decentralized Grid Scheduling Using Genetic Algorithms   227

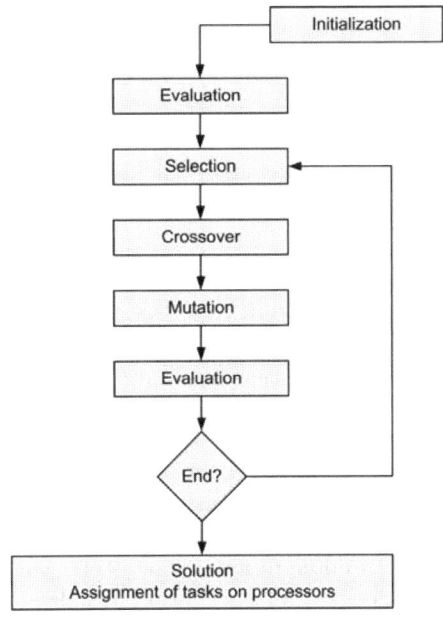

Fig. 8.5. Genetic Algorithms' main steps

The termination condition can be either:

1. No improvement in the solution after a certain number of generations.
2. The solution converges to a predetermined threshold.

8.4.1 Chromosome Encoding

In Genetic Algorithms, each chromosome (individual in the population) represents a possible solution to a problem. In the case of scheduling, each chromosome represents a schedule of a group (batch) of tasks on a group of processors. A chromosome can be represented as a sequence of individual schedules (one for each processor in the group) separated by a special value. Each individual schedule is a queue of tasks assigned to that processor. In another representation, each gene is a pair of values (T_j, P_i), indicating that task T_j is assigned to processor P_i. The execution order of tasks allocated to the same processor is given by the positions of the corresponding genes in the chromosome. Tasks allocated to different processors can be executed in parallel. A third representation adopts a matrix structure with processors represented on one dimension and queues represented on the second dimension.

In our chromosome encoding, we adopted the second representation, in which each gene is a pair of values (T_j, P_i), indicating that task T_j is assigned to processor P_i, where j is the index of the task in the batch of tasks and i is the processor id. For example, in the chromosome representation of Fig. 8.6, tasks

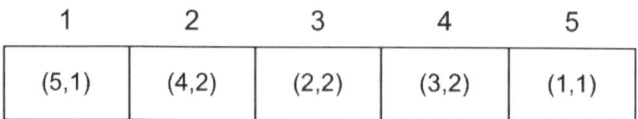

Fig. 8.6. Chromosome representation

5 and 1 are allocated to processor 1, while tasks 2, 3 and 4 are allocated to processor 2. Also task 5 will be executed before task 1.

This representation has been regarded in literature [1, 2] as efficient and compact, with reduced computational costs (crossover and mutation are easier to implement on this type of representation).

8.4.2 Population Initialization

The initial population is initialized by randomly placing each task on a processor. To ensure that the search space is thoroughly explored, the chromosomes are created using different random number generators. Various probabilistic distributions are used by each agent for population initialization (Poisson, Normal, Uniform, Laplace). This random generation of chromosomes can lead to configurations that do not correspond to valid schedules (e.g. the tasks' priorities do not correspond to the specification). The problem can be solved by applying corrections to make the generated chromosome compliant with the specification or by including penalties in the fitness function (see later).

8.4.3 Genetic Operators

A thorough analysis of the genetic operators is essential for an efficient inspection of the solution space. The current section presents design features of the genetic operators that we have used during the experiments.

Crossover

We have experimentally tested three types of crossover: single-point crossover, two-point crossover and uniform crossover. Single-point crossover functioned better in most of the cases and delivered the best scheduling solutions. For the selected chromosome representation, the single point crossover is performed as follows (see example in Fig. 8.7 for chromosomes of length 6):

- a cut position is randomly selected (for our example, the cut position is between genes three and four); for each chromosome, this produces a head segment and a trail segment;
- the first offspring is generated by retaining the head segment from the Parent 1 and combining it with a segment obtained by combining the trail segments of the parents; more specific, for each gene, the first value of the pair is taken from Parent 1, while the second value is taken from Parent 2;

Parent 1	(1, 2)	(5, 4)	(4, 2)	(3, 3)	(6, 2)	(2, 3)
Parent 2	(4, 1)	(6, 3)	(3, 2)	(1, 1)	(2, 4)	(5, 2)
Offspring 1	(1, 2)	(5, 4)	(4, 2)	(3, 1)	(6, 4)	(2, 2)
Offspring 2	(4, 1)	(6, 3)	(3, 2)	(1, 3)	(2, 2)	(5, 3)

Fig. 8.7. Single-point crossover between two chromosomes in GA scheduling

- the second offspring retains the first segment from parent 2; for each gene of the new trail segment, the first value of the pair is taken from parent 2, while the second value is taken from Parent 1.

With these rules, the tasks identities and order are inherited by the first offspring from the first parent, and by the second offspring from the second parent. Similar considerations can be made for the other forms of crossover.

The likelihood of crossover being applied is typically between 0.6 and 1.0[8]. We have adopted the same probability interval for our experiments. If crossover is not applied, offspring are produced simply by duplicating the parents. This gives each individual a chance of passing on its genes without the disruption of crossover.

Mutation

All new chromosomes have a certain probability of being affected by mutation. The search space expands to the vicinity of the population by randomly altering certain genes. The result is the tendency to converge to a global rather than to a local optimum[8, 10, 22, 33]. We have opted for an adaptive mutation operator, which gave better results than the usual static ones. Research has been carried out in the field of dynamic operators, but such studies usually focus on either increasing or decreasing the probability of mutation continuously during the entire run[8]. The adaptive operator that we introduced is more flexible. The novelty of our mutation operator is to dynamically adjust the mutation rate, depending on the fitness variation. In our experiments, we modeled a linear increase in mutation rate when the population stagnates, and a decrease towards a predefined threshold when population fitness varies and the search space has moved to a vicinity.

Based on the mutation rate, we first decide whether the current chromosome will be affected by mutation. If mutation has been decided, the next step is to modify the chromosome. We have conducted experiments with three different mutation operators, which are further described:

Partial-gene Mutation randomly selects a chromosome and change a randomly selected gene assigning the task to a new processor where it has earliest start time [10];

Order-based (Swap) Mutation randomly selects two processors, and then randomly selects a task on each processor. The tasks are interchanged between processors if they have similar properties (for example, the same priority); otherwise, the search continues [4];

Additive Mutation randomly selects two processors, then randomly selects a task on the first processor. Next, a starting point is randomly selected, after which the processor substring is searched for an insertion position. The task must be inserted such that the property rules (for example, priority order) still hold. After that, it is erased from the first processor string.

Selection

Genetic operators apply to chromosomes that are selected from the actual population and produce a new generation. Our algorithm implements the *roulette wheel* selection method, which proved to work well in similar studies [3, 5, 11]. According to the roulette wheel selection technique, the chance $p(i)$ of an individual to become a parent is directly proportional with its fitness value $F(i)$ (see Section 8.4.3):

$$p(i) = \frac{F(i)}{F_t} \qquad (8.2)$$

F_t is the total fitness of the population and p represents the population size.

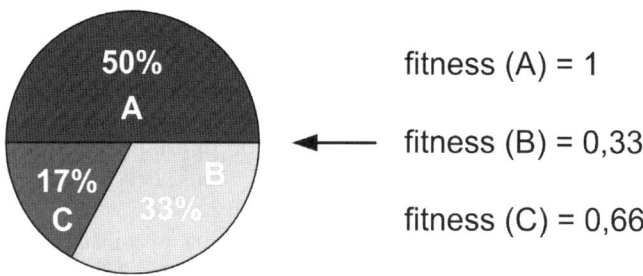

Fig. 8.8. The roulette wheel method

The three steps of a roulette wheel selection method are (see Fig. 8.8):

1. Sum the fitness of all the population members. Call this F_t (total fitness)

$$F_t = \sum_{i=1}^{p} F(i) \qquad (8.3)$$

2. Generate a random number n, between 0 and F_t.
3. Return the first individual whose fitness added to the preceding population members is greater than or equal to n.

8.4.4 Fitness Function

The fitness (or objective) function measures the quality of each individual in the population according to some criteria. For the scheduling problem, the goal is to obtain task assignments that ensure minimum execution time, maximum processor utilization, a well balanced load across all processors, or a combination of these.

The objective selected for our research was to obtain a well-balanced assignment of tasks to processors. This has, as a side effect, reduction of the overall completion time of tasks assigned to processors, which means minimization of makespan [5, 21]. Makespan is defined as:

$$t_M = \max_{1 \leq i \leq n} \{t_i\} \quad (8.4)$$

In this formula, n is the number of processors and t_i is the total execution time for processor i, computed as the sum of processing times for all tasks assigned to this processor in the current or previous schedules:

$$t_i = t_i^p + t_i^c = \sum_{j=1}^{T_i} (t_{i,j}) \quad (8.5)$$

We have considered T_i to be the total number of tasks assigned to processor i for execution and $t_{i,j}$ the running time of task j on processor i. t_i^p is the execution time for previously assigned tasks to processor i, and t_i^c is the processing time for currently assigned tasks.

A mapping of makespan in the $[0, 1]$ interval leads to the factor $\frac{1}{t_M}$ that is often considered for fitness computing [5, 21]. We optimized this factor for a more efficient search of well-balanced schedules. In our approach, we determined that reducing the difference between the minimum and the maximum processing times is a factor worth considering for fitness, in terms of load-balancing optimization. Therefore, one factor introduced for fitness computation is:

$$f_1 = \frac{t_m}{t_M} = \frac{\min_{1 \leq i \leq n}\{t_i\}}{\max_{1 \leq j \leq n}\{t_j\}}, 0 \leq f_1 \leq 1 \quad (8.6)$$

The factor converges to 1 when t_m approaches t_M, and the schedule is perfectly balanced.

The second factor considered for fitness computation is the average utilization of processors:

$$f_2 = \frac{1}{n} \sum_{i=1}^{n} \frac{t_i}{t_M}, 0 \leq f_2 \leq 1 \quad (8.7)$$

Zomaya [5] pointed out its utility of reducing idle times by keeping processors busy. Division by makespan is pursued in order to map the fitness values to the interval $[0, 1]$. In the ideal case, the total execution times on the processors are equal and equal to makespan, which leads to a value of 1 for average processor utilization.

Another factor considered in our research represents meeting the imposed restrictions. In a realistic scenario, task scheduling must meet both deadline and resource limitations (in terms of memory, cpu power). In deadline computation for a task t, we must consider the execution times for each of the tasks assigned to run before task t on the same processor. These tasks occupy a previous slot on the respective processor in our encoding of a chromosome [2, 13]. The fitness factor is subsequently defined as:

$$f_3 = \frac{T_s}{T}, 0 \leq f_3 \leq 1 \qquad (8.8)$$

We consider that T_s denotes the number of tasks which satisfy deadline and computation resource requirements, and T represents the total number of tasks in the current schedule.

This factor acts like a contract penalty on the fitness. Its value varies reaching 1 when all the requirements are satisfied and proportionally decreases with each requirement that is not met. The chance of being selected for future generations is reduced due to the penalty introduced by this factor. Still, the schedule is not dismissed, but may be used in subsequent reproduction stages that lead to valid chromosomes.

The fitness function applied in our research consists of the contribution of the factors presented:

$$F = f_1 \times f_2 \times f_3 = \left(\frac{t_m}{t_M}\right) \times \left(\frac{1}{n}\sum_{i=1}^{n}\frac{t_i}{t_M}\right) \times \left(\frac{T_s}{T}\right), 0 \leq F \leq 1 \qquad (8.9)$$

8.4.5 Algorithm Description

The description of the scheduling algorithm in pseudo-code is given below:

```
L = length of the chromosome;
S = number of steps for the GA;
T = waiting time;

Broker:
    Receive user request (input_file);
    Parse the input file and obtain a TaskBatch object;
    Broadcast the TaskBatch object to all the Agents in the
        Scheduling Group;

Agent:
    Queue taskQueue;
    Actively listens to the Brokers for new tasks and insert
        them sorted in the taskQueue;
```

```
While (NOT((length(taskQueue)>0 AND curr_waiting_time==T)
        OR taskQueue.length>=L)) do
    Wait for new tasks to come;
Endwhile

current length = min{taskQueue.length, L);

If current length < L then
    Fill chromosome with padding;

Interrogate the Grid Monitoring Service about the status
    of the processors in the Execution Group;
Initialize the GA algorithm;

While current step <= S do
    If current step == S then
        Determine the optimal individual (schedule);
        Save the optimal schedule in the history file;
        break;
    Endif

    Run the current step of the GA;
    Exchange the fittest individuals in the current
        generations (computed on every Agent) and insert
        them in the next population;
Endwhile
```

On the Brokers side, requests for the scheduling of a group of tasks are received. The input file from the user is parsed, and a TaskBatch object (which contains objects of the type Task) is obtained and sent to all the active Agents in the Scheduling Group. More than one Broker at a time may receive requests for task scheduling.

On the Agents side, the Tasks are stored in a local queue, on each of them. If the queue is not empty and either a given interval of time has passed or the length of the queue exceeds the length of the chromosome, the Agents start to run the genetic algorithm, otherwise wait for new tasks to come. Before GA running, the Agents interrogate the Grid Monitoring Service about the status of processors in the Execution Group, and find out the idle cpu and free memory on each of them. At each step of the GA or at a predefine interval, the fittest individuals computed by the Agents are exchanged, in order to improve the population. At the final step, the schedule is saved in a history file on the agents.

Here is a description of the scheduling algorithm in a logical flow of activities:

Step 1. A user requests that one or more tasks are scheduled. His request has as a parameter the name of the file containing a description of the tasks. The file has a standard XML format and presents requirements for each task relative to memory, cpu usage, execution time, etc.

Step 2. The input file is processed and a batch of tasks (group of tasks) object is constructed.

Step 3. The batch of tasks is broadcast to all the nodes in the cluster.

Step 4. The nodes receive the group of tasks to be scheduled. The tasks are inserted sorted in a queue according to a sorting criteria like arriving time or scheduling priority. If the number of tasks in the queue is less than a predefined length of the chromosome, they wait for T units of time before starting the genetic algorithm. If the chromosome is still not complete at the end of the waiting period, a non-influential padding is added. On the contrary, if the length of an arriving group of tasks exceeds the predefined dimension of the chromosome, some tasks are saved in the waiting queue and will be scheduled at the next time.

Step 5. On each node, a daemon keeps an up-to-date status of the computers in the Grid on which tasks are sent for execution, by constantly interrogating a monitoring system (we have used the Mon*ALISA* service). The Grid is a dynamic environment in which nodes' characteristics such as free memory or cpu utilization vary over time. The current configuration information is necessary for an accurate scheduling process. The nodes interrogate the daemon for at the beginning of the algorithm, to find out the current status of the Grid.

Step 6. The nodes in the cluster run the GA. Each node starts with a different, specific initialization of the genetic algorithm. The subsequent steps of the GA are similar for all the nodes in the cluster, and so is the fitness formula. In this way, the clients will compute different optima from which the best one will be chosen.

Step 7. The migration of the best current solutions is performed after each step of the GA, thus ensuring that the population finds a better optima. The nodes exchange the fittest individuals and insert them in the next generation.

Step 8. The generation of populations stops after a finite, predefined number of steps. At this point, each client in the cluster computes its optimal individual.

Step 9. The same communication procedure as above is used for the final step of the GA. Each node sends its optimum to all the other nodes in the cluster and the final optimal individual is decided by each one. The fittest chromosome is selected from the optimal individuals only. The result is the same on every node, because the computing procedure and the individuals at the last step of the GA are the same.

Step 10. The scheduling obtained is saved in a history file on each node in the cluster.

8.5 Experimental Results

The experimental cluster was configured with 11 computers, named P_1, P_2, ..., P_{11}, which represent heterogeneous computing resources with various processing capabilities and initial loads. The input tasks represent typical cpu-intensive computing programs. The processing time (in seconds), use for graphics, represents the execution time of scheduled tasks.

Default parameters for the genetic algorithm were established at 0.9 for crossover rate and 0.005 for mutation rate threshold. The chromosome length is 50. The values were experimentally determined, in order to widely and thoroughly explore the search space.

The reproduction operators applied were single point crossover and adaptive mutation, as described in section 8.4.3. We used the roulette wheel selection method to choose individuals who will survive in the next generation.

8.5.1 Algorithm Convergence

We have studied the improvement achieved by the algorithm throughout generations. The metric used for performance measurements is load-balancing, a performance attribute of high demand in distributed environments.

In this work, the load-balancing of the system was investigated according to the following definition:

$$L = 1 - \frac{\Delta}{\bar{u}}, 0 \leq L \leq 1 \tag{8.10}$$

$\bar{u}$ is the average processor utilization of the system, where u_i represents the utilization rate of processor i:

$$\bar{u} = \frac{1}{n} \sum_{i=1}^{n} u_i \tag{8.11}$$

We consider n to denote the number of processors in the system. The processor utilization rate is computed as:

$$u_i = \frac{t_i}{t_M} = \frac{t_i}{\max_{1 \leq i \leq n} \{t_i\}} \tag{8.12}$$

In the formula above, t_i and t_M have the same significance as described in section 8.4.4.

Furthermore, Δ in the load-balancing formula denotes the square deviation of from the mean $\bar{u}$:

$$\Delta = \sqrt{\frac{1}{n} \sum_{i=1}^{n} (u_i - \bar{u})^2} \tag{8.13}$$

Load-balancing in the system converges to 1 when the tasks are equitable disposed on processors, so that the processing times are approximately equal relative to one another and equal to makespan. In this case, the dispersion of processor utilization rates from the average tends to zero:

$$u_i \to \bar{u} \Rightarrow \Delta \to 0 \Rightarrow L \to 1 \tag{8.14}$$

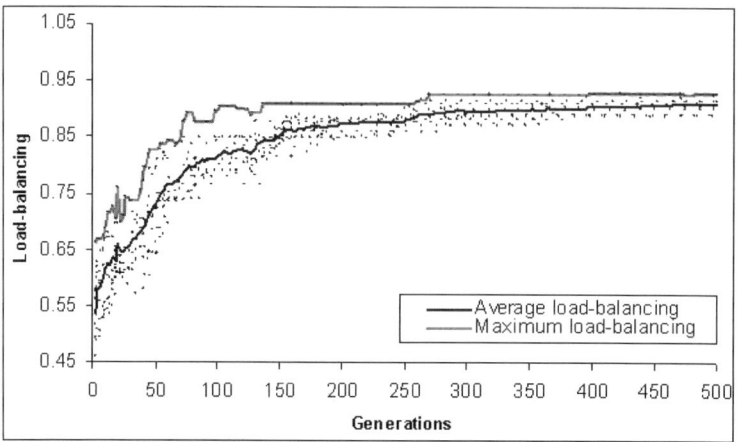

Fig. 8.9. Convergence of load-balancing in the system over 500 generations

For the purpose of this experiment, we use a GA scheduler with one 3. Fig. 8.9 shows the dispersion of load-balancing in the system over 500 generations, as well as the average and maximum load-balancing.

A number of 10 distinct experiments were pursued during the convergence study. The points in the figure represent values obtained at different generations. At reduced numbers of generations, the results are spread over a larger interval and usually achieve low values, as the algorithm does not have enough time to cover a larger search space. The algorithm converges for higher numbers of generations. For the set of tasks considered, little improvement is obtained over 200 generations.

8.5.2 Decentralization

Convergence speed-up

The influence of decentralization on the load-balancing performance metric was subsequently analyzed. Experiments were pursued with one, two, three and four agents, averaged over ten runs in each case. Fig. 8.10 shows the results obtained over 200 generations.

The increase in the number of agents gives best results in terms of convergence speed up when we are running the genetic algorithm with fewer generations, less than 100 on the tasks set used. This was expected since the GA is a stochastic algorithm, in which every run is initialized randomly. More agents mean more start points for the genetic algorithm in order to find a near optimal solution. While incrementing the number of agents, the improvement rate decreases. In our case, the benefit of four agents over three is almost negligible because the convergence is already established al low generations, below which the chances of finding optimal individuals are very low, although positive. At higher

8 Decentralized Grid Scheduling Using Genetic Algorithms   237

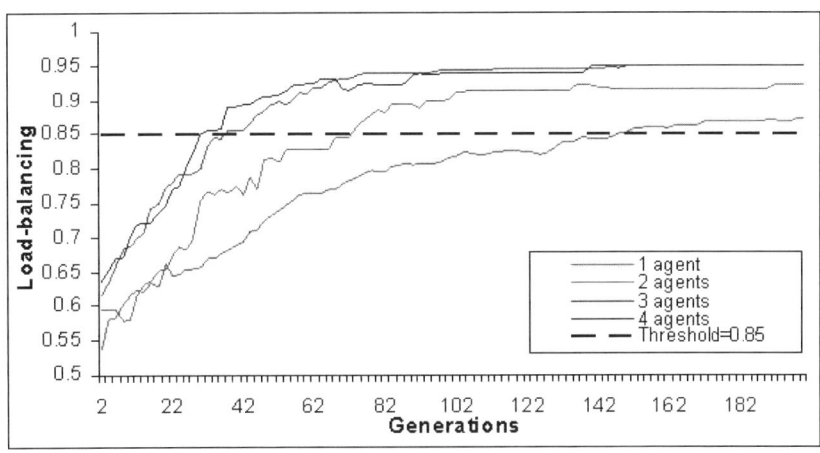

Fig. 8.10. Load-balancing with various number of agents

generations, the cost of employing multiple agents is not justified, because the algorithm has already converged to an optimum.

The $\alpha = 0.85$ threshold is used to illustrate speed-up for each situation. The one agent experiment reaches this threshold at approximately 150 generations, while the two agents experiment finds a solution of the same quality at 74 generations. Furthermore, three agents deliver an equally optimal solution at 38 generations and four agents at 30 generations.

Table 8.2 shows the speed-up increase rate for achieving an optimal solution at the 0.85 load-balancing threshold for different numbers of agents. The speed-up relative to the previous experiment (first column) is computed as:

$$s_i^\alpha = \frac{g_{i-1}^\alpha - g_i^\alpha}{g_{i-1}^\alpha} \times 100\% \qquad (8.15)$$

Table 8.2. Speed-up increase rates for various decentralization levels at $\alpha = 0.85$ load-balancing threshold

Experiment no.	Speed-up increase rate relative to previous exp. (%)	Speed-up increase rate relative to centralized alg. (%)	Relative execution time decrease (%)
Exp1: 1 agent	-	-	-
Exp2: 2 agents	50.66	50.66	5.46
Exp3: 3 agents	48.64	74.66	14.10
Exp4: 4 agents	21.05	80	-0.55

with g_i^α representing the number of generations at which the threshold is reached in the experiment with i agents, and g_{i-1}^α is the number of generations at which the threshold is reached in the experiment with $i-1$ agents.

The increase rate is high (50.66%) with 2 agents, meaning that the generations were reduced to less than half. In the case of three agents, a rate almost as high (48.64%) is obtained, while a low rate increase of 21.05% for the four agents experiment shows that the improvement is extremely reduced.

The speed-up relative to the experiment with one agent (that mean pseudo-centralized method) was computed as:

$$spc_i^\alpha = \frac{g_{pc}^\alpha - g_i^\alpha}{g_{pc}^\alpha} \times 100\% \qquad (8.16)$$

The g_{pc}^α term is the number of generations at which the α threshold is achieved in the experiment with 1 agent.

The method provides a substantial decrease as regards the execution time for the scheduling algorithm. For our experimental configuration, an optimum is achieved with 3 agents, when a decrease of over 14% execution time is registered comparative to the 1-agent experiment. The communication time costs are higher when running the algorithm with four agents, and therefore the performance is diminished. It is also worth noticing that not only execution performance, but also the probability to obtain a global optimum is improved by employing multiple agents with various start points in the solution space.

Scalability and robustness

We also pursued experiments for studying the capability of the system to resist to failure situations. Fig. 8.11 illustrates a scenario in which the system started to function with two agents and one of them dysfunctions after running the algorithm for 40 generations. For the purpose of comparison, we also represented the estimated evolution averaged over 10 runs with one agent and two agents.

Although in the initial phases the algorithm has performed well with two agents, achieving load-balancing values similar to those obtained during normal functioning, the improvement is reduced after one agent ceases to function. The convergence rate decreases, leading to slightly better results than in the case of a normal functioning with one agent. The performance is visible decreased, but the system resists failure and continues to perform scheduling.

8.5.3 Estimated Times Versus Real Execution Times

The quality of estimated times is essential for the quality of the schedule. We want the computed processing times to closely approximate the real execution times. That is especially important in hard real-time scheduling problems, in which missing deadlines is extremely problematic [2].

Fig. 8.12 provides a comparison between estimated time and real processing time achieved during the experiment. A configuration of the algorithm with 200 generations was used for computation of estimated times. Job execution was

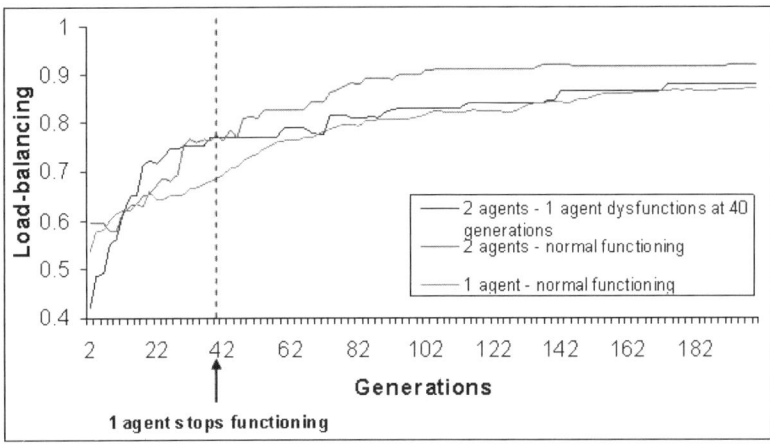

Fig. 8.11. Reliability: 1 agent out of 2 dysfunctions after 40 generations

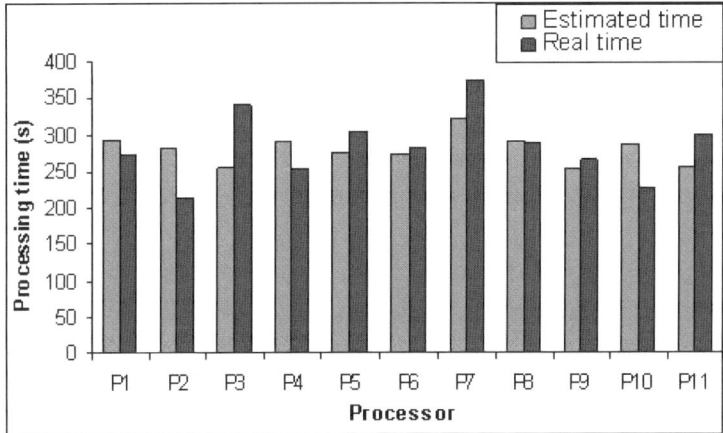

Fig. 8.12. Estimated and real processing times on each processor

pursued with PBS and job monitoring information was achieved by means of Mon*ALISA* Service and its extensions (ApMon and Mon*ALISA* Client).

The error of approximation is computed according to the method subsequently described.

The relative deviation of the estimated time from the real time for processor i is noted δ_i and determined as:

$$\delta_i = \frac{t_i^r - t_i^e}{t_i^r} \tag{8.17}$$

In this formula, t_i^r represents the real processing time obtained by running the jobs in the execution system, and t_i^e is the estimated processing time provided by the genetic algorithm.

The approximation error is further determined as:

$$\epsilon = \sqrt{\frac{1}{n} \sum_{i=1}^{n} \delta_i^2}, 0 \leq \epsilon \leq 1 \tag{8.18}$$

We obtained an average error value over 10 runs of approximately $\epsilon = 0.16$. The scheduling policy employing real monitoring data from the Grid environment is therefore a viable one, providing very good estimation results, with an accuracy of about 84%.

8.5.4 Comparison of Various Scheduling Methods

Four different methods were compared with respect to load-balancing achievement. In the first stage, we use the First Come First Served algorithm in order to assign 100 tasks on the eleven computation resources. The following experiments test heuristic strategies based on genetic algorithms. The experiments were pursued with 100 generations. The purpose was to demonstrate that very good results can be obtained even at reduced numbers of generations, which implies low computation costs.

First Come First Served Algorithm

In the First Come First Served policy, each task is assigned at its arrival on the processor with the minimum expected start time.

In order to determine the fittest processor for task allocation, we must first estimate the time at which already assigned tasks will finish execution on each processor. The expected completion time for processor i is computed as follows, where T_i and $t_{i,j}$ have the same significance as described in section 8.4.4:

$$tc_i = \sum_{j=1}^{T_i} t_{i,j}, 1 \leq i \leq n \tag{8.19}$$

We can now compute the estimated start time for a newly arrived task k:

$$ts_k = \min_{1 \leq i \leq n} \{tc_i\} = \min_{1 \leq i \leq n} \left\{ \sum_{j=1}^{T_i} t_{i,j} \right\} \tag{8.20}$$

The arriving times of the tasks influence the balancing level in the system, so that early arrival times for the larger tasks would lead to better results. Improved solutions are obtained if we use a heuristic, which would reorder the task set on the expenses on algorithm complexity [21]. For the task set considered, the

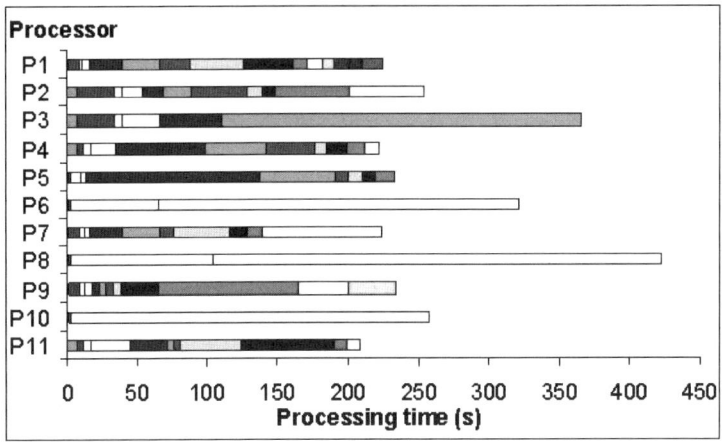

Fig. 8.13. Task assignment with First Come First Served algorithm

strategy obtains an average processor utilization of about 0.64 on a [0, 1] interval and a load-balancing of about 0.754 (Fig. 8.13). The least loaded processor (P1), has over 200 seconds (about half the total execution time) additional idle time relative to the overloaded processor $P8$.

It is clear that heterogeneity of tasks is a highly influential factor and leads to low values for load-balancing. The large tasks disposed on processors $P3$ and $P8$ heavily overload these computation resources.

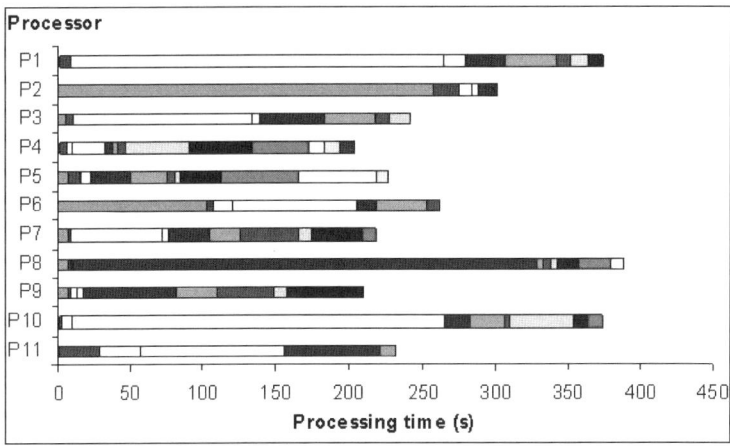

Fig. 8.14. Task assignment with the centralized genetic algorithm

Centralized GA-based Scheduling Algorithm

This heuristic method searches for an optimal assignment of tasks in order to achieve total execution time reduction and highly homogeneous loads on processors.

In the second experiment, we have tested the centralized genetic algorithm on the same set of input tasks with a stop point at 100 generations (see Fig. 8.14). The load-balancing obtained is with $\Delta_l \approx 0.2\%$ lower in comparison with the First Come First Served policy. The result is indeed satisfactory considering the reduced number of generations run by the algorithm. Moreover, an important makespan reduction of $\Delta_t = 33s$ was achieved, as well as an increase in average processor utilization of $\Delta_u = 6\%$.

In a First Come First Served strategy, monitoring information must be collected at the arrival of each job, in order to determine the processor with the earliest start time. Although the complexity of this method is reduced comparative to genetic algorithms, high execution times are induced by the need of monitoring data. The genetic algorithms, on the other hand, have the advantage of scheduling a whole group of tasks at a time, without the necessity of status interrogation for every task assigned.

Decentralized Non-Cooperative Genetic Algorithm

In experiment 3, the decentralized non-cooperative genetic algorithm was studied. Based on the experimental results described in section 8.5.2, the decentralized algorithm was pursued with a number of three agents. No cooperation mechanism is applied among agents, therefore, optimal individuals are not interchanged at different stages during algorithm run. The genetic algorithm starts

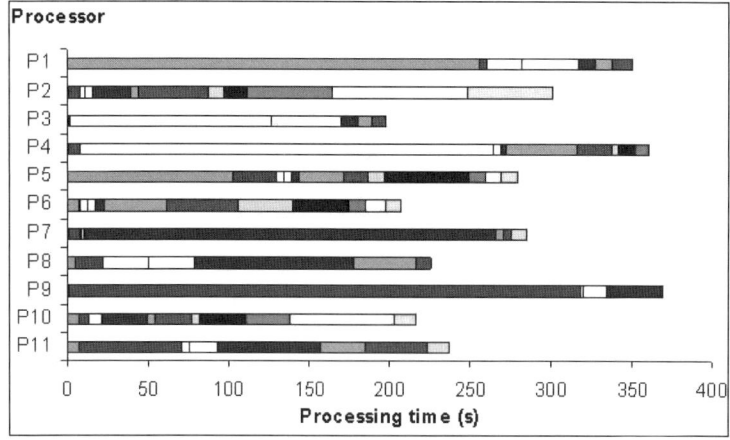

Fig. 8.15. Task assignment with the decentralized non-cooperative genetic algorithm

from different search points, and the fittest individual is chosen at the final step from the optimal results determined by all agents.

The disposal of tasks after 100 generations can be seen in Fig. 8.15. Compared to the previous experiments, all load-balancing, average processor utilization and makespan performance metrics are improved. The load-balancing has a $\Delta_l \approx 3\%$ increase relative to the centralized algorithm, and an average processor utilization improvement of $\Delta_u = 4\%$. The maximum execution time has been reduced by $\Delta_t = 18.5s$.

All metrics indicate that resources are better utilized, although processors like P3 or P6 are still idle 0.47% and 0.44% respectively of the total execution time, to the detriment of overloaded processors like P9, P4 and P1.

Decentralized Cooperative Genetic Algorithm

The fourth experiment analysis the metrics previously discussed in a decentralized scheduling scenario, in which a cooperative genetic strategy has been employed. The cooperative characteristic implies optimal individuals interchange in order to speed up convergence. The input task set is the same as previously, as well as the level of decentralization (3), and the number of generations (100).

Fig. 8.16 illustrates the schedule obtained. An essential improvement of all metrics has been achieved, with a load-balancing of 0.94, in comparison with 0.78 determined by the non-cooperative strategy ($\Delta_l \approx 16\%$). The average processor utilization has increased with $\Delta_u = 12\%$ (from 0.74 to 0.86), and the makespan has been reduced by $\Delta_t = 52.5s$, with processor P3 executing only the largest task in the system.

It is clear that a substantial improvement could hardly be achieved. It could only be possible if the largest task was assigned alone on one other processor

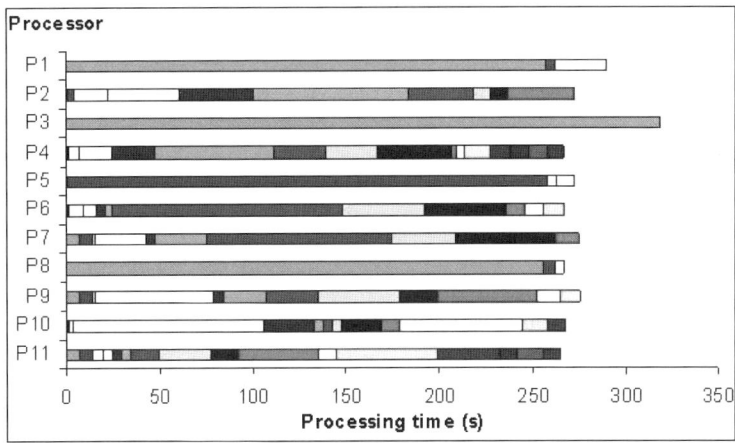

Fig. 8.16. Task assignment with the decentralized cooperative genetic algorithm

with better computation capabilities. Therefore, we can appreciate that the result obtained was very close to the best possible distribution. The cooperative algorithm represents a viable solution, with significant improvement of the solution quality.

Previous work on distributed strategies usually develops scheduling scenarios in which the execution resources or the set of tasks are divided among agents. In a distributed multi-agent mechanism, as described in [21], each agent has only up-to-date information on its neighboring agents, which limits the scheduling effect. Therefore, load-balancing at the global level of Grid is reduced, comparative to a centralized strategy. As the experimental results show, our approach achieves global load-balancing while also ensuring scalability and reliability.

8.6 Conclusions

In Grid environments, various real-time applications require dynamic scheduling for optimized assignment of tasks. This paper describes a genetic scheduling approach, which features a decentralized strategy for the problem of task allocation. We carry out our experiments with complex scheduling scenarios and with heterogeneous input tasks and computation resources. We improve upon centralized genetic approaches with respect to scalability and robustness. Our experimental results show that the system continues to work well even when agents dysfunction. The use of lookup services also facilitates rapid integration of new agents that arise in the system. Moreover, the strategy of starting the search from multiple initial points in the problem space is favorable for obtaining global convergence and avoiding premature blocking in a local optimum. Also, significant convergence speed-up is achieved by means of the cooperative scheduling strategy, although there is a trade-off in terms of implied communication costs.

We compare the performances of the Decentralized Cooperative Genetic Algorithm with three other strategies: OLB, Centralized GA and Decentralized Non-cooperative GA. It is shown that the algorithm clearly outperforms these methods. Decentralization and cooperation provide significantly better results of load-balancing and average processor utilization increase, as well as of total execution time minimization.

Furthermore, instead of employing simulated scenarios, we have validated our research in real-time environments by utilizing existing monitoring and job execution systems. The experiments show a high level of accuracy in the results obtained.

Future investigation would involve the extension of the algorithm towards classes of dependent tasks, as well as the incorporation of new features into the current framework (e.g. Recovery Service for task backup and migration). Also, the slow nature of the GA method and node dynamics in a Grid may lead to less suitable results for estimates of processing times. The solution would combine grid monitoring with prediction of status for the Grid nodes [25].

References

1. Wu, A., Yu, H., Jin, S., Lin, K.-C., Schiavone, G.: An incremental genetic algorithm approach to multiprocessor scheduling. IEEE Trans. on Parallel and Distributed Systems 15(9), 824–834 (2004)
2. Mahmood, A.: A Hybrid Genetic Algorithm for Task Scheduling in Multiprocessor Real-Time Systems. Journal of Studies in Informatics and Control 9(3) (2000)
3. Page, A.J., Naughton, T.J.: Dynamic task scheduling using genetic algorithms for heterogeneous distributed computing. In: Proceedings of the 19th International Parallel and Distributed Processing Symposium, Denver, Colorado, USA, April 2005, pp. 189a.1–189a.8 (2005)
4. Zomaya, A.Y., Ward, C., Macey, B.: Genetic Scheduling for Parallel Processor Systems: Comparative Studies and Performance Issues
5. Zomaya, A.Y., Teh, Y.-H.: Observations on using genetic algorithms for dynamic load-balancing. IEEE Transactions on Parallel and Distributed Systems 12(9), 899–911 (2001)
6. Aggarwal, M., Kent, R.D., Ngom, A.: Genetic algorithm based scheduler for computational grids. In: 19th International Symposium on High Performance Computing Systems and Applications 2005. HPCS 2005, May 15-18, pp. 209–215 (2005)
7. Csaji, B.C., Monostori, L., Kadar, B.: Learning and Cooperation in a Distributed Market-Based Production Control System. In: Proceedings of the 5th International Workshop on Emergent Synthesis, pp. 109–116 (2004)
8. Beasley, D., Bull, D., Martin, R.: An overview of genetic algorithms: Part 2, research topics. University Computing 15(4), 170–181 (1993)
9. Thain, D., Tannenbaum, T., Livny, M.: Condor and the Grid. In: Berman, F., Hey, A.J.G., Fox, G. (eds.) Grid Computing: Making The Global Infrastructure a Reality. John Wiley, Chichester (2003)
10. Goldberg, D.E.: Genetic Algorithms in Search, Optimization and Machine Learning. Addison-Wesley, Reading (1989)
11. Hou, E., Ansari, N., Ren, H.: A genetic algorithm for multiprocessor scheduling. IEEE Transactions on Parallel and Distributed Systems 5(2), 113–120 (1994)
12. Seredynski, F., Koronacki, J., Janikow, C.Z.: Distributed Scheduling with Decomposed Optimization Criterion: Genetic Programming Approach. In: Rolim, J.D.P. (ed.) IPPS-WS 1999 and SPDP-WS 1999. LNCS, vol. 1586. Springer, Heidelberg (1999)
13. Manimaram, G., Murthy, C.S.R.: An Efficient Dynamic Scheduling Algorithm for Multiprocessor Real-time Systems. IEEE Transactions on Parallel and Distributed Systems 9(3), 312–319 (1998)
14. Syswerda, G.: Uniform crossover in genetic algorithms. In: Schafer, J.D. (ed.) Proceedings of the Third International Conference on Genetic Algorithms. Morgan Kaufmann, San Francisco (1989)
15. Weichhart, G., Affenzeller, M., Reitbauer, A., Wagner, S.: Modelling of an Agent-Based Schedule Optimisation System. In: Proceedings of the IMS International Forum (2004)
16. Iordache, G., Boboila, M., Pop, F., Stratan, C., Cristea, V.: A Decentralized Strategy for Genetic Scheduling in Heterogeneous Environments. In: Meersman, R., Tari, Z. (eds.) OTM 2006. LNCS, vol. 4276, pp. 1234–1251. Springer, Heidelberg (2006)

17. Ghosh, S., Melhem, R., Mosse, D.: Fault-tolerance Through Scheduling of Aperiodic Tasks in Hard Real-time Multiprocessor Systems. IEEE Transactions on Parallel and Distributed Systems 8(3), 272–284 (1997)
18. Newman, H.B., Legrand, I.C., Galvez, P., Voicu, R., Cirstoiu, C.: MonALISA: A Distributed Monitoring Service. In: CHEP 2003, La Jolla, California (2003)
19. Yin, H., Wu, H., Zhou, J.: An Improved Genetic Algorithm with Limited Iteration for Grid Scheduling, gcc. In: Sixth International Conference on Grid and Cooperative Computing (GCC 2007), pp. 221–227 (2007)
20. Legrand, I.C.: End User Agents: extending the intelligence to the edge in Distributed Service Systems. In: Fall 2005 Internet2 Member Meeting, Philadelphia (2005)
21. Cao, J., Spooner, D.P., Jarvis, S.A., Saini, S., Nudd, G.R.: Grid load balancing using intelligent agents, Future Generation Computer Systems special issue on Intelligent Grid Environments: Principles and Applications (2004)
22. Schaffer, J.D., Eshelman, L.J.: On crossover as an evolutionarily viable strategy. In: Belew, R.K., Booker, L.B. (eds.) Proceedings of the Fourth International Conference on Genetic Algorithms, pp. 61–68. Morgan Kaufmann, San Francisco (1991)
23. Maheswaran, M., Ali, S., Siegel, H.J., Hensgen, D., Freund, R.F.: Dynamic Mapping of a Class of Independent Tasks onto Heterogeneous Computing Systems. JPDC, 107–131 (1999)
24. Theys, M.D., Braun, T.D., Siegal, H.J., Maciejewski, A.A., Kwok, Y.K.: Mapping Tasks onto Distributed Heterogeneous Computing Systems Using a Genetic Algorithm Approach. John Wiley, Chichester (2001)
25. Phinjareonphan, P., Bevinakoppa, S., Zeephongsekul, P.: An Algorithm to Predict Reliability of a Grid Node. In: 11th ISSAT International Conference on Reliability and Quality in Design, pp. 37–41 (2005)
26. Prodan, R., Fahringer, T.: Dynamic scheduling of scientific workflow applications on the grid: a case study, Symposium on Applied Computing. In: Proceedings of the 2005 ACM symposium on Applied computing, Santa Fe, New Mexico, USA, pp. 687–694 (2005)
27. Henderson, R.L.: Job scheduling under the Portable Batch System. In: Feitelson, D.G., Rudolph, L. (eds.) IPPS-WS 1995 and JSSPP 1995. LNCS, vol. 949, pp. 279–294. Springer, Berlin (1995)
28. Ramamritham, K.: Allocation and scheduling of precedence related periodic tasks. IEEE TPDS, 382–397 (1993)
29. Baghavathi Priya, S., Prakash, M., Dhawan, K.K.: Fault Tolerance-Genetic Algorithm for Grid Task Scheduling using Check Point, gcc. In: Sixth International Conference on Grid and Cooperative Computing (GCC 2007), pp. 676–680 (2007)
30. Zhou, S.: LSF: load sharing in large-scale heterogeneous distributed systems. In: Proceedings of the Cluster Computing (1992)
31. Gentzsch, W.: Sun Grid Engine: Towards Creating a Compute Power Grid. In: Proceedings of the 1st International Symposium on Cluster Computing and the Grid, pp. 35–36 (2001)
32. Greene, W.A.: Dynamic Load-Balancing via a Genetic Algorithm. In: 13th IEEE International Conference on Tools with Artificial Intelligence, Dallas, Texas, USA, pp. 121–129 (2001)
33. Spears, W.M.: Crossover or mutation? In: Whitley, L.D. (ed.) Foundations of Genetic Algorithms, pp. 221–237. Morgan Kaufmann, San Francisco (1993)

9

Nature Inspired Meta-heuristics for Grid Scheduling: Single and Multi-objective Optimization Approaches

Ajith Abraham[1], Hongbo Liu[2,3], Crina Grosan[4], and Fatos Xhafa[5]

[1] Centre for Quantifiable Quality of Service in Communication Systems,
    Norwegian University of Science and Technology, NO-7491 Trondheim, Norway
    ajith.abraham@ieee.org
    http://www.softcomputing.net
[2] School of Computer Science and Engineering, Dalian Maritime University,
    116026 Dalian, China
[3] Department of Computer, Dalian University of Technology, 116023 Dalian, China
    lhb@dlut.edu.cn
[4] Department of Computer Science, Faculty of Mathematics and Computer Science,
    Babeş Bolyai University, Kogalniceanu 1, Cluj-Napoca, 3400, Romania
    cgrosan@cs.ubbcluj.ro
[5] Dept. de Llenguatges i Sistemes Informàtics
    Universitat Politècnica de Catalunya
    C/Jordi Girona 1-3, 08034 Barcelona, Spain
    fatos@lsi.upc.edu

Summary. In this chapter, we introduce several nature inspired meta-heuristics for scheduling jobs on computational grids. Our approach is to dynamically generate an optimal schedule so as to complete the tasks in a minimum period of time as well as utilizing the resources in an efficient way. We evaluate the performance of Genetic Algorithm (GA), Simulated Annealing (SA), Ant Colony optimization (ACO) and Particle Swarm Optimization (PSO) Algorithm. Finally, the usage of Multi-objective Evolutionary Algorithm (MOEA) for two scheduling problems are also illustrated.

Keywords: Nature Inspired Meta-heuristics, Multi-objective Optimization, Job Scheduling, Grid Computing, Genetic Algorithms, Simulated Annealing, Ant Colony, Particle Swarm Optimization.

9.1 Introduction

A computational grid is a large scale, heterogeneous collection of autonomous systems, geographically distributed and interconnected by low latency and high bandwidth networks [1]. The sharing of computational jobs is a major application of grids. Grid resource management provides functionality for discovery and publishing of resources as well as scheduling, submission and monitoring of jobs. However, computing resources are geographically distributed under different ownerships each having their own access policy, cost and various constraints. Every resource owners will have a unique way of managing and scheduling resources and the grid schedulers are to ensure that they do not conflict with resource owner's policies. In the worst-case situation, the resource owners might charge different prices to different grid users for their resource usage and it might vary from time to time. The job schedule problem is known to be NP-complete [2]. Recently several metaheuristics were introduced to minimize the average completion time of jobs through optimal job allocation on each grid node in application-level scheduling [3], [4]. Because of the intractable nature of the problem and its importance in grid computing, it is desirable to explore other avenues for developing good heuristic algorithms for the problem.

Particularly, with its sound exploration ability both global and local, some new search techniques, nature inspired meta-heuristics, has become the new focus of research. In this chapter, we introduce several nature inspired meta-heuristics for scheduling jobs on computational grids. The nature inspired meta-heuristics involved are Genetic Algorithm (GA), Simulated Annealing (SA), Ant Colony optimization (ACO) and Particle Swarm Optimization (PSO) Algorithm. The PSO approach for scheduling jobs on computational grids is based on fuzzy matrices to represent the position and velocity of the particles in PSO [5], in which a new mapping between the job scheduling problem and the particle is constructed [13]. The approach is to dynamically generate an optimal schedule so as to complete the tasks in a minimum period of time as well as utilizing the resources in an efficient way. We also illustrate the use of Multi-objective evolutionary algorithms for job scheduling [7].

The Chapter is organized as follows. Section 9.2 deals with some theoretical foundations related to job scheduling. Various nature inspired heuristics are introduced in Section 9.3. In Section 9.4, experiment results and discussions are provided. Finally we conclude our work in Section 9.5.

9.2 Scheduling Problem Formulation

In the grid environment, there is usually a general framework focusing on the interaction between grid resource broker, domain resource manager and the grid information server [8]. Usually it is easy for the grid to get information about the speed of the available grid nodes but quite complicated to know the computational processing time requirements from the user. To conceptualize the problem as an algorithm, we need to dynamically estimate the job lengths from user application specifications or historical data. For clarity purposes, some key terminologies are defined as follows:

- Grid Node (computing unit)
  Grid node is a set of computational resources with limited capacities. It may be a simple personal machine, a workstation, a super-computer, or a cluster of workstations in the grid environment. The computational capacity of the node depends on its number of CPUs, amount of memory, basic storage space and other specializations. In other words, each node has its own processing speed, which can be expressed in number of Cycles Per Unit Time (CPUT).

- **Jobs and Operations**
  A job is considered as a single set of multiple atomic operations/tasks. Each operation will be typically allocated to execute on one single node without preemption. It has input and output data, and processing requirements in order to complete its task. The operation has the processing length expressed in number of cycles.
- **Schedule and Scheduling Problem**
  A schedule is the mapping of the tasks to specific time intervals of Grid nodes. A scheduling problem is specified by a set of machines, a set of jobs/operations, optimality criteria, environmental specifications, and by other constraints.

To formulate the problem, we consider J_j ($j \in \{1, 2, \cdots, n\}$) independent user jobs on G_i ($i \in \{1, 2, \cdots, m\}$) heterogeneous grid nodes with an objective of minimizing the completion time and utilizing the nodes effectively. The speed of each node is expressed in number of CPUT, and the length of each job in number of cycles. Each job J_j has its processing requirement (cycles) and the node G_i has its calculating speed (cycles/second). Any job J_j has to be processed in the one of grid nodes G_i, until completion. Since all nodes at each stage are identical and preemptions are not allowed, to define a schedule it suffices to specify the completion time for all tasks comprising each job.

To formulate our objective, define $C_{i,j}$ ($i \in \{1, 2, \cdots, m\}$, $j \in \{1, 2, \cdots, n\}$) as the completion time that the grid node G_i finishes the job J_j, $\sum C_i$ represents the time that the grid node G_i finishes all the jobs scheduling to itself. Define $C_{max} = max\{\sum C_i\}$ as the makespan, and $\sum_{i=1}^{m}(\sum C_i)$ as the flowtime.

An optimal schedule will be the one that optimizes the flowtime and makespan. The conceptually obvious rule to minimize $\sum_{i=1}^{m}(\sum C_i)$ is to schedule Shortest Job on the Fastest Node (SJFN). The simplest rule to minimize C_{max} is to schedule the Longest Job on the Fastest Node (LJFN). Minimizing $\sum_{i=1}^{m}(\sum C_i)$ asks the average job finishes quickly, at the expense of the largest job taking a long time, whereas minimizing C_{max}, asks that no job takes too long, at the expense of most jobs taking a long time. Minimization of C_{max} will result in maximization of $\sum_{i=1}^{m}(\sum C_i)$.

9.3 Nature Inspired Meta-heuristics

Combinatorial optimizationproblems are important in many real life applications and recently, the area has attracted much research with the advances in nature inspired heuristics and multi-agent systems. For scheduling problems, the dramatic increase in the size of the search space and the need for real-time solutions motivated research ideas into solving scheduling problems using nature inspired heuristic techniques. In this Chapter, we included evolutionary algorithms, simulated annealing, ant colony optimization and particle swarm optimization algorithm. The generic pseudo-code for the algorithms is illustrated in Algorithm 9.1.

Algorithm 9.1. General Description for Nature Inspired Algorithm
01. Initialize the solution vectors randomly and other parameters.
02. Evaluate the candidate solution(s);
03. Repeat
04.   Generate new candidate solutions following the nature or social behaviors;
05.   Evaluate the candidate solution;
06. Until terminating criteria.

The termination criteria are usually one of the following:

- Maximum number of iterations: the optimization process is terminated after a fixed number of iterations, for example, 1000 iterations.
- Number of iterations without improvement: the optimization process is terminated after some fixed number of iterations without any improvement.
- Minimum objective function error: the error between the obtained objective function value and the best fitness value is less than a pre-fixed anticipated threshold.
- Cost threshold: allocated budget (computation time/cost) reached.
- Manual inspection: the process is executed by human-computer interactively.
- Combinations of the above.

9.3.1 Evolutionary Algorithms

In nature, evolution is mostly determined by natural selection, where individuals that are better are more likely to survive and propagate their genetic material. The encoding of genetic information (genome) is done in a way that admits asexual reproduction which results in offspring's that are genetically identical to the parent. Sexual reproduction allows some exchange and re-ordering of chromosomes, producing offspring that contain a combination of information from each parent. This is the recombination operation, which is often referred to as crossover because of the way strands of chromosomes crossover during the exchange. Diversity in the population is achieved by mutation. A typical evolutionary (genetic) algorithm procedure takes the following steps: A population of candidate solutions (for the optimization task to be solved) is initialized. New

Algorithm 9.2. Evolutionary Algorithm
01. Initialize the population randomly, and other parameters.
02. Evaluate the fitness of each individual in the population.
03. Repeat
04.   Select best-ranking individuals to reproduce;
05.   Breed new generation through crossover operator and give birth to offspring;
06.   Breed new generation through mutation operator and give birth to offspring;
07.   Evaluate the individual fitness of the offspring;
08.   Replace worst ranked part of population with offspring;
09. Until terminating criteria.

solutions are created by applying genetic operators (mutation and/or crossover). The fitness (how good the solutions are) of the resulting solutions are evaluated and suitable selection strategy is then applied to determine which solutions will be maintained into the next generation. The procedure is then iterated [9]. A canonical version of the pseudo-code for the evolutionary algorithm is illustrated in Algorithm 9.2.

9.3.2 Evolutionary Multi-objective Optimization

Even though some real world problems can be reduced to a matter of single objective very often it is hard to define all the aspects in terms of a single objective. Defining multiple objectives often gives a better idea of the task. In single objective optimization, the search space is often well defined. As soon as there are several possibly contradicting objectives to be optimized simultaneously, there is no longer a single optimal solution but rather a whole set of possible solutions of equivalent quality. When we try to optimize several objectives at the same time the search space also becomes partially ordered. To obtain the optimal solution, there will be a set of optimal trade-offs between the conflicting objectives. A multiobjective optimization problem is defined by a function f which maps a set of constraint variables to a set of objective values.

A solution could be best, worst and also indifferent to other solutions (neither dominating or dominated) with respect to the objective values. Best solution means a solution not worst in any of the objectives and at least better in one objective than the other. An optimal solution is the solution that is not dominated by any other solution in the search space. Such an optimal solution is called Pareto optimal and the entire set of such optimal trade-offs solutions is called Pareto optimal set. As evident, in a real world situation a decision making (trade-off) process is required to obtain the optimal solution. Even though there are several ways to approach a multiobjective optimization problem, most work is concentrated on the approximation of the Pareto set.

Evolutionary algorithm is characterized by a population of solution candidates and the reproduction process enables to combine existing solutions to generate new solutions. Finally, natural selection determines which individuals of the current population participate in the new population. Multi-objective Evolutionary Algorithms (MOEA) can yield a whole set of potential solutions, which are all optimal in some sense. After the first pioneering work on multiobjective evolutionary optimization in the eighties [10], several different algorithms have been proposed and successfully applied to various problems. For comprehensive overviews and discussions, the reader is referred to [11].

9.3.3 Simulated Annealing

Simulated Annealing (SA) exploits an analogy between the way in which a metal cools and freezes into a minimum energy crystalline structure (the annealing process) and the search for a minimum in a more general system. SA's major advantage over other methods is an ability to avoid becoming trapped at local minima [12]. The annealing schedule, i.e., the temperature-decreasing rate used in SA is an important factor, which affects SA's rate of convergence. The algorithm employs a random search, which not only accepts changes that decrease objective function "f", but also some changes that increase it. The latter are accepted with a probability $p = \exp\left(-\frac{\delta f}{T}\right)$, where δf is the increase in objective function, and "f" and T are control parameters. Several SAs have been developed with annealing schedule inversely linear in time (Fast SA), exponential function of time (Very Fast SA) etc. We explain a SA algorithm [13], which is exponentially faster than Very Fast SA whose annealing schedule is given by $T(k) = \frac{T_0}{\exp(e^k)}$, where T_0 is the initial temperature, $T(k)$ is the temperature we wish to approach to zero for $k = 1, 2, \ldots$. If the generation function of the simulated annealing algorithm is represented as:

$$g_k(Z) = \prod_{i=1}^{D} g_k(z_i) = \prod_{i=1}^{D} \frac{1}{2(|z_i| + \frac{1}{\ln(1/T_i(k))}) \ln(1 + \ln(1/T_i(k)))} \quad (9.1)$$

where $T_i(k)$ is the temperature in dimension i at time k. The generation probability will be represented by

$$G_k(Z) = \int_{-1}^{z_1} \int_{-1}^{z_2} \ldots \int_{-1}^{z_D} g_k(Z) dz_1 dz_2 \ldots dz_D = \prod_{i=1}^{D} G_{ki}(z_i) \quad (9.2)$$

where $G_{ki}(z_i) = \frac{1}{2} + \frac{sgn(z_i) \ln(1+|z_i| \ln(1/T_i(k)))}{2\ln(1+\ln(1/T_i(k)))}$

It is straightforward to prove that an annealing schedule for

$$T_i(k) = T_{0i} \exp(-\exp(b_i k^{1/D})) \quad (9.3)$$

A global minimum, statistically, can be obtained. That is,

$$\sum_{k=k_o}^{\infty} g_k = \infty \quad (9.4)$$

Algorithm 9.3. Simulated Annealing
---
01. Set initial temperature T_0, and other parameters.
02. Initialize the solution vectors randomly.
03. Repeat
04.   Counter = 0;
05.   Repeat
06.   Evaluate the candidate solution;
07.   Generate a neighbor and evaluate the cost of the neighbor solution;
08.   Accept or reject the neighbor with a probability p;
09.   Counter++;
10.   Until (Counter = Number of Iterations at T_i);
11. $T_{i+1} = c * T_i$ (temperature reduction);
12. Until terminating criteria.

where $b_i > 0$ is a constant parameter and k_0 is a sufficiently large constant to satisfy Eq.(9.4), if the generation function in Eq.(9.1) is adopted. The pseudo-code for simulated annealing is illustrated in Algorithm 9.3.

9.3.4 Ant Colony Optimization

In nature, ants usually wander randomly, and upon finding food return to their nest while laying down pheromone trails. If other ants find such a path (pheromone trail), they are likely not to keep traveling at random, but to instead follow the trail, returning and reinforcing it if they eventually find food. However, as time passes, the pheromone starts to evaporate. The more time it takes for an ant to travel down the path and back again, the more time the pheromone has to evaporate (and the path to become less prominent). A shorter path, in comparison will be visited by more ants (can be described as a loop of positive feedback) and thus the pheromone density remains high for a longer time.

ACO is implemented as a team of intelligent agents which simulate the ants behavior, walking around the graph representing the problem to solve using mechanisms of cooperation and adaptation. ACO algorithm requires to define the following [14], [15]:

- The problem needs to be represented appropriately, which would allow the ants to incrementally update the solutions through the use of a probabilistic transition rules, based on the amount of pheromone in the trail and other problem specific knowledge. It is also important to enforce a strategy to construct only valid solutions corresponding to the problem definition.
- A problem-dependent heuristic function η that measures the quality of components that can be added to the current partial solution.
- A rule set for pheromone updating, which specifies how to modify the pheromone value τ.
- A probabilistic transition rule based on the value of the heuristic function η and the pheromone value τ that is used to iteratively construct a solution.

ACO was first introduced using the Traveling Salesman Problem (TSP). Starting from its start node, an ant iteratively moves from one node to another. When being at a node, an ant chooses to go to a unvisited node at time t with a probability given by

$$p_{i,j}^k(t) = \frac{[\tau_{i,j}(t)]^\alpha [\eta_{i,j}(t)]^\beta}{\sum_{l \in N_i^k} [\tau_{i,j}(t)]^\alpha [\eta_{i,j}(t)]^\beta} \quad j \in N_i^k \quad (9.5)$$

where N_i^k is the feasible neighborhood of the ant_k, that is, the set of cities which ant_k has not yet visited; $\tau_{i,j}(t)$ is the pheromone value on the edge (i,j) at the time t, α is the weight of pheromone; $\eta_{i,j}(t)$ is a priori available heuristic information on the edge (i,j) at the time t, β is the weight of heuristic information. Two parameters α and β determine the relative influence of pheromone trail and heuristic information. $\tau_{i,j}(t)$ is determined by

$$\tau_{i,j}(t) = \rho\tau_{i,j}(t-1) + \sum_{k=1}^{n} \Delta\tau_{i,j}^{k}(t) \quad \forall(i,j) \tag{9.6}$$

$$\Delta\tau_{i,j}^{k}(t) = \begin{cases} \frac{Q}{L_k(t)} & \text{if the edge } (i,j) \text{ chosen by the } ant_k \\ 0 & \text{otherwise} \end{cases} \tag{9.7}$$

where ρ is the pheromone trail evaporation rate $(0 < \rho < 1)$, n is the number of ants, Q is a constant for pheromone updating.

Reader is advised to consult [16], [17], [15] for more technical details and other applications of ACO. A generalized version of the pseudo-code for the ACO algorithm with reference to the TSP is illustrated in Algorithm 9.4.

Algorithm 9.4. Ant Colony Optimization Algorithm
---
01. Initialize the number of ants n, and other parameters.
02. Repeat
03. $t++$;
04.   For $k= 1$ to n
05. ant_k is positioned on a starting node;
06.   For $m= 2$ to *problem_size*
07.   Choose the state to move into
07.   according to the probabilistic transition rules;
08.   Append the chosen move into $tabu_k(t)$ for the ant_k;
09.   Next m
10.   Compute the length $L_k(t)$ of the tour $T_k(t)$ chosen by the ant_k;
11.   Compute $\Delta\tau_{i,j}(t)$ for every edge (i,j) in $T_k(t)$ according to Eq.(9.7);
12.   Next k
13.   Update the trail pheromone intensity for every edge (i,j) according to Eq.(9.6);
14.   Compare and update the best solution;
15. Until terminating criteria.

9.3.5 Particle Swarm Optimization

Particle swarm algorithm is inspired by social behavior patterns of organisms that live and interact within large groups. In particular, it incorporates swarming behaviors observed in flocks of birds, schools of fish, or swarms of bees, and even human social behavior, from which the Swarm Intelligence (SI) paradigm has emerged [11], [12]. It could be implemented and applied easily to solve various function optimization problems, or the problems that can be transformed to function optimization problems.

As an algorithm, its main strength is its fast convergence, which compares favorably with many global optimization algorithms [9], [12]. The canonical PSO model consists of a swarm of particles, which are initialized with a population of random candidate solutions. They move iteratively through the d-dimension problem space to search the new solutions, where the fitness, f, can be calculated as the certain qualities measure.

Each particle has a position represented by a position-vector $\mathbf{x}_i$ (i is the index of the particle), and a velocity represented by a velocity-vector $\mathbf{v}_i$. Each particle remembers its own best position so far in a vector $\mathbf{x}_i^{\#}$, and its j-th dimensional value is $x_{ij}^{\#}$. The best position-vector among the swarm so far is then stored in a vector $\mathbf{x}^*$, and its j-th dimensional value is x_j^*. During the iteration time t, the update of the velocity from the previous velocity to the new velocity is determined by Eq.(9.8). The new position is then determined by the sum of the previous position and the new velocity by Eq.(9.9).

$$v_{ij}(t+1) = wv_{ij}(t) + c_1 r_1(x_{ij}^{\#}(t) - x_{ij}(t)) + c_2 r_2(x_j^*(t) - x_{ij}(t)). \qquad (9.8)$$

$$x_{ij}(t+1) = x_{ij}(t) + v_{ij}(t+1). \qquad (9.9)$$

where w is called as the inertia factor, r_1 and r_2 are the random numbers, which are used to maintain the diversity of the population, and are uniformly distributed in the interval [0,1] for the j-th dimension of the i-th particle. c_1 is a positive constant, called as coefficient of the self-recognition component, c_2 is a positive constant, called as coefficient of the social component.

From Eq.(9.8), a particle decides where to move next, considering its own experience, which is the memory of its best past position, and the experience of its most successful particle in the swarm. In the particle swarm model, the particle searches the solutions in the problem space with a range $[-s, s]$ (If the range is not symmetrical, it can be translated to the corresponding symmetrical range.) In order to guide the particles effectively in the search space, the maximum moving distance during one iteration must be clamped in between the maximum velocity $[-v_{max}, v_{max}]$ given in Eq.(9.10):

$$v_{ij} = sign(v_{ij}) min(|v_{ij}|, v_{max}). \qquad (9.10)$$

$$x_{i,j} = sign(x_{i,j}) min(|x_{i,j}|, x_{max}). \qquad (9.11)$$

The value of v_{max} is $p \times s$, with $0.1 \leq p \leq 1.0$ and is usually chosen to be s, i.e. $p = 1$. The pseudo-code for particle swarm optimization algorithm is illustrated in Algorithm 9.5.

The role of inertia weight w, in Eq.(9.8), is considered critical for the convergence behavior of PSO. The inertia weight is employed to control the impact of the previous history of velocities on the current one. Accordingly, the parameter w regulates the trade-off between the global (wide-ranging) and local (nearby) exploration abilities of the swarm. A large inertia weight facilitates global exploration (searching new areas), while a small one tends to facilitate local exploration, i.e. fine-tuning the current search area. A suitable value for the inertia weight w usually provides balance between global and local exploration abilities and consequently results in a reduction of the number of iterations required to locate the optimum solution. Initially, the inertia weight is set as a constant. However, some experiment results indicates that it is better to initially set the inertia to a large value, in order to promote global exploration of the search space, and gradually decrease it to get more refined solutions [20], [21]. Thus, an

Algorithm 9.5. Particle Swarm Optimization Algorithm

01. Initialize the size of the particle swarm n, and other parameters.
02. Initialize the positions and the velocities for all the particles randomly.
03. Repeat
04. $t++$;
05.   Calculate the fitness value of each particle;
06. $\mathbf{x}^* = argmin_{i=1}^{n}(f(\mathbf{x}^*(t-1)), f(\mathbf{x}_1(t)), f(\mathbf{x}_2(t)), \cdots, f(\mathbf{x}_i(t)), \cdots, f(\mathbf{x}_n(t)))$;
07.   For $i= 1$ to n
08. $\mathbf{x}_i^\#(t) = argmin_{i=1}^{n}(f(\mathbf{x}_i^\#(t-1)), f(\mathbf{x}_i(t))$;
09.   For $j = 1$ to $Dimension$
10.   Update the j-th dimension value of $\mathbf{x}_i$ and $\mathbf{v}_i$
10.   according to Eqs.(9.8), (9.9), (9.10), (9.11);
12.   Next j
13.   Next i
14. Until terminating criteria.

initial value around 1.2 and gradually reducing towards 0 can be considered as a good choice for w. A better method is to use some adaptive approaches (example: fuzzy controller), in which the parameters can be adaptively fine tuned according to the problem under consideration [22], [16].

The parameters c_1 and c_2, in Eq.(9.8), are not critical for the convergence of PSO. However, proper fine-tuning may result in faster convergence and alleviation of local minima. As default values, usually, $c_1 = c_2 = 2$ are used, but some experiment results indicate that $c_1 = c_2 = 1.49$ might provide even better results. Recent work reports that it might be even better to choose a larger cognitive parameter, c_1, than a social parameter, c_2, but with $c_1 + c_2 \leq 4$ [23].

The particle swarm algorithm can be described generally as a population of vectors whose trajectories oscillate around a region which is defined by each individual's previous best success and the success of some other particle. Various methods have been used to identify some other particle to influence the individual. Eberhart and Kennedy called the two basic methods as "gbest model" and "lbest model" [11]. In the lbest model, particles have information only of their own and their nearest array neighbors' best (lbest), rather than that of the entire group.

In the gbest model, the trajectory for each particle's search is influenced by the best point found by any member of the entire population. The best particle acts as an attractor, pulling all the particles towards it. Eventually all particles will converge to this position. The lbest model allows each individual to be influenced by some smaller number of adjacent members of the population array. The particles selected to be in one subset of the swarm have no direct relationship to the other particles in the other neighborhood.

Typically lbest neighborhoods comprise exactly two neighbors. When the number of neighbors increases to all but itself in the lbest model, the case is equivalent to the gbest model. Some experiment results testified that gbest model converges quickly on problem solutions but has a weakness for becoming trapped in local optima, while lbest model converges slowly on problem solutions but is able to "flow around" local optima, as the individuals explore different regions. The gbest model has faster convergence. But very often for multi-modal problems involving high dimensions it tends to suffer from premature convergence.

9.3.6 A Fuzzy Scheme Based on Particle Swarm Optimization

In this section, we design a fuzzy scheme based on discrete particle swarm optimization to solve the job scheduling problem on computational grids. The vectors to fuzzy matrices are extended to represent the position and velocity of the particles for computational grid job scheduling.

Suppose $G = \{G_1, G_2, \cdots, G_m\}, J = \{J_1, J_2, \cdots, J_n\}$, then the fuzzy scheduling relation from G to J can be expressed as follows:

$$S = \begin{bmatrix} s_{11} & s_{12} & \cdots & s_{1n} \\ s_{21} & s_{22} & \cdots & s_{2n} \\ \vdots & \vdots & \ddots & \vdots \\ s_{m1} & s_{m2} & \cdots & s_{mn} \end{bmatrix}$$

Here s_{ij} represents the degree of membership of the i-th element G_i in domain G and the j-th element J_j in domain J to relation S. The fuzzy relation S between G and J has the following meaning: for each element in the matrix S, the element

$$s_{ij} = \mu_R(G_i, J_j), i \in \{1, 2, \cdots, m\}, j \in \{1, 2, \cdots, n\}. \tag{9.12}$$

μ_R is the membership function, the value of s_{ij} means the degree of membership that the grid node G_j would process the job J_i in the feasible schedule solution. In the grid job scheduling problem, the elements of the solution must satisfy the following conditions:

$$s_{ij} \in [0, 1], i \in \{1, 2, \cdots, m\}, j \in \{1, 2, \cdots, n\}. \tag{9.13}$$

$$\sum_{i=1}^{m} s_{ij} = 1, i \in \{1, 2, \cdots, m\}, j \in \{1, 2, \cdots, n\}. \tag{9.14}$$

For applying PSO successfully, one of the key issues is finding how to map the the problem solution into the PSO particle, which directly affects its feasibility and performance. We assume that the jobs and grid nodes are arranged in an ascending order according to the job lengths and the node processing speeds. The information related job lengths may be derived from historical data, some kind of strategy defined by the user or through load profiling.

$$X = \begin{bmatrix} x_{11} & x_{12} & \cdots & x_{1n} \\ x_{21} & x_{22} & \cdots & x_{2n} \\ \vdots & \vdots & \ddots & \vdots \\ x_{m1} & x_{m2} & \cdots & x_{mn} \end{bmatrix}$$

Accordingly, the elements of the matrix X must satisfy the following conditions:

$$x_{ij} \in [0,1], i \in \{1,2,\cdots,m\}, j \in \{1,2,\cdots,n\}. \tag{9.15}$$

$$\sum_{i=1}^{m} x_{ij} = 1, i \in \{1,2,\cdots,m\}, j \in \{1,2,\cdots,n\}. \tag{9.16}$$

We define similarly the velocity of the particle as:

$$V = \begin{bmatrix} v_{11} & v_{12} & \cdots & v_{1n} \\ v_{21} & v_{22} & \cdots & v_{2n} \\ \vdots & \vdots & \ddots & \vdots \\ v_{m1} & v_{m2} & \cdots & v_{mn} \end{bmatrix}$$

The symbol "$\otimes$" is used to denote the modified multiplication. Let α be a real number, $\alpha \otimes V$ or $\alpha \otimes X$ means all the elements in the matrix V or X are multiplied by α. The symbol "$\oplus$" and symbol "$\ominus$" denote the addition and subtraction between matrices respectively. Suppose A and B are two matrices which denote position or velocity, then $A \oplus B$ and $A \ominus B$ are regular addition and subtraction operation between matrices.

Then we get the equations (9.8) and (9.9) for updating the positions and velocities of the particles in the fuzzy discrete PSO:

$$V(t+1) = w \otimes V(t) \oplus (c_1 * r_1) \otimes X^{\#}(t) \ominus X(t)) \oplus (c_2 * r_2) \otimes (X^{*}(t) \ominus X(t)). \tag{9.17}$$

$$X(t+1) = X(t) \oplus V(t+1)). \tag{9.18}$$

The position matrix may violate the constraints (9.15) and (9.16) after some iterations, so it is necessary to normalize the position matrix. First we make all the negative elements in the matrix become zero. If all elements in a column of the matrix are zero, they need be re-evaluated using a series of random numbers with the interval [0,1]. Then the matrix undergoes the following transformation without violating the constraints:

$$Xnormal = \begin{bmatrix} x_{11}/\sum_{i=1}^{m} x_{i1} & x_{12}/\sum_{i=1}^{m} x_{i2} & \cdots & x_{1n}/\sum_{i=1}^{m} x_{in} \\ x_{21}/\sum_{i=1}^{m} x_{i1} & x_{22}/\sum_{i=1}^{m} x_{i2} & \cdots & x_{2n}/\sum_{i=1}^{m} x_{in} \\ \vdots & \vdots & \ddots & \vdots \\ x_{m1}/\sum_{i=1}^{m} x_{i1} & x_{m2}/\sum_{i=1}^{m} x_{i2} & \cdots & x_{mn}/\sum_{i=1}^{m} x_{in} \end{bmatrix}$$

Since the position matrix indicates the potential scheduling solution, we should "decode" the fuzzy matrix and get the feasible solution. We use a flag array to record whether we have selected the columns of the matrix and a scheduling array to record the scheduling solution. First all the columns are not selected, then for each columns of the matrix, we choose the element which has the max value, then mark the column of the max element "selected", and the column number are recorded to the scheduling array. After all the columns have been processed, we get the scheduling solution from the scheduling array and the makespan of the scheduling solution.

To optimize the makespan and flowtime we propose to swap the usage of LJFN and SJFN heuristic alternatively every time the new jobs are allocated to the grid nodes. If the number of jobs is less than the number of grid nodes, we propose to allocate the jobs based on a First-Come-First-Serve basis and LJFN heuristic (if possible). In a grid environment, a scheduler might have to make a multi-criteria decision analysis (access policy, access cost, resource requirements, processing speed, etc.) for selecting an optimal solution. To formulate the algorithm, we propose the following job lists and grid node lists. $JList_1$ and $GList_1$ are to be dynamically updated through load profiling, grid node health status, and forecasted load status, etc. along with grid information services. The entire job and the grid node lists are to be arranged in the ascending order of the job lengths and processing speeds/access-cost (based on multi-criteria decision analysis). Frequency of updating the lists will very much depend on the grid condition, availability of grid nodes and jobs.

- $JList_1 =$ Job list maintaining the list of all the jobs to be processed.
- $JList_2 =$ Job list maintaining only the list of jobs being scheduled.
- $JList_3 =$ Job list maintaining only the list of jobs already allocated ($JList_3 = JList_1 - JList_2$).
- $GList_1 =$ List of available grid nodes (including time frame).
- $GList_2 =$ List of grid nodes already allocated to jobs.
- $GList_3 =$ List of free grid nodes ($GList_3 = GList_1 - GList_2$).

A scheme based on fuzzy discrete PSO for job scheduling is depicted in Algorithm 9.6.

9.4 Experimental Illustrations

The scheduling problem is to determine both an assignment and a sequence of the operations on all machines that minimize some criteria. The following optimality criteria are to be minimized:

1. the maximum completion time (makespan): C_{max};
2. the sum of the completion times (flowtime): C_{sum}.

The weighted aggregation is the most common approach to the problems. According to this approach, the objectives, $F_1 = min\{C_{max}\}$ and $F_2 = min\{C_{sum}\}$, are aggregated as a weighted combination:

$$F = w_1 min\{F_1\} + w_2 min\{F_2\} \qquad (9.19)$$

where w_1 and w_2 are non-negative weights, and $w_1 + w_2 = 1$. These weights can be either fixed or adapt dynamically during the optimization. The fixed weights, $w_1 = w_2 = 0.5$, are used in this article. In fact, the dynamic weighted aggregation mainly takes C_{max} into account [25] because C_{sum} is commonly much larger

Algorithm 9.6. A scheduling scheme based on fuzzy discrete PSO

0 If the grid is active and ($JList_1 = 0$) and no new jobs have been submitted, wait for new jobs to be submitted. Otherwise, update $GList_1$ and $JList_1$.
1 If ($GList_1 = 0$), wait until grid nodes are available. If $JList_1 > 0$, update $JList_2$. If $JList_2 < GList_1$ allocate the jobs on a first-come-first-serve basis and if possible allocate the longest job on the fastest grid node according to the LJFN heuristic. If $JList_1 > GList_1$, job allocation is to be made by following the fuzzy discrete PSO algorithm detailed below. Take jobs and available grid nodes from $JList_2$ and $GList_3$. If $m*n$ (m is the number of the grid nodes, n is the number of the jobs) is larger than the dimension threshold D_T, the jobs and the grid nodes are grouped into the fuzzy discrete PSO algorithm loop, and the single node flowtime is accumulated. The LJFN-SJFN heuristic is applied alternatively after a batch of jobs and nodes are allocated.
2 At $t = 0$, represent the jobs and the nodes using fuzzy matrix.
3 Begin fuzzy discrete PSO Loop
3.0 Initialize the size of the particle swarm n and other parameters.
3.1 Initialize a random position matrix and a random velocity matrix for each particle, and then normalize the matrices.
3.2 Repeat
3.2.0 $t++$;
3.2.1   Defuzzify the position, and calculate the makespan and total flowtime for each particle (the feasible solution);
3.2.2 $X^* = argmin_{i=1}^{n}(f(X^*(t-1)), f(X_1(t)), f(X_2(t)), \cdots, f(X_i(t)), \cdots, f(X_n(t)))$;
3.2.3   For each particle, $X_i^{\#}(t) = argmin_{i=1}^{n}(f(X_i^{\#}(t-1)), f(X_i(t))$
3.2.4   For each particle, update each element in its position matrix and its velocity matrix according to equations (9.17, 9.11, 9.18 and 9.10);
3.2.5   Normalize the position matrix for each particle;
3.3 Until terminating criteria.
4 End of the fuzzy discrete PSO Loop.
5 Check the feasibility of the generated schedule with respect to grid node availability and user specified requirements. Then allocate the jobs to the grid nodes and update $JList_2$, $JList_3$, $GList_2$ and $GList_3$. Un-allocated jobs (infeasible schedules or grid node non-availability) shall be transferred to $JList_1$ for re-scheduling or dealt with separately.
6 Repeat steps 0-5 as long as the grid is active.

than C_{max} and the solution has a large weight on C_{sum} during minimization of the objective. Alternatively, the weights can be changed gradually according to the Eqs. (9.20) and (9.21). The changes in the dynamic weights ($R = 200$) are illustrated in Fig. 9.1.

$$w_1(t) = |sin(2\pi t/R)| \quad (9.20)$$

$$w_2(t) = 1 - w_1(t) \quad (9.21)$$

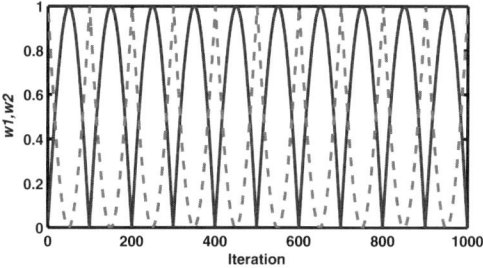

Fig. 9.1. Dynamic weight adaptation

9.4.1 Scheduling Using Fuzzy Particle Swarm Optimization Algorithm

Since the position matrix indicates the potential scheduling solution, we choose the element which has the max value, then tag it as "1", and other numbers in the column are set as "0" in the scheduling array. After all the columns have been processed, we get the scheduling solution from the scheduling array and the makespan (solution). In the experiments, genetic algorithm and simulated annealing were used to compare the performance with PSO. Specific parameter settings of all the considered algorithms are described in Table 9.1.

Each experiment (for each algorithm) was repeated 10 times with different random seeds. Each trial had a fixed number of $50 * m * n$ iterations (m is the number of the grid nodes, n is the number of the jobs). The makespan values of the best solutions throughout the optimization run were recorded and the averages and the standard deviations were calculated from the 10 different trials. In a grid environment, the main emphasis is to generate the schedules as fast as possible. So the completion time for 10 trials were used as one of the criteria to improve their performance.

To illustrate, we start with a small scale job scheduling problem involving 3 nodes and 13 jobs represented as (3, 13). The node speeds are 4, 3, 2 CPUT, and the job lengths of 13 jobs are 6, 12, 16, 20, 24, 28, 30, 36, 40, 42, 48, 52, 60 cycles, respectively.

Fig. 9.2 illustrates the performance of GA, SA and PSO algorithms. The empirical results (makespan) for 10 GA runs were {47, 46, 47, 47.3333, 46, 47, 47, 47, 47.3333, 49}, with an average value of 47.1167. The results of 10 SA runs were {46.5, 46.5, 46, 46,46, 46.6667, 47, 47.3333, 47, 47}with an average value of 46.6. The results of 10 PSO runs were {46, 46, 46, 46, 46.5, 46.5, 46.5, 46, 46.5, 46.6667}, with an average value of 46.2667. The optimal result is supposed to be 46. While GA provided the best results twice, SA and PSO provided the best results three and five times respectively. Table 9.2 depicts one of the best job scheduling results for (3,13), in which "1" means the job is scheduled to the respective grid node.

Table 9.1. Parameter settings for the algorithms

Algorithm	Parameter name	Parameter value
GA	Size of the population	20
	Probability of crossover	0.8
	Probability of mutation	0.02
	Scale for mutations	0.1
SA	Number operations before temperature adjustment	20
	Number of cycles	10
	Temperature reduction factor	0.85
	Vector for control step of length adjustment	2
	Initial temperature	50
PSO	Swarm size	20
	Self-recognition coefficient c_1	1.49
	Social coefficient c_2	1.49
	Inertia weight w	$0.9 \to 0.1$

Table 9.2. An optimal schedule for (3,13)

Grid Node	Job												
	J_1	J_2	J_3	J_4	J_5	J_6	J_7	J_8	J_9	J_{10}	J_{11}	J_{12}	J_{13}
G_1	0	0	1	0	0	0	1	1	0	1	0	0	1
G_2	1	0	0	1	1	0	0	0	1	0	1	0	0
G_3	0	1	0	0	0	1	0	0	0	0	0	1	0

Table 9.3. Performance comparison between GA, PSO and SA

Algorithm	Item	Instance			
		(3,13)	(5,100)	(8,60)	(10,50)
GA	Average makespan	47.1167	85.7431	42.9270	38.0428
	Standard Deviation	±0.7700	±0.6217	±0.4150	±0.6613
	Time	302.9210	2415.9	2263.0	2628.1
SA	Average makespan	46.6000	90.7338	55.4594	41.7889
	Standard Deviation	±0.4856	±6.3833	±2.0605	±8.0773
	Time	332.5000	6567.8	6094.9	6926.4
PSO	Average makespan	46.2667	84.0544	41.9489	37.6668
	Standard Deviation	±0.2854	±0.5030	±0.6944	±0.6068
	Time	106.2030	1485.6	1521.0	1585.7

Table 9.4. Run time performance comparison for large dimension problems

(G,J)	PSO	GA
(60,100)	1721.1	1880.6
100,1000)	3970.80	5249.80

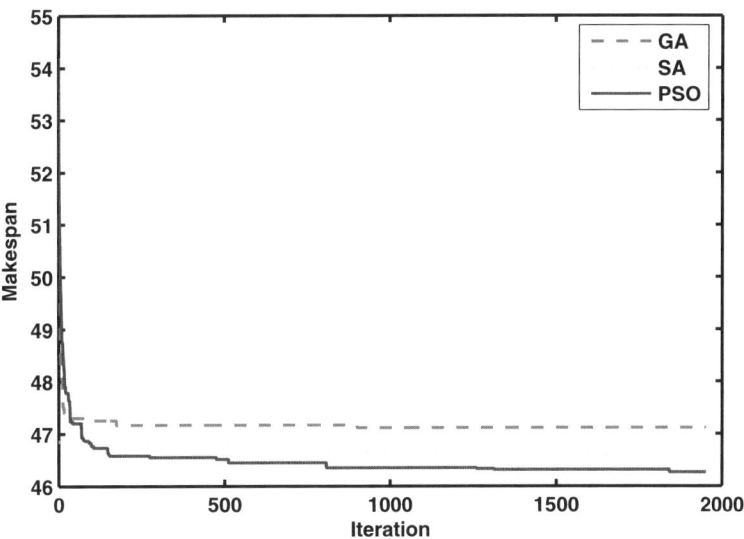

Fig. 9.2. Performance for job scheduling (3,13)

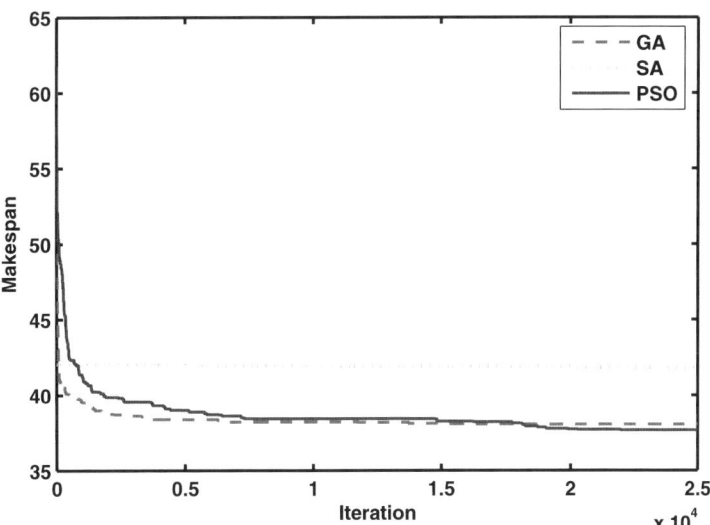

Fig. 9.3. Performance for job scheduling (5,100)

Further, we considered the three algorithms for other three (G, J) pairs, i.e. (5,100), (8,60) and (10,50). All the jobs and the nodes were submitted at one time. Figs. 9.2, 9.3 illustrate the performance for GA, SA and PSO algorithms during the search process for (3, 13), (5,100) respectively. The average makespan

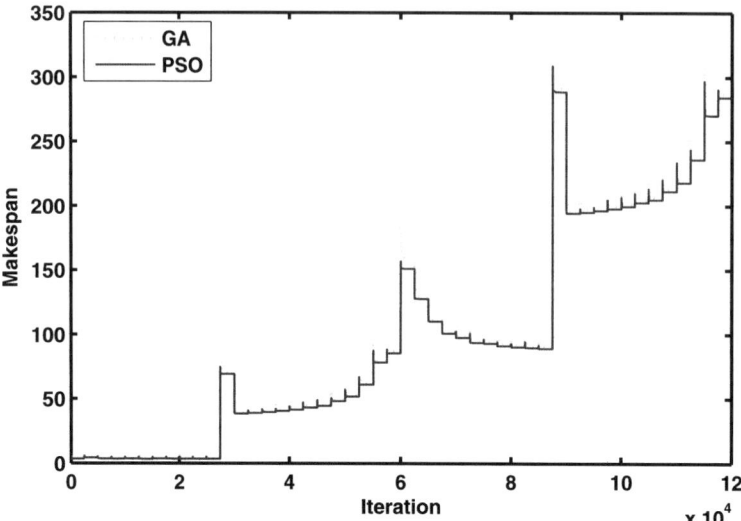

Fig. 9.4. Performance for job scheduling (60,500)

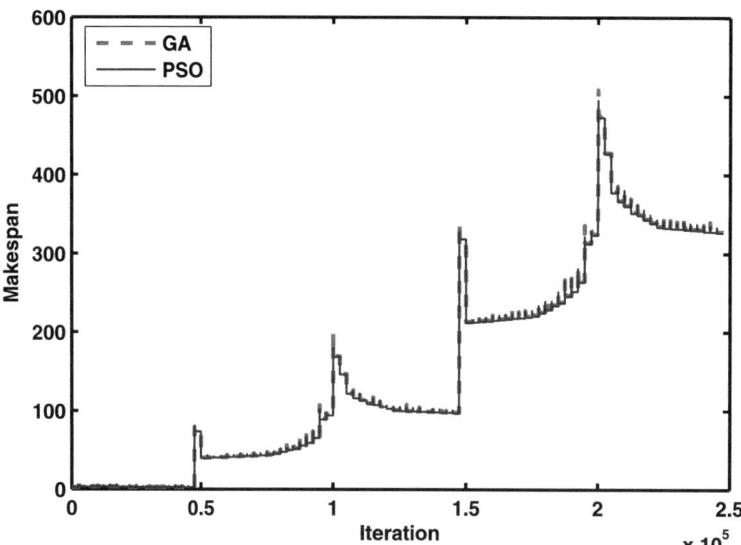

Fig. 9.5. Performance for job scheduling (100,1000)

values, the standard deviations and the time for 10 trials are illustrated in Table 9.3. Although the average makespan value of SA was better than that of GA for (3,13), the case was reversed for bigger problem sizes. PSO usually had better average makespan values than the other two algorithms. The makespan results of SA seemed to depend on the initial solutions extremely. Although the best values in the ten trials for SA were not worse than other algorithms, it had larger standard deviations. For SA, there were some "bad" results in the ten trials, so the averages were the largest. In general, for larger (G, J) pairs, the time was much longer. PSO usually spent the least time to allocate all the jobs on the grid node, GA was the second, and SA had to spent more time to complete the scheduling. It is to be noted that PSO usually spent the shortest time to accomplish the various job scheduling tasks and had the best results among all the considered three algorithms.

It is possible that (G, J) is larger than the dimension threshold D_T. We considered two large-dimensions of (G, J), (60, 500) and (100, 1000) by submitting the jobs and the nodes in multi-stages consecutively. In each stage, 10 jobs were allocated to 5 nodes, and the single node flowtime was accumulated. The LJFN-SJFN heuristic was applied alternatively after a batch of jobs and nodes were allocated. Figs. 9.4, 9.5 and Table 9.4 illustrate the performance of GA and PSO during the search process for the considered (G, J) pairs. As evident, even though the performance were close enough, PSO generated the schedules much faster than GA as illustrated in Table 9.4.

9.4.2 Job Scheduling Using ACO

For illustration, we considered two problem instances: (3,13) and (5,100) [26]. The parameters used for GA, SA and PSO were the same as depicted in Table 9.1 and the ACO algorithm parameters are as follows:

Parameter	Value
Number of ants	5
Weight of pheromone trail α	1
Weight of heuristic information β	5
Pheromone evaporation parameter ρ	0.8
Constant for pheromone updating Q	10

Each experiment (for each algorithm) was repeated 10 times with different random seeds. Each trial had a fixed number of $50 * m * n$ iterations (m is the number of the grid nodes, n is the number of the jobs). The makespan values of the best solutions throughout the optimization run were recorded and the averages and the standard deviations were calculated from the 10 different trials.

Fig. 9.6 illustrates the performance of GA, SA, PSO and ACO algorithms for (3,13). The empirical results for 10 ACO runs were {46, 46, 46, 46, 46.5, 46.5, 46.5, 46, 46, 46.5}, with an average value of 46.2667. The optimal result is supposed to be 46. While GA provided the best results twice, SA, PSO, ACO provided the best results three, five and six times respectively. Empirical results are summarized in Table 9.5 for (3,13) and (5,100). As evident, ACO algorithm seems to work well but as the problem dimensions got bigger, the computational time also increased drastically.

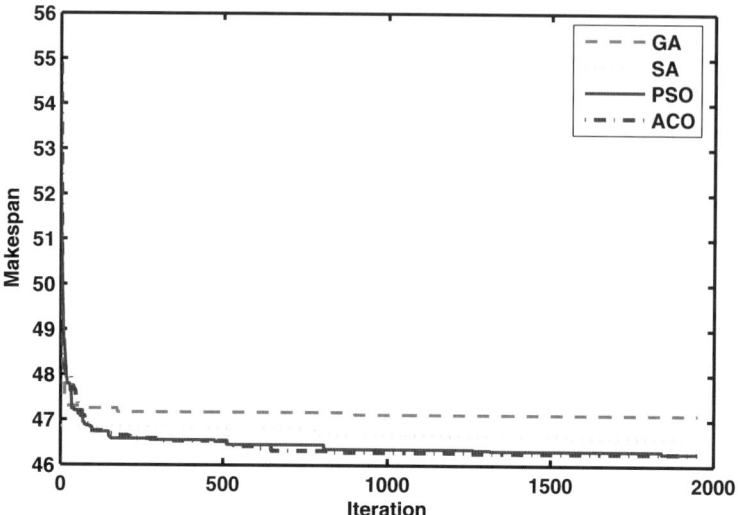

Fig. 9.6. ACO algorithm performance for (3,13)

Table 9.5. Comparing the performance of the considered algorithms

Algorithm	Item	Instance	
		(3,13)	(5,100)
GA	Average makespan	47.1167	85.7431
	Standard Deviation	±0.7700	±0.6217
	Time	302.9210	2415.9
SA	Average makespan	46.6000	90.7338
	Standard Deviation	±0.4856	±6.3833
	Time	332.5000	6567.8
PSO	Average makespan	46.2667	84.0544
	Standard Deviation	±0.2854	±0.5030
	Time	106.2030	1485.6
ACO	Average makespan	46.2667	88.1575
	Standard Deviation	±0.2854	±0.6423
	Time	340.3750	6758.3

9.4.3 Scheduling Using Evolutionary Multi-objective Optimization Approach

Instead of considering the objectives involved by using techniques which combines objectives and reduce the problem to a single objective one (as illustrated in the previous Experiment sections), in this Section, we illustrate the use of Pareto dominance concept and all the objectives are considered as independent.

Even though several optimization criteria can be considered, we considered a bi-objective minimization problem with the task of minimization of *makespan* and *flowtime*. The most common approaches of a multiobjective optimization problem use the concept of Pareto dominance as defined below:

Pareto Dominance Concept

Consider a maximization problem. Let x, y be two decision vectors (solutions) from the definition domain. Solution x *dominate* y (also written as $x \succ y$), if and only if the following conditions are fulfilled:
  (i) $f_i(x) \geq f_i(y); \forall i = 1, 2, \ldots, n$;
  (ii) $\exists j \in \{1, 2, \ldots, n\} : f_j(x) > f_j(y)$.

That is, a feasible vector x is Pareto optimal if no feasible vector y can increase some criterion without causing a simultaneous decrease in at least one other criterion.

Multi-objective Evolutionary Algorithms (MOEA) can yield a whole set of potential solutions, which are all optimal in some sense. The main challenge in a multiobjective optimization environment is to minimize the distance of the generated solutions to the Pareto set and to maximize the diversity of the developed Pareto set. A good Pareto set may be obtained by appropriate guiding of the search process through careful design of reproduction operators and fitness assignment strategies. To obtain diversification special care has to be taken in the selection process. Special care is also to be taken care to prevent non-dominated solutions from being lost.

Solution Representation and Genetic Operators

The solution is represented as a string of length equal to the number of jobs. The value corresponding to each position i in the string represent the machine to which job i was allocated. Consider we have 10 jobs and 3 machines. Then a chromosome and the job allocation is represented as follows:

1	2	3	2	1	1	3	2	1	3

Machine 1: Job1, Job 5, Job 6, Job 9
Machine 2: Job 2, Job 4, Job 8
Machine 3: Job 3, Job 7, Job 10

Mutation and crossover are used as operators and binary tournament selection was used in the implementation. The Pareto dominance concept is used in order to compare 2 solutions. The one which dominates is preferred. In case of nondominance, the solution whose jobs allocation between machines is uniform is preferred. This means, there will not be idle machines as well as overloaded machines. The evolution process is similar to the evolution scheme of a standard evolutionary algorithm for multiobjective optimization. Reader is advised to consult [11] more details about MOEA approach.

Experiment Illustrations Using MOEA

We considered two scheduling instances (3,13) and (10,50). Specific parameter settings for MOEA, SA, PSO and GA are depicted in Table 9.6. Each experiment was repeated 10 times with different random seeds. Each trial (except for MOEA) had a fixed number of $50*m*n$ iterations (m is the number of the grid nodes, n is the number of the jobs). The makespan values of the best solutions throughout the optimization run were recorded. First we tested a small scale job scheduling problem involving 3 nodes and 13 jobs represented as (3,13). The node speeds of the 3 nodes are 4, 3, 2 CPUT, and the job length of 13 jobs are 6, 12, 16, 20, 24, 28, 30, 36, 40, 42, 48, 52, 60 cycles, respectively. The results (makespan) for 10 runs are as follows:

Genetic Algorithm: {47, 46, 47, 47.3333, 46, 47, 47, 47, 47.3333, 49}, average value = 47.1167

Table 9.6. Parameter settings for the different algorithms

Algorithm	Parameter name	Parameter value
GA	Population size	20
	Probability of crossover	0.8
	Probability of mutation	0.02
	Scale for mutations	0.1
SA	Number operations before temperature adjustment	20
	Number of cycles	10
	Temperature reduction factor	0.85
	Vector for control step of length adjustment	2
	Initial temperature	50
PSO	Swarm size	20
	Self-recognition coefficient c_1	1.49
	Social coefficient c_2	1.49
	Inertia weight w	$0.9 \rightarrow 0.1$
MOEA	Population size	100 (500 for the second experiment)
	Number of generations	200 (1000 for the second experiment)
	Mutation probability	1 (0.9 for the second experiment)
	Crossover probability	1 (0.9 for the second experiment)

Table 9.7. Performance comparison for (10, 50)

Algorithm	Average makespan
GA	38.04
SA	41.78
PSO	37.66
MOEA	36.68

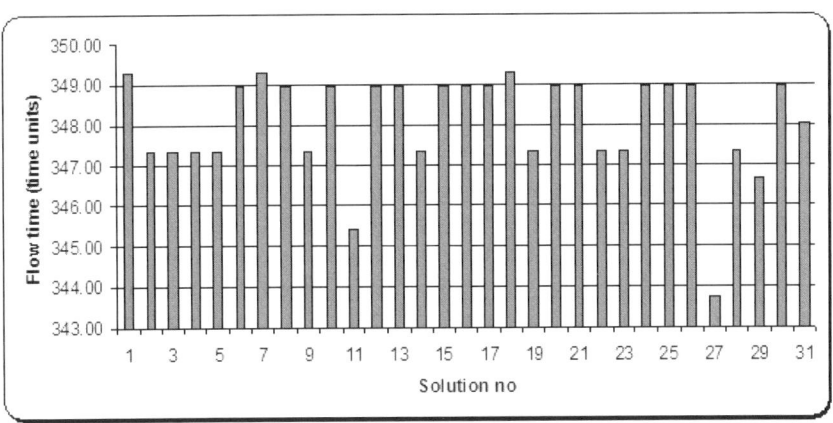

Fig. 9.7. Makespan from 31 non-dominated solutions in the final population for (10, 50)

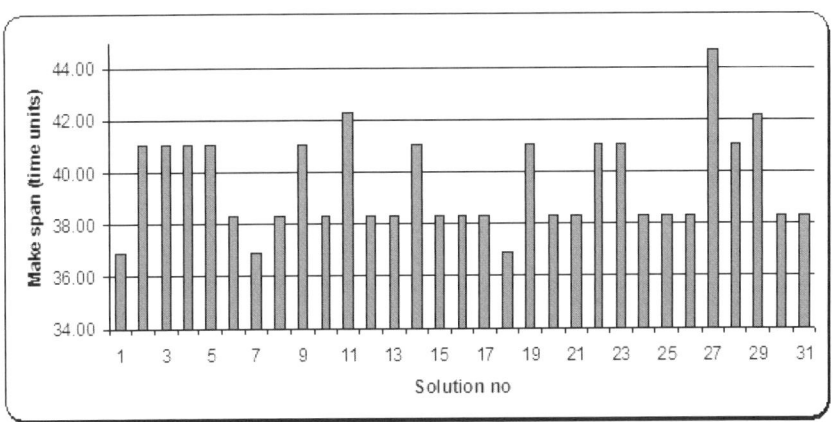

Fig. 9.8. Flowtime from 31 non-dominated solutions in the final population for (10, 50)

Simulated Annealing: {46.5, 46.5, 46, 46, 46, 46.6667, 47, 47.3333, 47, 47} average value = 46.6

Particle Swarm Optimization Algorithm: {46, 46, 46, 46, 46.5, 46.5, 46.5, 46, 46.5, 46.6667}, average value = 46.2667

Multi-objective Optimization Algorithm: 46, 46, 46, 46, 46, 46, 46, 46, 46, 46, average value = 46

The optimal result for (3,13) makespan is supposed to be 46 and the MOEA approach gave 46. It is to be noted that the MOEA approach obtained the best results in each of the considered runs.

Further, we tested the MOEA approach for (10, 50). The average makespan values for 10 trials are illustrated in Table 9.7. Although the average makespan value of SA was better than that of GA for (3,13), the case was reversed for this second case. Using the MOEA approach, the total average flow time obtained is = 348.07. Figs. 9.7 and 9.4.3 illustrate the makespan and flow time given by 31 non-dominated solutions from the final population. The user would have the option to go for a better flow time solution at the expense of a non-optimal makespan. As evident from the figure, the lowest flow time was 343.72 with the makespan of 44.75 for solution no. 27.

As evident from the empirical results, MOEA have given excellent results when compared to other techniques modeled using a single objective approach. Figs.9.7 and 9.4.3 illustrate the makespan and flow time given by 31 non-dominated solutions from the final population. The user would have the option to go for a better flow time solution at the expense of a non-optimal makespan. As evident from the Figs. 9.7 and 9.4.3, the lowest flow time was 343.72 with the makespan of 44.75 for solution no. 27. By seeing the population of solutions as illustrated in Figs. 9.7 and 9.4.3, the user will have the option to choose a particular schedule depending on the importance of the objectives. For example, the user can give more preference to a schedule which could offer a minimal flowtime but not an optimal makespan, etc.

9.5 Conclusions

In this Chapter, we illustrated the usage of several nature inspired metaheuristics for scheduling jobs. Our approach was to dynamically generate an optimal schedule so as to complete the tasks in a minimum period of time as well as utilizing the resources in an efficient way. We evaluated the performance of the heuristic approaches using a single and multi-objective optimization approaches.

Empirical results clearly illustrate the success of nature inspired heuristics in providing real-time good solutions especially when the search space is very huge. Our experiments also illustrate the importance and benefits of considering the objectives separately (multi-objective optimization approach) rather than combining them for the sake of simplicity.

Acknowledgments

F. Xhafa acknowledges partial support by Projects ASCE TIN2005-09198-C02-02, FP6-2004-ISO-FETPI (AEOLUS) and MEC TIN2005-25859-E and FORMALISM TIN2007-66523.

References

1. Foster, I., Kesselman, C.: The Grid: Blueprint For A New Computing Infrastructure. Morgan Kaufmann, USA (2004)
2. Garey, M.R., Johnson, D.S.: Computers and Intractability: A Guide to the Theory of NP-Completeness. Freeman, CA (1979)
3. Martino, V.D., Mililotti, M.: Sub optimal scheduling in a grid using genetic algorithms. Parallel Computing 30, 553–565 (2004)
4. Gao, Y., Rong, H.Q., Huang, J.Z.: Adaptive Grid Job Scheduling With Genetic Algorithms. Future Generation Computer Systems 21, 151–161 (2005)
5. Pang, W., Wang, K.P., Zhou, C.G., et al.: Fuzzy discrete particle swarm optimization for solving traveling salesman problem. In: Proceedings of the 4th International Conference on Computer and Information Technology. IEEE CS Press, Los Alamitos (2004)
6. Abraham, A., Liu, H., Zhang, W., Chang, T.G.: Job Scheduling on Computational Grids Using Fuzzy Particle Swarm Algorithm. In: Gabrys, B., Howlett, R.J., Jain, L.C. (eds.) KES 2006. LNCS (LNAI), vol. 4252, pp. 500–507. Springer, Heidelberg (2006)
7. Grosan, C., Abraham, A., Helvik, B.: Multi-objective Evolutionary Algorithms for Scheduling Jobs on Computational Grids. In: Guimaraes, N., Isaias, P. (eds.) International Conference on Applied Computing 2007, Salamanca, Spain, pp. 459–463 (2007) ISBN 978-972-8924-30-0
8. Abraham, A., Buyya, R., Nath, B.: Nature's Heuristics For Scheduling Jobs on Computational Grids. In: Proceedings of the 8th International Conference on Advanced Computing and Communications, pp. 45–52. Tata McGraw-Hill, India (2000)
9. Goldberg, D.E.: Genetic Algorithms in search, optimization, and machine learning. Addison-Wesley Publishing Corporation, Inc., Reading (1989)
10. Schaffer, J.D.: Multiple Objective Optimization with Vector Evaluated Genetic Algorithms, Ph. D. Thesis, Vanderbilt University, Nashville, TN (1984)
11. Abraham, A., Jain, L., Goldberg, R. (eds.): Evolutionary Multi-objective Optimization: Theoretical Advances and Applications, ch. 12, p. 315. Springer, London (2005)
12. Kirkpatrick, S., Gelatt, C.D., Vecchi, M.P.: Optimization by Simulated Annealing. Science 220(4598), 671–680 (1983)
13. Yao, X.: A New Simulated Annealing Algorithm. International Journal of Computer Mathematics 56, 161–168 (1995)
14. Bonabeau, E., Dorigo, M., Theraulaz, G.: Swarm Intelligence: From Natural to Artificial Systems. Oxford University Press, New York (1999)
15. Dorigo, M., Stützle, T.: Ant Colony Optimization. MIT Press, Cambridge (2004)
16. Gambardella, L.M., Dorigo, M.: Ant-Q: A reinforcement learning approach to the traveling salesman problem. In: Proceedings of the 11th International Conference on Machine Learning, pp. 252–260 (1995)
17. Stützle, T., Hoo, H.H.: MAX-MIN ant system. Future Generation Computer Systems 16, 889–914 (2000)
18. Kennedy, J., Eberhart, R.: Swarm Intelligence. Morgan Kaufmann, San Francisco (2001)
19. Clerc, M.: Particle Swarm Optimization. ISTE Publishing Company, London (2006)

20. Kennedy, J., Mendes, R.: Population structure and particle swarm performance. In: Proceeding of IEEE conference on Evolutionary Computation, pp. 1671–1676 (2002)
21. Abraham, A., Liu, H., Chang, T.G.: Variable neighborhood particle swarm optimization algorithm. In: Genetic and Evolutionary Computation Conference (GECCO 2006), Seattle, USA (2006)
22. Shi, Y.H., Eberhart, R.C.: Fuzzy adaptive particle swarm optimization. In: Proceedings of IEEE International Conference on Evolutionary Computation, pp. 101–106 (2001)
23. Liu, H., Abraham, A.: Fuzzy Adaptive Turbulent Particle Swarm Optimization. In: Proceedings of the Fifth International conference on Hybrid Intelligent Systems, pp. 445–450 (2005)
24. Clerc, M., Kennedy, J.: The Particle Swarm-explosion, Stability, and Convergence in A Multidimensional Complex Space. IEEE Transactions on Evolutionary Computation 6, 58–73 (2002)
25. Parsopoulos, K.E., Vrahatis, M.N.: Recent Approaches to Global Optimization Problems through Particle Swarm Optimization. Natural Computing 1, 235–306 (2002)
26. Abraham, A., Guo, H., Liu, H.: Swarm Intelligence: Foundations, Perspectives and Applications. In: Nedjah, N., Mourelle, L. (eds.) Swarm Intelligent Systems. Studies in Computational Intelligence, pp. 3–25. Springer, Germany (2006)

10

Efficient Batch Job Scheduling in Grids Using Cellular Memetic Algorithms

Fatos Xhafa[1], Enrique Alba[2], Bernabé Dorronsoro[3], Bernat Duran[1], and Ajith Abraham[3]

[1] Dept. de Llenguatges i Sistemes Informàtics
   Universitat Politècnica de Catalunya
   C/Jordi Girona 1-3, 08034 Barcelona, Spain
   fatos@lsi.upc.edu
[2] Dpto. de Lenguajes y Ciencias de la Computación
   E.T.S.I. Informática, Campus de Teatinos
   29071 Málaga, Spain
   eat@lcc.uma.es
[3] Faculty of Science, Technology and Communication University of Luxembourg
   6, rue Richard Coudenhove-Kalergi L-1359 Luxembourg
   bernabe.dorronsoro@uni.lu
[4] Centre for Quantifiable Quality of Service in Communication Systems, Norwegian
   University of Science and Technology, NO-7491 Trondheim, Norway
   ajith.abraham@ieee.org
   http://www.softcomputing.net

Summary. Due to the complex nature of Grid systems, the design of efficient Grid schedulers is challenging since such schedulers have to be able to optimize many conflicting criteria in very short periods of time. This problem has been tackled in the literature by several different meta-heuristics, and our main focus in this work is to develop a new highly competitive technique with respect to the existing ones. For that, we exploit the capabilities of Cellular Memetic Algorithms, a kind of Memetic Algorithm with structured population, for obtaining efficient batch schedulers for Grid systems, and the resulting scheduler is experimentally tested through a Grid simulator.

Keywords: Cellular Memetic Algorithms, Job Scheduling, Grid Computing, ETC model, Makespan, Dynamic computing environment, Simulation.

10.1 Introduction

One of the main motivations of the Grid computing paradigm has been the computational need for solving many complex problems from science, engineering, and business such as hard combinatorial optimization problems, protein folding, financial modelling, etc. [19, 21, 22]. One key issue in Computational Grids is the allocation of jobs (applications) to Grid resources. The resource allocation problem is known to be computationally hard as it is a generalization of the standard scheduling problem. Some of the features of the Computational Grids that make the problem challenging are the high degree of heterogeneity of resources, their connection with heterogenous networks, the high degree of dynamics, the large scale of the problem regarding number of jobs and resources, and other features related to existing local schedulers, policies on resources, etc. (see Chapter 1, this volume).

Meta-heuristic approaches have shown their effectiveness for a wide variety of hard combinatorial problems and also for multi-objective optimization problems. In this work we address the use of Cellular Memetic Algorithms (cMAs) [3, 4, 5, 6, 16] for efficiently scheduling jobs to Grid resources. cMAs are population-based algorithms that maintain a structured population as opposed to GAs or MAs of unstructured population. Research on cMAs has shown that, due to the structured population, this family of algorithms is able to better control the tradeoff between the exploitation and exploration of the solution space with respect to other non-structured algorithms [3, 4, 5]. It should be noted that this feature is very important if high quality solutions are to be found in a very short time. This is precisely the case of the job scheduling in Computational Grids whose highly dynamic nature makes indispensable the use of schedulers that would be able to deliver high quality planning of jobs to resources very fast in order to deal with the changes of the Grid. On the other hand, population-based heuristics are potentially good also for solving complex problems in the long run providing, for many problems, near optimal solutions. This is another interesting feature to explore regarding the use of cMAs for the job scheduling problem. The evidence reported in the literature that cMAs are capable to maintain a high diversity of the population in many generations suggests that cMAs could be appropriate for scheduling jobs that periodically arrive in the Grid system since in this case the Grid scheduler would dispose longer intervals of time to compute the planning of jobs to Grid resources. Finally, cMAs are used here to solve the bi-objective case of the job scheduling, namely makespan and flowtime are simultaneously optimized.

Many different cMA configurations have been developed and compared in this study on a benchmark of static instances of the problem (proposed by Braun et al. [9]). After that, we have also studied the behavior of the best obtained configuration in a more realistic benchmark of dynamic instances. Our algorithms will be validated by comparing the obtained results versus other results in the literature for the same studied benchmarks (both the static and the dynamic ones). Moreover, we studied the robustness of our cMA implementation since robustness is a desired property of Grid schedulers, which are very changing in nature. Because the cMA scheduler is able to deliver very high quality planning of jobs to Grid nodes, it can be used to design efficient dynamic schedulers for real Grid systems. Such dynamic schedulers are obtained by running the cMA-based scheduler in batch mode for a very short time to schedule jobs arriving in the systems since the last activation of the cMA scheduler.

This chapter is organized as follows. We give in Section 10.2 a description of the job scheduling in computational grids. The cMAs and their particularization for job scheduling in Grids together with the tuning process for the values of the parameters of the algorithm are given in Section 10.3. Some computational results as well as their evaluation for a benchmark of static instances are presented in Section 10.4. In Section 10.5, the best of the tested cMA configurations are evaluated in the more realistic case of dynamic instances, and the results are compared versus those of other algorithms found in the literature. Finally, we end in Section 10.6 with some conclusions.

10.2 The Batch Job Scheduling on Grids

In this work we consider the version of the problem[1] that arises quite frequently in parameter sweep applications, such as Monte-Carlo simulations [11]. In these applications, many jobs with almost no interdependencies are generated and submitted to the Grid system. In fact, more generally, the scenario in which the submission of independent jobs to a Grid system is quite natural given that Grid users independently submit their jobs or applications to the Grid system and expect an efficient allocation of their jobs/applications. We notice that the efficiency means that we are interested to allocate jobs as fast as possible and to optimize two conflicting criteria: *makespan* and *flowtime*.

In our scenario, jobs are originated from different users/applications, have to be completed in unique resource unless it drops from the Grid due to its dynamic environment (*non-preemptive* mode), are independent of each other and could have their hardware and/or software requirements over resources. On the other hand, resources could dynamically be added/dropped from the Grid, can process one job at a time, and have their own computing characteristics regarding consistency of computing. More precisely, assuming that the computing time needed to perform a task is known (assumption that is usually made in the literature [9, 15, 18]), we use the Expected Time to Compute (ETC) model by Braun et al. [9] to formalize the instance definition of the problem as follows:

- A *number* of independent (user/application) *jobs* to be scheduled.
- A *number* of heterogeneous *machines* candidates to participate in the planning.
- The *workload* of each job (in millions of instructions).
- The *computing capacity* of each machine (in *mips*).
- Ready time $ready_m$ indicates when machine m will have finished the previously assigned jobs.
- The Expected Time to Compute (ETC) matrix ($nb\_jobs \times nb\_machines$) in which $ETC[i][j]$ is the expected execution time of job i in machine j.

10.2.1 Optimization Criteria

We consider the job scheduling as a bi-objective optimization problem, in which both makespan and flowtime are simultaneously minimized. These criteria are defined as follows:

---
[1] The problem description and simultaneous optimization criteria are given in Chapter 1 and are reproduced here for completeness.

- *Makespan* (the finishing time of latest job) defined as $\min_S \max\{F_j : j \in Jobs\}$,
- *Flowtime* (the sum of finishing times of jobs), that is, $\min_S \sum_{j\in Jobs} F_j$,

where F_j is the finishing time of job j in schedule S.

For a given schedule, it is quite useful to define the *completion time* of a machine, which indicates the time in which the machine will finalize the processing of the previous assigned jobs as well as of those already planned for the machine. Formally, for a machine m and a schedule S, the completion time of m is defined as follows:

$$completion[m] = ready_m + \sum_{j\in S^{-1}(m)} ETC[j][m] \ . \tag{10.1}$$

We can then use the values of completion times to compute the makespan as follows:

$$\min_S \max\{completion[i] \mid i \in Machines'\} \ . \tag{10.2}$$

In order to deal with the simultaneous optimization of the two objectives we have used a simple weighted sum function of makespan and flowtime, which is possible since both parameters are measured in the same unit (time units). This way of tackling multiobjective optimization problems is widely accepted in the literature [12, 13], and its drawbacks are well known: only a single solution from the Pareto front (a set containing the best non-dominated solutions to the problem) is found in each run, and only solutions located in the convex region of the Pareto front will be found. However, the use of a weighted function is justified in our case by the convex search space of the considered problem and also by the need of providing a unique solution to the grid system, since there is not any decision maker to select the most suitable solution from a set of non-dominated ones.

The makespan and flowtime values are in incomparable ranges, since flowtime has a higher magnitude order over makespan, and its difference increases with the number of jobs and machines to be considered. For this reason, the value of mean flowtime, $flowtime/nb\_machines$, is used instead of flowtime. Additionally, both values are weighted in order to balance their importance. Fitness value is thus calculated as:

$$fitness = \lambda \cdot makespan + (1-\lambda) \cdot mean\_flowtime \ , \tag{10.3}$$

where λ has been *a priori* fixed after a preliminary tuning process to the value $\lambda = 0.75$ for the studies made in this work. Hence, we are considering in this work the makespan as the most important objective to optimize, while we give less importance to the total flowtime obtained in our solutions.

10.3 A cMA for Resource Allocation in Grid Systems

We present in this section a description of the cMA we are proposing in this work (Section 10.3.1) and its application to the batch job scheduling problem (Section 10.3.2).

10.3.1 Cellular Memetic Algorithms

In Memetic Algorithms (MAs) the population of individuals could be unstructured or structured. In the former, there is no relation between the individuals of the population while in the latter individuals can be related to only some other specific individuals of the population. The structured MAs are usually classified into *coarse-grained* model and *fine-grained* (Cellular MAs) model [4,5,6,16]. In Cellular MAs the individuals of the population are spatially distributed forming neighborhoods and the evolutionary operators are applied to neighbor individuals making thus cMAs a new family of evolutionary algorithms. As in the case of other evolutionary algorithms, cMAs are high level algorithms whose description is independent of the problem being solved. Thus, for the purposes of this work, we have considered the cMA template given in Algorithm 10.1.

As it can be seen, this template is quite different from the canonical cGA approximation [4,5], in which individuals are updated in a given order by applying the recombination operator to the two parents and the mutation operator to the obtained offspring. In the case of the proposed algorithm in this work, mutation and recombination operators are applied to individuals independently of each other, and in different orders. This model was adopted after a previous experimentation, in which it performed better than the cMA following the canonical model for the studied problems. After each recombination (or mutation), a local search step is applied to the newly obtained solution, which is then evaluated. If this new solution is better than the current one, it replaces the latter in the population. This process is repeated until a termination condition is met.

10.3.2 Application of the cMA to job Scheduling

Given the generic template showed in Algorithm 10.1, we proceed in this section to define the different parameters and operators we will use for solving the problem of batch job scheduling in grids. In order to efficiently solve the problem, we have to particularize the template with operators incorporating some specific knowledge of the problem at hand. The objective is to design an efficient algorithm for optimizing the QoS and productivity of grid systems. For that, we will use genetic operators focussed in balancing the load of all the available machines, and taking into account the presence of heterogeneous computers. We give next the description of the cMA particularization for job scheduling.

Regarding the problem representation, a feasible solution, *schedule*, is considered as a vector of size the number of jobs (nb_jobs) in which its jth position (an integer value) indicates the machine where job j is assigned: $schedule[j] = m, m \in \{1, \ldots, nb\_machines\}$.

Algorithm 10.1. A Cellular MA template

```
Initialize the mesh of n individuals P(t=0);
Initialize permutations rec_order and mut_order;
For each i ∈ P, LocalSearch(i);
Evaluate(P);
while not stopping condition do
  for j = 1...#recombinations do
    SelectToRecombine S ⊆ N_{P[rec_order.current]};
    i' = Recombine(S);
    LocalSearch(i'); Evaluate(i');
    Replace P[rec_order.current] by i';
    rec_order.next();
  end for
  for j = 1...#mutations do
    i = P[mut_order.current()];
    i' = Mutate(i);
    LocalSearch(i'); Evaluate(i');
    Replace P[rec_order.current] by i';
    rec_order.next();
  end for
  Update rec_order and mut_order;
end while
```

As it can be seen in Algorithm 10.1, many parameters are involved in the cMA template. Tuning these parameters is a crucial step in order to achieve a good performance, since they influence in a straightforward way on the search process. The tuning process was done by using randomly generated instances of the problem according to the ETC matrix model. This way we would expect a robust performance of our cMA implementation since no specific instance knowledge is used in fixing the values of the parameters. An extensive experimental study was done in order to identify the best configuration for the cMA. Thus, we experimentally studied the choice of the local search method, the neighborhood pattern, the selection, recombination and mutation operators, and the cell update orders. The tuning process was made step by step, starting from an initial configuration set by hand, and adding in each step the tuned parameters of the previous ones. We give in Figs. 10.2 to 10.8 the graphical representation for the makespan reduction of the cMA with the considered parameters. The results are obtained after making 20 independent runs in standard configuration computer.

Population's topology and neighborhood structure

Both the topology of the population and the neighborhood pattern are very important parameters in deciding the selective pressure of the algorithm and, therefore, they have a direct influence on the tradeoff between exploration and exploitation of the algorithm [2, 7]. The topology of the population is a two-dimensional toroidal grid of pop_height × pop_width size. Regarding the neighborhood patterns, several well-known patterns are used for this work: L5 (5 individuals), L9 (9 individuals), C9 (9 individuals) and C13 (13 individuals) (see Fig. 10.1). Additionally, in our quest for efficiency, we have considered the case in which the neighborhood is equal to the whole population, so an individual can interact with any other one in the population. Using this boundary

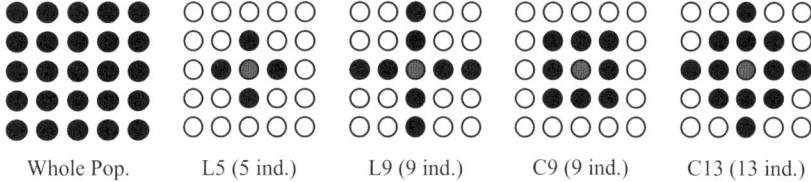

Fig. 10.1. Neighborhood patterns

neighborhood we remove a typical feature of cellular populations from our cMA, namely, the isolation by distance. The pursued effect is to accelerate the convergence of the algorithm up to the limit in order to check if it is profitable for the cMA.

We study in Fig. 10.2 the effects of using the different neighborhood structures previously proposed in our cMA in order to identify the pattern that leads to the best performance for the job scheduling problem. As it can be seen, we obtain from this study that the obtained makespan worsens when increasing the radius of the neighborhood (refer to [7] for a definition of the neighborhood radius). Among the tested neighborhoods, L5 and C9 (those with the smallest radii) perform the best exploration/exploitation tradeoffs of the algorithm for this problem. Between them, we can see that L5 yields a very fast reduction, although C9 performs better in the "long run" (see Fig. 10.2).

Finally, the case of considering the whole population as the neighborhood throws the worst performance (slowest convergence) of the algorithm. This is probably because the diversity in the population is quickly lost and thus the speed of the population evolution becomes very slow.

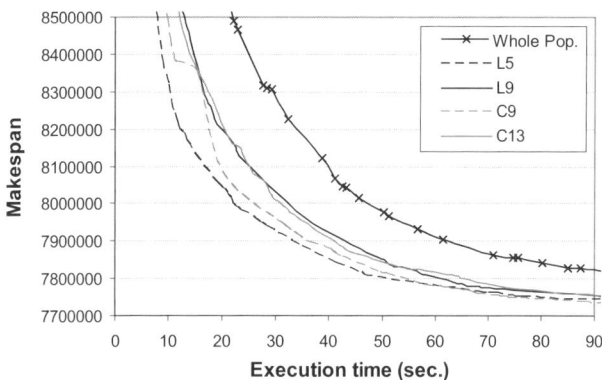

Fig. 10.2. Makespan reduction obtained with different neighborhood patterns (Makespan values are in arbitrary time units)

Population initialization

In this work we make use of some problem knowledge for creating the initial population of individuals. This way, one individual is generated using the *Longest Job to Fastest Resource - Shortest Job to Fastest Resource (LJFR-SJFR)* heuristic [1], while the rest are randomly obtained from the first individual by large perturbations. The LJFR-SJFR method has been chosen because it tries to simultaneously minimize both makespan and flowtime. LJFR minimizes the makespan and it is alternated with the SJFR which minimizes the flowtime. The method starts by increasingly sorting jobs with respect to their workload. At the beginning, the first *nb_machines* longest jobs are assigned to the *nb_machines* idle machines (the longest job to the fastest machine and so on). For the remaining jobs, at each step the fastest machine (that has finished its jobs) is chosen to which is assigned alternatively either the shortest job (SJFR) or the longest job (LJFR).

Cell updating

Unlike many unstructured MAs, in cMAs the population is kept constant by applying cell updating mechanisms by which an individual of the population is updated with a new offspring obtained by either recombination or mutation process (see later for the definition of these two operators). Two well-known methods of cell updating are the synchronous and asynchronous updating. For the purpose of this work, we have considered the asynchronous updating since it is less computationally expensive and usually shows a good performance in a very short time [8], which is interesting for the scheduling problem given the dynamic nature of Grid systems. In the asynchronous mode, cell updating is done sequentially (an individual is aware of other neighbor individual updates during the same iteration). The following asynchronous mechanisms have been implemented and experimentally studied for our job scheduling problem:

- **Fixed Line Sweep (FLS):** The individuals of the *grid* are updated in a sequential order row by row.
- **Fixed Random Sweep (FRS):** The sequence of cell updates is at random. This sequence is defined at the beginning of the algorithm and it is the same during all the cMA iterations.
- **New Random Sweep (NRS):** At each iteration, a new cell update sequence (at random) is applied.

It should be noted that recombination and mutation are independent processes in our cMAs (cf. `rec_order` and `mut_order` in the cMAs template) and therefore different update orders are used for them. Next, we study some different update policies and probabilities for applying them for the recombination and mutation steps.

In Fig. 10.3 we provide a study of the three proposed update policies for the recombination operator applied with two different probabilities. As regards to the cell updating for the recombination operator, the three considered mechanisms

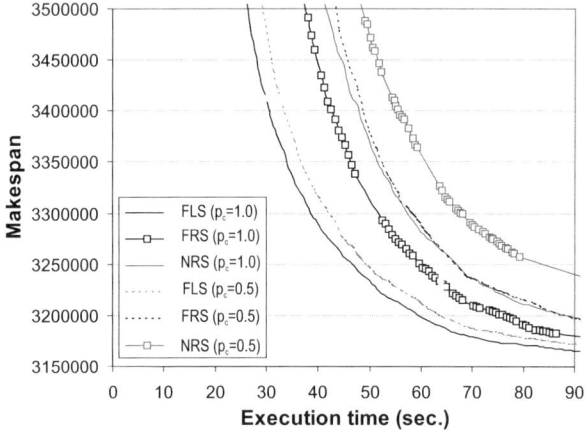

Fig. 10.3. Makespan reduction with different recombination orders and probabilities (p_c)

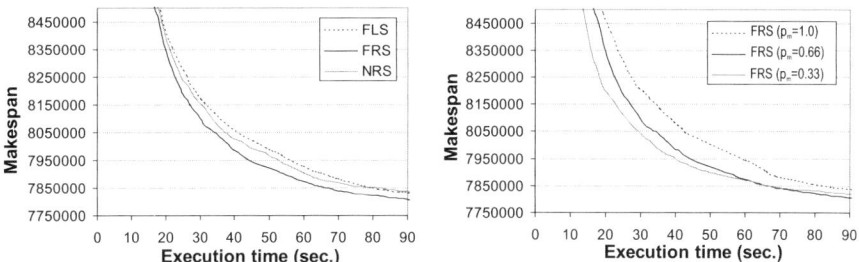

Fig. 10.4. Makespan reduction obtained with different mutation orders –left– and probabilities (p_m) –right

performed similarly, the FLS being the best performer (see Fig. 10.3). For the three update policies, the case of always recombining the individuals ($p_c = 1.0$) is advantageous versus applying the operator with probability $p_c = 0.5$.

Regarding the mutation operator, it can be seen in Fig. 10.4 (left hand plot) that, like in the case of the recombination operator, the three update policies perform in a similar way, being FRS slightly better than the other two ones. In the right hand plot of this same figure we study three different probabilities of applying the mutation operator when using FRS. The main result that can be drawn from this study is that the two lower probabilities ($p_m = 0.66$ and $p_m = 0.33$) perform better than the highest one ($p_m = 1.0$). When comparing these two lowest probabilities between them, one can notice that the case $p_m = 0.33$ converges faster, but after the 90 seconds allowed for the execution using $p_m = 0.66$ seems to be beneficial.

Selection operator for recombination

We have considered in this work the use of six different well-known selection policies in our cMA, namely linear ranking (LR), N-tournament (Ntour) with $N = 2, 3, 5$, and 7, and selecting the best individual in the neighborhood (Best). The results of our experiments are given in Fig. 10.5. As it can be seen, the slowest convergence is given by both linear ranking and binary tournament (Ntour with $N = 2$), although at the end of the run the makespan found using these two selection methods is close to that of the other compared ones, for which the convergence is faster at the beginning of the run, although its speed is drastically slowed after a few seconds. From all the compared selection methods, the one reporting the best makespan at the end of the run is Ntour with $N = 3$.

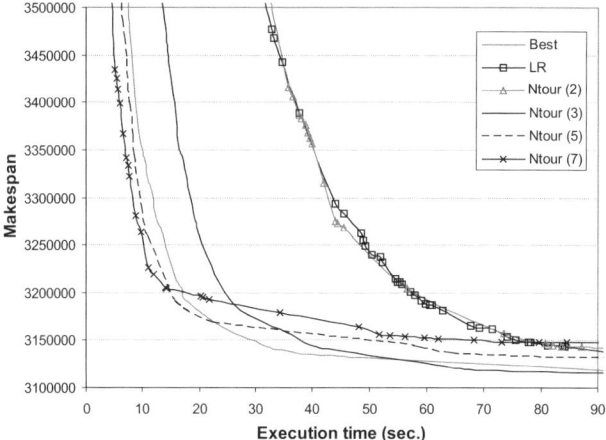

Fig. 10.5. Makespan reduction obtained with different selection methods

Recombination operator

Three recombination operators, very well known in the literature, were tested in this study for tuning our cMA. They are the *one-point* (OP2), the *uniform* (Uni2), and the *fitness-based* (FB2) recombination. The *one-point* operator lies in splitting the two chromosomes into two parts (in a randomly selected point), and joining each part of one parent chromosome with the other part of the chromosome of the second parent. In the case of the *uniform* recombination, an offspring is constructed by selecting for each gene the value of the corresponding gene of one of the two parents with equal probability. Finally, in the case of the *fitness-based* recombination both the structure and the relative fitness of the two parents are taken into account. The offspring is obtained as follows. Let us suppose that the two parents are P_1 and P_2, being $P_1[i]$ the ith gene of P_1 and f_{P_1} its fitness. The offspring is noted as C. If the two parents have the same value for a given gene i ($P_1[i] = P_2[i]$) this value is adopted for the same gene

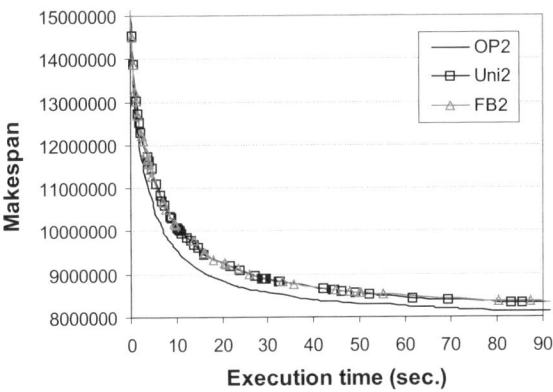

Fig. 10.6. Makespan reduction obtained with different recombination operators

of the offspring $C[i] = P_1[i]$. In other case, when $P_1[i] \neq P_2[i]$, $C[i] = P_1[i]$ with probability $p = f_{P_2}/(f_{P_1} + f_{P_2})$, while $C[i] = P_2[i]$ with probability $1 - p$.

From the results showed in Fig. 10.6, the *one-point* method has been chosen as the one reporting the best performance from the three compared recombination operators. The other two tested recombination operators (Uni2 and FB2) perform in a similar way, and slightly worse than OP2.

Mutation operator

We have tested four different mutation operators in our cMA. They are *move*, *swap*, *both*, and *rebalance*:

- *Move* is a simple operator that lies in changing the location of a given job in the chromosome of the individual, i.e. it assigns the machine of job i to job j.
- *Swap* exchanges the value of two genes. In our problem, this means that we are exchanging the machines assigned to two different jobs.
- *Both*. In this case we are applying one of the two previously explained operators (*move* and *swap*) with equal probability.
- *Rebalance*. The mutation is done by *rebalancing* of machine loads of the given schedule. The load factor of a machine m is defined as load_factor(m) = completion$[m]$/makespan (load_factor$(m) \in (0, 1]$). The idea behind this choice is that in a schedule, some machines could be overloaded (when its completion time is equal to the current makespan –load_factor$(m) = 1$–) and some others less overloaded (regarding the overloaded machines, we sort the machines in increasing order of their completion times and 25% first machines are considered less overloaded), in terms of their completion times. It is useful then to mutate the schedule by a load balancing mechanism, which transfers a job assigned to an overloaded machine to another less loaded machine.

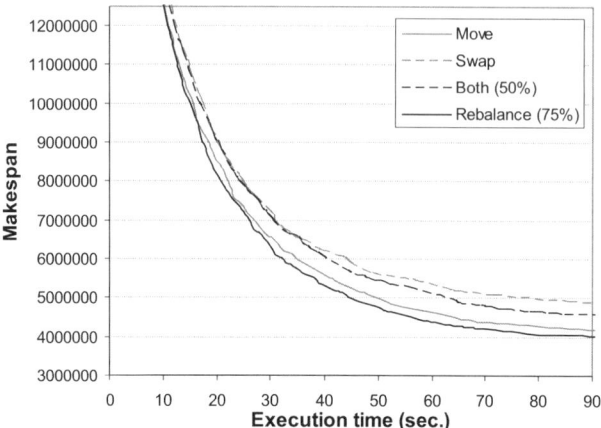

Fig. 10.7. Makespan reduction obtained with different mutation operators

Three of the studied mutation operators (*move*, *swap*, and *both*) are generic ones, while *rebalance* was specifically designed for this problem. They all are compared in Fig. 10.7. As it can be seen, the best performance is given by the *rebalance* operator (the unique specific method of the studied ones). Comparing the generic operators, *swap* is the worst one, and *move* is the best, being the results obtained by *both* between they two.

Local search methods

Local search is a proper feature of Memetic Algorithms. As it can be seen from the template of Algorithm 10.1, each individual is improved by a local search both after being generated by the recombination operator and after being mutated. Improvement of the descendants is thus done not only by means of genetic information but also by local improvements. The presence of this local search method in the algorithm does not increase selection pressure too much due to the exploration capabilities intrinsic to the cellular model. Four local search methods have been implemented and experimentally studied. These are the Local Move (LM), Steepest Local Move (SLM), Local Minimum Completion Time Swap (LMCTS), and Local Tabu Hop (LTH).

- LM is similar to the mutation operator (a randomly chosen job is transferred to a new randomly chosen machine).
- In SLM method, the job transfer is done to the machine that yields the best improvement in terms of the reduction of the completion time.
- In LMCTS method, two jobs assigned to different machines are swapped; the pair of jobs that yields the best reduction in the completion time is applied.
- Local Tabu Hop is a local search method based on the Tabu Search (TS) meta-heuristic. The main feature of TS [17] is that it maintains an

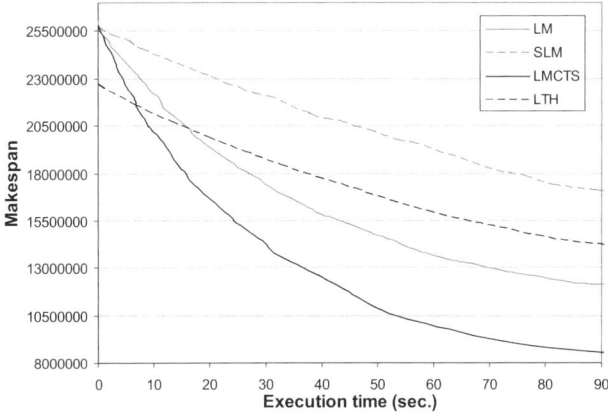

Fig. 10.8. Makespan reduction obtained with four local search methods

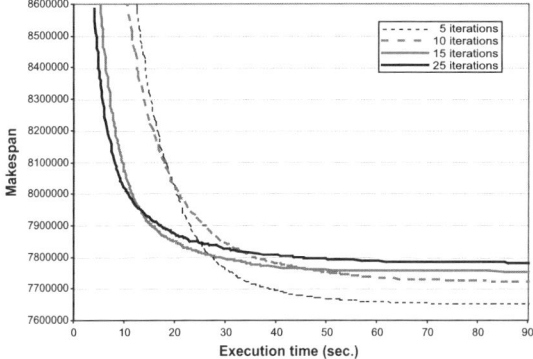

Fig. 10.9. Makespan reduction obtained with different intensities of the local search method

adaptive memory of forbidden (tabu) movements in order to avoid cycling among already visited solutions and thus escape from local optimal solutions. In the LTH algorithm for job scheduling, the implemented neighborhood relationship is based on the idea of the load balancing. The neighborhood of solutions consists of all those solutions to which we can reach via *swap* of the tasks of an overloaded resource with those of the less overloaded ones, or via *move* of tasks of an overloaded resource to the less overloaded resources. LTH is essentially a *phase* of Tabu Search and is taken from the Tabu Search implementation for the problem by Xhafa et al. [23].

In Fig. 10.8 we compare the behavior of our cMAs implementing the four proposed local search methods. From that graphical representation we can easily observe that the LMCTS method performs best among the four considered local

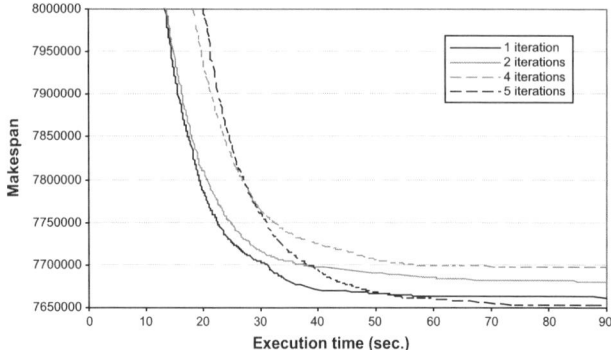

Fig. 10.10. Makespan reduction obtained with different maximum allowed iterations of the local search method when the solution is not improved

search methods. In fact, a clear difference in the behavior of the considered local search methods is observed, though all of them provide an accentuated reduction in the makespan value (see Fig. 10.8).

The bad behavior of the cMA using LTH is probably because this local search method is very heavy (computationally speaking) with respect to the other compared ones, and also the termination condition of the cMA is very hard (only 90 seconds of execution). Thus, the cMA only has time for making a few generations before the termination condition is met. Hence, it should be interesting to try some other parameters in order to reduce the number of LTH steps made by cMA+LTH in each generation, what hopefully should lead us to better results.

We present in Fig. 10.9 a study of the influence of the number of iterations of the LMCTS local search algorithm in the behavior of the cMA. Specifically, we study the cases of performing 5, 10, 15, and 25 iterations. As it can be seen in the figure, the smaller the number of iterations is the slower the convergence of the algorithm, and also the better the resulting makespan. Hence, the use of a strong local search provokes a premature convergence of the population, and this fast lost of diversity induces a bad behavior into the algorithm.

Once the number of iterations of the local search step is set, there is still one parameter to be tuned for the local search. This parameter is the number of iterations of the local search to perform even if no improvements were obtained in the previous ones. We present in Fig. 10.10 a study in which the cases of performing 1, 2, 4, and 5 iterations without any improvement are analyzed (recall that the maximum number of iterations was previously set to 5). From the results shown in Fig. 10.10 we decided to set the number of iterations of the local search to 5 even if no improvements are found.

Population Size and Shape and Replacement Policy

In this final step of our tuning procedure, we set the population size and shape as well as the replacement policy that we will use in our experiments. We compare

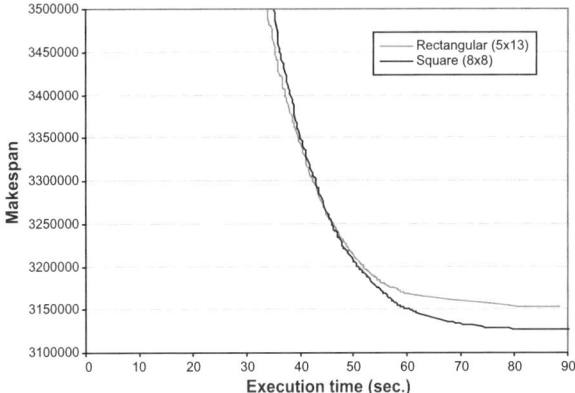

Fig. 10.11. Makespan reduction obtained with two different population shapes

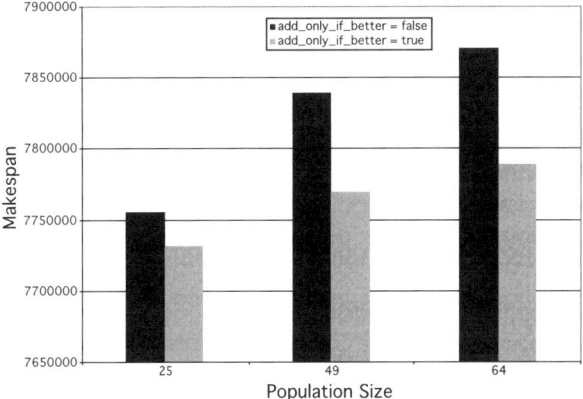

Fig. 10.12. Makespan obtained with two different replacement methods and three (square) population sizes

in Fig. 10.11 the behavior of our cMA with two populations of different shapes but having approximately the same size. The reason for this study is that the shape of the population markedly influences the behavior of the search in cellular evolutionary algorithms [3, 2]. The populations compared in Fig. 10.11 are a rectangular ones composed by 65 individuals arranged in a 5×13 mesh and a square 8×8 individuals population. As it can be seen in the figure, the latter performs better than the former for the studied instance.

We now study the influence of the replacement policy of new individuals into the population. Specifically, we considered two different options, namely, allowing that worse individuals can replace the current ones in the population ($add\_only\_if\_better = false$) or not ($add\_only\_if\_better = true$). As it can be seen in

Fig. 10.12, we always obtained better (lower) makespan values when individuals in the population can only be replaced by offsprings having better fitness values.

Additionally, we can see in Fig. 10.12 that the smallest of the three tested populations was the one providing the best makespan value (the three populations have square shape). The reason is that the use of a larger population allows to maintain the diversity for longer, but as a consequence the convergence speed is slowed down, so the algorithm generally requires a longer time to converge. This property is desirable for very difficult problems and large computation times. However, the computational time is fixed and very limited in our case of study, so it is desirable to enhance the exploitation capabilities of our algorithm.

10.4 Computational Results on Static Instances

After tuning our cMA on a set of random instances of the problem according to the ETC matrix model in Section 10.3.2, we present in this section some computational results obtained with our tuned cMAs for the benchmark of instances by Braun et al. [9] for distributed heterogenous systems. This benchmark is described in the next section, while the results of our algorithm are discussed and compared versus those obtained by other algorithms in Section 10.4.2.

10.4.1 Benchmark Description

The instances of this benchmark are classified into 12 different types of ETC matrices, each of them consisting of 100 instances, according to three parameters: job heterogeneity, machine heterogeneity and consistency. Instances are labelled as $u\_x\_yyzz.k$ where:

- u stands for uniform distribution (used in generating the matrix).
- x stands for the type of consistency (c–consistent, i–inconsistent, and s means semi-consistent). An ETC matrix is considered consistent when, if a machine m_i executes job j faster than machine m_j, then m_i executes all the jobs faster than m_j. Inconsistency means that a machine is faster for some jobs and slower for some others. An ETC matrix is considered semi-consistent if it contains a consistent sub-matrix.
- yy indicates the heterogeneity of the jobs (hi means high, and lo means low).
- zz indicates the heterogeneity of the resources (hi means high, and lo means low).

Note that all instances consist of 512 jobs and 16 machines. We report computational results for 12 instances, which are made up of three groups of four instances each. These three groups represent different Grid scenarios regarding the computing capacity. The first group corresponds to consistent ETC matrices (for each of them combinations between *low* and *high* are considered), the second represent instances of inconsistent computing capacity and the third one to semi-consistent computing capacity.

10.4.2 Evaluation and Discussion

In this section we present and discuss the results obtained by our algorithms, and compare them versus some other algorithms in the literature. Specifically, we propose two different cMAs: cMA+LMCTS and cMA+LTH. First, we compare the results obtained with the two proposed cMAs. Since one of these versions uses as a local search the Local Tabu Hop, we also compare the obtained results with those obtained by Tabu Search implementation by Xhafa et al. [23]. The algorithms run for 90 seconds (a single run) and 10 runs per instance are made. These decisions are the same than those adopted for the compared algorithms in order to make fair comparisons, since the compared results are directly taken from the original papers.

Table 10.1. Parameterization of cMA+LMCTS

Termination condition	Maximum of 90 seconds running
Population size	5×5
Probability of recombination	$p_c = 1.0$
Probability of mutation	$p_m = 0.5$
Population initialization	LJFR-SJFR (Longest / Shortest Job to Fastest Resource)
Neighborhood pattern	C9
Recombination order	FLS (Fixed Line Sweep)
Mutation order	NRS (New Random Sweep)
Selection method	3-Tournament
Recombination operator	One-Point recombination
Mutation operator	Rebalance
Local search method	LMCTS (Local Minimum Completion Time Swap)
Number of iterations of the local search	5
Replacement policy	Replace if better

The resulting configuration for cMA+LMCTS we decided to use after the initial tuning step made in Section 10.3.2 is given in Table 10.1. The parameterizations for cMA+LTH is similar to the one shown in Table 10.1, but in this case the population was set to 3×3 in order to reduce the number of local search steps due to the high computational requirements of LTH (see Section 10.3.2). Because of the small population used in this case, we adopt the L5 neighborhood pattern for cMA+LTH.

We give in Table 10.2 the computational results[2] for the makespan objective, where the first column indicates the name of the instance, and the other three ones present the average makespan with standard deviation (in %) obtained by the two proposed CMA algorithms (cMA+LMCTS and cMA+LTH) and TS [23]. Again, the results are averaged over 10 independent runs of the algorithms for every instance. The algorithm cMA+LMCTS provides the worst results in terms of average makespan, while the other proposed cellular memetic algorithm, cMA+LTH, is the best one for all the consistent instances, and it is the best performing algorithm if we do not take into account the inconsistent instances. This observation is interesting if the Grid characteristics were known in advance, since cMA+LTH seems to be more appropriate for consistent

---
[2] Values are in arbitrary time units.

Table 10.2. Comparison of the three proposed algorithms. Average makespan values.

Instance	cMA+LMCTS	TS	cMA+LTH
u_c_hihi.0	7700929.751 ±0.73%	7690958.935 ±0.28%	**7554119.350** ±0.47%
u_c_hilo.0	155334.805 ±0.13%	154874.145 ±0.41%	**154057.577** ±0.10%
u_c_lohi.0	251360.202 ±0.62%	250534.874 ±0.59%	**247421.276** ±0.47%
u_c_lolo.0	5218.18 ±0.30%	5198.430 ±0.52%	**5184.787** ±0.07%
u_i_hihi.0	3186664.713 ±1.80%	**3010245.600** ±0.26%	3054137.654 ±0.83%
u_i_hilo.0	75856.623 ±0.79%	**74312.232** ±0.35%	75005.486 ±0.31%
u_i_lohi.0	110620.786 ±1.72%	**103247.354** ±0.42%	106158.733 ±0.54%
u_i_lolo.0	2624.211 ±0.83%	**2573.735** ±0.39%	2597.019 ±0.39%
u_s_hihi.0	4424540.894 ±0.85%	**4318465.107** ±0.49%	4337494.586 ±0.71%
u_s_hilo.0	98283.742 ±0.47%	**97201.014** ±0.56%	97426.208 ±0.21%
u_s_lohi.0	130014.529 ±1.11%	**125933.775** ±0.38%	128216.071 ±0.83%
u_s_lolo.0	3522.099 ±0.55%	3503.044 ±1.52%	**3488.296** ±0.19%

Table 10.3. Comparison versus other algorithms in the literature. Average makespan values.

Instance	Braun et al. GA	GA (Carretero&Xhafa)	Struggle GA (Xhafa)	cMA+LTH
u_c_hihi.0	8050844.50	7700929.75	7752349.37	**7554119.35**
u_c_hilo.0	156249.20	155334.85	155571.80	**154057.58**
u_c_lohi.0	258756.77	251360.20	250550.86	**247421.28**
u_c_lolo.0	5272.25	5218.18	5240.14	**5184.79**
u_i_hihi.0	3104762.50	3186664.71	3080025.77	**3054137.65**
u_i_hilo.0	75816.13	75856.62	76307.90	**75005.49**
u_i_lohi.0	107500.72	110620.79	107294.23	**106158.73**
u_i_lolo.0	2614.39	2624.21	2610.23	**2597.02**
u_s_hihi.0	4566206.00	4424540.89	4371324.45	**4337494.59**
u_s_hilo.0	98519.40	98283.74	983334.64	**97426.21**
u_s_lohi.0	130616.53	130014.53	**127762.53**	128216.07
u_s_lolo.0	3583.44	3522.10	3539.43	**3488.30**

and semi-consistent Grid scenarios. Moreover, we consider that cMA+LTH is a more robust algorithm with respect to TS because the standard deviation values of the results obtained by the former are lower than those of the latter, in general.

We believe that it is possible to improve the results of cMA+LTH if we apply longer steps of the LTH method. Additionally, as it happened in [4,5] for the case of the satisfiability problem, we believe that the memetic algorithm (cMA+LTH) should outperform the local search by itself (TS) for larger instances of the problem. Moreover, it makes sense in our case to solve much larger instances of the problem, since we are tackling grids composed by only 16 machines in this preliminary study, and it is desirable to solve instances including hundreds or even thousands of processors.

The comparison of our best cMA (cMA+LTH) with three other versions of GAs taken from the literature is given in Table 10.3. Like in the case of Tables 10.2 and 10.3, values are the average makespan and standard deviation obtained after 10 independent runs. The compared algorithms are the Braun et al. GA [9], the GA by Carretero and Xhafa [10], the Struggle GA [26], and our best memetic algorithm cMA+LTH. For all the compared algorithms, the termination condition is set to a 90 seconds runtime. As it can be seen, cMA+LTH is

Table 10.4. Comparison versus other algorithms in the literature. Average flowtime values.

Instance	LJFR-SJFR	Struggle GA (Xhafa)	TS	cMA+LMCTS	cMA+LTH
u_c_hihi.0	2025822398.7	1039048563.0	1043010031.4	**1037049914.2**	1048630695.5
u_c_hilo.0	35565379.6	27620519.9	27634886.7	**27487998.9**	27684456.0
u_c_lohi.0	66300486.3	34566883.8	34641216.8	**34454029.4**	34812809.9
u_c_lolo.0	1175661.4	917647.31	919214.3	**913976.2**	922378.0
u_i_hihi.0	3665062510.4	379768078.0	**357818309.3**	361613627.3	370506405.1
u_i_hilo.0	41345273.2	12674329.1	**12542316.2**	12572126.6	12754803.6
u_i_lohi.0	118925453.0	13417596.7	**12441857.7**	12707611.5	12975406.6
u_i_lolo.0	1385846.2	440729.0	**437956.9**	439073.7	445529.3
u_s_hihi.0	2631459406.5	524874694.0	515743097.6	**513769399.1**	532276376.7
u_s_hilo.0	35745658.3	16372763.2	16385458.2	**16300484.9**	16628576.7
u_s_lohi.0	86390552.3	15639622.5	15255911.2	**15179363.5**	15863842.1
u_s_lolo.0	1389828.8	598332.7	597263.2	**594666.0**	605053.4

the best one of the four compared algorithms for all the studied instances, with the exception of the semi-consistent instance with low heterogeneity of jobs and high heterogeneity of the resources (u_s_lohi.0), for which cMA+LTH is the second best algorithm, just after the Struggle GA.

Additionally, when comparing cMA+LMCTS against the three other versions of GAs shown in Table 10.3 (Braun et al. GA, Carretero&Xhafa's GA [10] and Xhafa's Struggle GA [26]), cMA+LMCTS obtains better schedules than the compared GAs for half of the considered instances, and for the rest of the instances, the solutions found by cMA+LMCTS have a similar quality than the best of the other three GAs.

Computational results for flowtime parameter are given in Table 10.4 wherein we compare the average flowtime value obtained after 10 independent runs by the *ad hoc* heuristic LJFR-SJFR, the Xhafa's Struggle GA [26], Xhafa et al. TS [23] and the two cMAs proposed in this work. As it can be seen, the improvement made by the two cMAs on the initially constructed solution (obtained by the LJFR-SJFR heuristic) is very important. Additionally, it is noticeable in this table the improvement obtained by cMA+LMCTS over the compared algorithms, since it outperforms the compared algorithms for all considered instances. The exception are the inconsistent instances, for which the TS algorithm is the best one. The other proposed cMA, cMA+LTH, which obtained the best results for the makespan value is worse than both cMA+LMCTS and the Struggle GA for the flowtime objective.

10.5 Computational Results on Dynamic Instances

The study made in Section 10.4 using static instances for the problem of resource allocation in grids allowed us to better know the behavior of the cMAs, showing their main differences in the resolution of the problem and the results we could expect from them for several different cases. However, even if we can define static instances with really complex features, we still need to analyze the behavior of the algorithms in a more realistic dynamic grid system environment. In this case, the algorithm typically has to schedule the tasks in very short time intervals, and in a dynamic scenario that is continuously changing with time (resources that join and leave the Grid system). Thus, we study in this section the behavior of our algorithms in a more realistic set of dynamic instances. These instances are obtained using a simulator of a grid environment proposed in [24] that allows us to simulate different grid environments with distinct parameterizations. This simulator is briefly described in Section 10.5.1.

10.5.1 Dynamic Grid Simulator

The dynamic grid simulator was built from the Hypersim [20] framework, which is at the same time based in the simulation of systems of discrete events. The dynamic grid simulator allows us to emulate a set of dynamic resources that appear and disappear along time simulating resources that are registered and unregistered in grids. These simulated resources have different computing capacities (by means of number of instructions per time unit), and there is no limit on the number of tasks that can be assigned to a given resource. Moreover, every resource could have its own local scheduling policy.

New tasks arrive to the system following different distributions. The modelled tasks have intensive computing requirements, and they differ each other only in the work load (number of instructions). Tasks are considered to be sequential and have no dependencies on the other ones, so they are not restricted by the order in which they are executed, and no communication is needed among them. Hence, tasks are run in one single resource, and cannot be interrupted unless there is some error during the run. The scheduling process of these tasks is centralized, allowing to compare the scheduling algorithms easier than in the case of a decentralized system, since in this case the result of the scheduling is highly dependent on the the structure defined by the schedulers. The design of this simulator allows to easily adapt different scheduling policies, and it offers already implemented some scheduling policies. Anyway, the simulator is compatible both with static and dynamic schedulers.

The scheduler in our simulated grid is dynamically adapted to the evolution of the grid through the re-scheduling of the tasks either with a given frequency or when a change in the grid resources is made. As it could be expected from a scheduler of a real grid. In our simulator (at least in the version we are using in this work), no possible dependencies are considered between tasks and resources, so tasks can be run in any resource, and the computation time depends on both the length of the task and the resource capacity.

Finally, this simulator provides a configurable environment that allows the user to define different grid scenarios simply by changing some parameters. The simulator provides a large number of statistical measures that allows the user to evaluate and compare different schedulers, as well as the influence of the different parameter values.

10.5.2 Dynamic Benchmark Description

In this section, we present the parametrization used for the simulator described in Section 10.5.1 in order to define the benchmark for testing our schedulers. This parametrization have been carefully set in order to have different kinds of real grids. This way, we have defined grids of different (random) sizes, that we have enclosed in four different sets called small, medium, large, and very large, having the resources composing these grids random computing capacities. The details on the parametrization of the used simulator are given in Table 10.5, and the meaning of every parameter in the table is explained next:

- *Init. hosts*: Number of resources initially in the environment.
- *Max. hosts*: Maximum number of resources in the grid system.
- *Min. hosts*: Minimum number of resources in the grid system.
- *MIPS*: Normal distribution modelling computing capacity of resources.
- *Add host*: Normal distribution modelling the frequency of new resources being added to the system.
- *Delete host*: Normal distribution modelling the frequency of resources being dropped from the system.
- *Total tasks*: Number of tasks to be scheduled.
- *Init. tasks*: Initial number of tasks in the system to be scheduled.
- *Workload*: Normal distribution modelling the workload of tasks.
- *Interarrival*: Frequency (given by an exponential distribution) of new tasks arriving to the system (it is ensured that each time the simulator is activated, there will be at least one new task per resource).
- *Activation*: Establishes the activation policy according to an exponential distribution.
- *Reschedule*: When the scheduler is activated, this parameter indicates whether the already assigned tasks, which have not yet started their execution, will be rescheduled.

Table 10.5. Settings for the dynamic grid simulator

	Small	Medium	Large	Very Large
Init. hosts	32	64	128	256
Max. hosts	37	70	135	264
Min. hosts	27	58	121	248
MIPS		$N(1000, 175)^*$		
Add host	$N(625000, 93750)$	$N(562500, 84375)$	$N(500000, 75000)$	$N(437500, 65625)$
Delete host		$N(625000, 93750)$		
Total tasks	512	1024	2048	4096
Init. tasks	384	768	1536	3072
Workload		$N(2.5 * 10^8, 4.375 * 10^7)$		
Interarrival	$E(7812.5)^\dagger$	$E(3906.25)$	$E(1953.125)$	$E(976.5625)$
Activation		Resource_and_time_interval(250000)		
Reschedule		True		
Host select		All		
Task select		All		
Number of runs		15		

$^*N(\mu, \sigma)$ is a uniform distribution with average value μ and standard deviation σ.
$^\dagger E(\mu)$ is an exponential distribution an average value μ.

- *Host selection*: Selection policy of resources (*all* means that all resources of the system are selected for scheduling purposes).
- *Task selection*: Selection policy of tasks (*all* means that all tasks in the system must be scheduled).
- *Number runs*: Number of simulations done with the same parameters. Reported results are then averaged over this number.

As it can be seen in Table 10.5, we have defined four different grid sizes for our studies. Small grids are composed of a maximum of 37 hosts and a minimum of 27. The initial value is set to 32 and then it dynamically changes in that interval. When the simulation starts, 384 tasks must be scheduled, and new tasks arrive along time until a total of 512 ones. Medium grids are composed by a number of hosts in the interval [58, 70], starting with 64, and the total number of tasks is 1024 (being 768 at the beginning of the simulation). The large grids are considered to have between 121 and 135 hosts (128 initially) and a number of 2048 tasks (starting from 1536). Finally, the largest grids studied in this section are composed by an average of 256 hosts (varying this value in the interval [248, 264]), and the number of tasks to schedule grows from 3072 (initial fixed value) up to 4096. In all the configurations, the computing capacity of resources, their frequency of appearing and disappearing, the length of tasks and their arrival frequency are randomly set parameters (with values in the specified intervals). In the rescheduling process, all non executed tasks are considered even if they were previously scheduled.

10.5.3 Evaluation and Discussion

We proceed in this section to evaluate our schedulers in the dynamic benchmark previously defined. In order to make more realistic simulations, we have reduced the run time for the algorithm from 90 to 25 seconds, and we run the algorithm in a a Pentium IV 3.5GHz with 1GB RAM under windows XP operating system without any other process in background. The parametrization of the algorithm is shown in Table 10.6. As it can be seen, there are some small differences between this configuration and the one used in Section 10.4. These changes were made in order to promote the exploration capabilities of the algorithm for this set of more complex instances, and improve its answer to the instance changes. Thus, we have increased the population size to 6×6 (instead of 5×5) for solving all the instances except the small ones, the neighborhood is changed to L5 instead of C9, and we also improved the generation of the initial population. In the case of Section 10.4, this was made by generating one first individual using the LJFR-SJFR method, and the other individuals of the population were obtained after applying strong mutations to this initial individual. In this case, two initial individuals are generated instead of one: one with LJFR-SJFR, and the other one using the minimum completion time method (MCT). Then, the rest of the population is generated by mutating one of these two initial individuals (selected with equal probability).

Table 10.6. Parameterization of cMA+LMCTS for the dynamic benchmark

Termination condition	25 seconds run or $2 \times nb\_tasks$ generations
Population size	5×5 (small grids)
	6×6 (medium, large, and very large grids)
Probability of recombination	$p_c = 1.0$
Probability of mutation	$p_m = 0.5$
Population initialization	MCT and LJFR-SJFR
Neighborhood pattern	L5
Recombination order	FLS (Fixed Line Sweep)
Mutation order	NRS (New Random Sweep)
Selection method	3-Tournament
Recombination operator	One-Point recombination
Mutation operator	Rebalance
Local search method	LMCTS
Number of iterations of the local search	5
Replacement policy	Replace if better

The MCT method assigns a job to the machine yielding the earliest completion time (the ready times of the machines are used). When a job arrives in the system, all available resources are examined to determine the resource that yields the smallest completion time for the job (note that a job could be assigned to a machine that does not have the smallest execution time for that job). This method is also known as Fast Greedy, originally proposed for SmartNet system [14].

The parametrization for cMA+LTH is the same one proposed for cMA+LMCTS in Table 10.6 with only one exception: the population is set to 3×3 as an attempt to reduce the computational overload introduced by the expensive LTH method.

In our tests, the algorithms were run for 30 independent runs. The results are given in tables 10.7 and 10.8 for the makespan and the flowtime, respectively. Specifically, we present the average in the 30 runs of the average value (for the makespan and the flowtime, respectively) during every run (this value is continuously changing during the run due to the grid dynamism), the standard deviation (with a 95% confidence interval –CI–), and the deviation between the current solution and the best one for the same instance size. A 95% CI means that we can be 95% sure that the range of makespan (flowtime) values are within the shown interval, if the experiment were to run again. We are comparing in these tables the results obtained by our two cMAs and the same algorithms but with a panmictic (non structured) population, which are the best results we found in the literature for the studied problems [25].

In Table 10.7 we can see that the best overall algorithm for the four kinds of instances is cMA+LTH, which is the best algorithm in three out of the four cases (best values for every instance size are **bolded**). Only in the case of the largest instances cMA+LTH is outperformed by another algorithm (namely MA+LTH), but it is the second best algorithm in this case. Regarding the local search method used, we obtain from the results that the algorithms using TS as a local search method (MA+LTH and cMA+LTH) clearly outperform the other two ones for the four different instance sizes, since these two algorithms are the best ones for the four instances.

Table 10.7. Makespan values for the dynamic instances

	Heuristic	Makespan	% CI (0.95)	Best Dev.
Small	MA+LMCTS	4161118.81	1.47%	0.34%
	MA+LTH	4157307.74	1.31%	0.25%
	cMA+LMCTS	4175334.61	1.45%	0.68%
	cMA+LTH	**4147071.06**	**1.33%**	**0.00%**
Medium	MA+LMCTS	4096566.76	0.94%	0.32%
	MA+LTH	4083956.30	0.70%	0.01%
	cMA+LMCTS	4093488.97	0.71%	0.25%
	cMA+LTH	**4083400.11**	**0.62%**	**0.00%**
Large	MA+LMCTS	4074842.81	0.69%	0.29%
	MA+LTH	4067825.95	0.77%	0.12%
	cMA+LMCTS	4087570.52	0.57%	0.60%
	cMA+LTH	**4063033.82**	**0.49%**	**0.00%**
Very Large	MA+LMCTS	4140542.54	0.80%	0.82%
	MA+LTH	**4106945.59**	**0.74%**	**0.00%**
	cMA+LMCTS	4139573.56	0.35%	0.79%
	cMA+LTH	4116276.64	0.72%	0.23%

Table 10.8. Flowtime values for the dynamic instances

	Heuristic	Flowtime	% CI (0.95)	Best Dev.
Small	MA+LMCTS	1045280118.16	0.93%	0.15%
	MA+LTH	1045797293.10	0.93%	0.20%
	cMA+LMCTS	**1044166223.64**	**0.92%**	**0.00%**
	cMA+LTH	1046029751.67	0.93%	0.22%
Medium	MA+LMCTS	2077936674.17	0.61%	0.07%
	MA+LTH	2080903152.40	0.62%	0.22%
	cMA+LMCTS	**2076432235.04**	**0.60%**	**0.00%**
	cMA+LTH	2080434282.38	0.61%	0.19%
Large	MA+LMCTS	4146872566.09	0.54%	0.02%
	MA+LTH	4153455636.89	0.53%	0.18%
	cMA+LMCTS	**4146149079.39**	**0.55%**	**0.00%**
	cMA+LTH	4150847781.82	0.53%	0.11%
Very Large	MA+LMCTS	**8328971557.96**	**0.35%**	**0.00%**
	MA+LTH	8341662800.11	0.35%	0.15%
	cMA+LMCTS	8338100602.75	0.34%	0.11%
	cMA+LTH	8337173763.88	0.35%	0.10%

The results obtained for the flowtime are given in Table 10.8. As it happened in the previous case, the best cMA (cMA+LMCTS in this case) outperforms the best MA (MA+LMCTS) algorithm for all the tested instance sizes, with the only exception of the very large one. We notice that in this case, the results are also somehow opposite to the ones obtained for the makespan, since the algorithms implementing the LMCTS local search method outperform those using LTH for all the instances. However, these results make sense, since both makespan and flowtime are conflictive objective values. This means that, for high quality solutions, it is not possible to improve one of the two objectives without decreasing the quality of the other. This is related to the concept of Pareto optimal front in multi-objective optimization (see [12,13]). In this paper we are tackling a multi-objective problem by weighting the two objectives into a single fitness function. Thus, in this work we are giving more importance to the makespan objective by weighting this value by 0.75 in the fitness function, while the weight of flowtime was set to 0.25. So we can consider that cMA+LTH is the algorithm obtaining the best overall results among the tested ones.

10.6 Conclusions and Future Work

In this work we have presented two implementations of Cellular Memetic Algorithms (cMAs) for the problem of job scheduling in Computational Grids when both makespan and flowtime are simultaneously minimized. cMAs are a family of population-based metaheuristics that have turned out to be an interesting approach due to their structured population, which allows to better control the tradeoff between the exploitation and exploration of the search space. We have implemented and experimentally studied several methods and operators of cMA for the job scheduling in Grid systems, which is a challenging problem in today's large-scale distributed applications.

The proposed cMAs were tested and compared versus other algorithms in the literature for benchmarks using both static and dynamic instances. Our experimental study showed that cMAs are a good choice for scheduling jobs in Computational Grids given that they are able to deliver high quality planning in a very short time. This last feature makes cMAs useful to design efficient dynamic schedulers for real Grid systems, which can be obtained by running the cMA-based scheduler in batch mode for a very short time to schedule jobs arriving in the systems since the last activation of the cMA scheduler. The use of the proposed cMA could highly improve the behavior of real clusters in which very simple methods (e.g., queuing systems or *ad hoc* schedulers using specific knowledge of the grid infrastructure) are used.

In our future work we would like to better understand some issues raised by the experimental study such as the good performance of the cMAs for consistent and semi-consistent Grid Computing environments and the not so good performance for inconsistent computing instances. Also, we plan to extend the experimental study by considering other operators and methods as well as studying the performance of cMA-based scheduler(s) in longer periods of time and considering larger grids. Additionally, we are studying different policies for applying the local search method in order to make this important step of the algorithm less computationally expensive. Other interesting line for future research is to tackle the problem with a multi-objective algorithm in order to find a set of non-dominated solutions to the problem.

Acknowledgments

F. Xhafa acknowledges partial support by Projects ASCE TIN2005-09198-C02-02, FP6-2004-ISO-FETPI (AEOLUS) and MEC TIN2005-25859-E and FORMALISM TIN2007-66523. E. Alba acknowledges that this work has been partially funded by the Spanish MEC and FEDER under contract TIN2005-08818-C04-01 (the OPLINK project).

References

1. Abraham, A., Buyya, R., Nath, B.: Nature's heuristics for scheduling jobs on computational grids. In: The 8th IEEE International Conference on Advanced Computing and Communications (ADCOM), India, pp. 45–52. IEEE Press, Los Alamitos (2000)
2. Alba, E., Dorronsoro, B.: The exploration/exploitation tradeoff in dynamic cellular evolutionary algorithms. IEEE Transactions on Evolutionary Computation 9(2), 126–142 (2005)
3. Alba, E., Dorronsoro, B.: Cellular Genetic Algorithms. In: Operations Research/Computer Science Interfaces. Springer, Heidelberg (to appear)
4. Alba, E., Dorronsoro, B., Alfonso, H.: Cellular memetic algorithms. Journal of Computer Science and Technology 5(4), 257–263 (2005)
5. Alba, E., Dorronsoro, B., Alfonso, H.: Cellular memetic algorithms evaluated on SAT. In: XI Congreso Argentino de Ciencias de la Computación (CACIC) (2005) DVD Edition
6. Alba, E., Tomassini, M.: Parallelism and evolutionary algorithms. IEEE Transactions on Evolutionary Computation 6(5), 443–462 (2002)
7. Alba, E., Troya, J.M.: Cellular evolutionary algorithms: Evaluating the influence of ratio. In: Deb, K., Rudolph, G., Lutton, E., Merelo, J.J., Schoenauer, M., Schwefel, H.-P., Yao, X. (eds.) PPSN 2000. LNCS, vol. 1917, pp. 29–38. Springer, Heidelberg (2000)
8. Alba, E., Dorronsoro, B., Giacobini, M., Tomassini, M.: Handbook of Bioinspired Algorithms and Applications. In: Decentralized Cellular Evolutionary Algorithms, ch. 7, pp. 103–120. CRC Press, Boca Raton (2006)
9. Braun, H., Siegel, T.D., Beck, N., Bölöni, L., Maheswaran, M., Reuther, A., Robertson, J., Theys, M., Yao, B.: A comparison of eleven static heuristics for mapping a class of independent tasks onto heterogeneous distributed computing systems. Journal of Parallel and Distributed Computing 61(6), 810–837 (2001)
10. Carretero, J., Xhafa, F.: Using genetic algorithms for scheduling jobs in large scale grid applications. Journal of Technological and Economic Development –A Research Journal of Vilnius Gediminas Technical University 12(1), 11–17 (2006)
11. Casanova, H., Legrand, A., Zagorodnov, D., Berman, F.: Heuristics for scheduling parameter sweep applications in grid environments. In: Heterogeneous Computing Workshop, pp. 349–363 (2000)
12. Coello, C.A., Van Veldhuizen, D.A., Lamont, G.B.: Evolutionary Algorithms for Solving Multi-Objective Problems. In: Genetic Algorithms and Evolutionary Computation. Kluwer Academic Pubishers, Dordrecht (2002)
13. Deb, K.: Multi-Objective Optimization using Evolutionary Algorithms. Wiley, Chichester (2001)
14. Freund, R., Gherrity, M., Ambrosius, S., Campbell, M., Halderman, M., Hensgen, D., Keith, E., Kidd, T., Kussow, M., Limaand, J., Mirabile, F., Moore, L., Rust, B., Siegel, H.J.: Scheduling resources in multi-user, heterogeneous, computing environments with smartnet. In: Seventh Heterogeneous Computing Workshop, pp. 184–199 (1998)
15. Ghafoor, A., Yang, J.: Distributed heterogeneous supercomputing management system. IEEE Comput. 26(6), 78–86 (1993)
16. Giacobini, M., Tomassini, M., Tettamanzi, A.G.B., Alba, E.: Selection intensity in cellular evolutionary algorithms for regular lattices. IEEE Transactions on Evolutionary Computation 9(5), 489–505 (2005)

17. Glover, F., Laguna, M.: Tabu Search. Kluwer Academic Publishers, Boston (1997)
18. Kafil, M., Ahmad, I.: Optimal task assignment in heterogeneous distributed computing systems. IEEE Concurrency 6(3), 42–51 (1998)
19. Luna, F., Nebro, A.J., Alba, E.: Observations in using grid-enabled technologies for solving multi-objective optimization problems. Parallel Computing 32, 377–393 (2006)
20. Phatanapherom, S., Kachitvichyaunukul, V.: Fast simulation model for grid scheduling using hypersim. In: Proceedings of the 2003 Winter Simulation Conference, pp. 1494–1500 (2003)
21. Talbi, E.-G.: Parallel Combinatorial Optimization. John Wiley & Sons, USA (2006)
22. Talbi, E.-G., Zomaya, A.: Grids for Bioinformatics and Computational Biology. John Wiley & Sons, USA (2007)
23. Xhafa, F., Carretero, J., Alba, E., Dorronsoro, B.: Design and Evaluation of Tabu Search Method for Job Scheduling in Distributed Environments. In: The 11th International Workshop on Nature Inspired Distributed Computing (NIDISC 2008) held in conjunction with The 22th IEEE/ACM International Parallel and Distributed Processing (NIDISC 2008), Florida, USA, April 14-18 (to appear, 2008)
24. Xhafa, F., Carretero, J., Barolli, L., Durresi, A.: Requirements for an event-based simulation package for grid systems. Journal of Interconnection Networks 8(2), 163–178 (2007)
25. Xhafa, F.: Hybrid Evolutionary Algorithms. In: A Hybrid Heuristic for Job Scheduling in Computational Grids, ch. 11. Studies in Computational Intelligence, vol. 75, pp. 269–311. Springer, Heidelberg (2007)
26. Xhafa, F.: An experimental study on GA replacement operators for scheduling on grids. In: The 2nd International Conference on Bioinspired Optimization Methods and their Applications (BIOMA), Ljubljana, Slovenia, October 2006, pp. 212–130 (2006)

11

P2P B&B and GA for the Flow-Shop Scheduling Problem

A. Bendjoudi[1], S. Guerdah[2], M. Mansoura[2], N. Melab[3], and E-G. Talbi[3]

[1] Université A/Mira de Béjaia, Département d'Informatique
   CEntre de Recherche sur l'Information Scientifique et Technique (CERIST),
   Laboratoire d'Ingénierie et Théories des Systèmes Informatiques, Algiers, Algeria
   ahcene.bendjoudi@gmail.com
[2] Université Mouloud Mammeri de Tizi Ouzou,
   Département Informatique, Tizi Ouzou, Algeria
   {samir.guerdah,madjid.mansoura}@gmail.com
[3] Université des Sciences et Technologies de Lille, France
   Laboratoire d'Informatique Fondamentale de Lille LIFL
   UMR CNRS 8022, Cité scientifique
   INRIA Futurs - DOLPHIN
   59655 - Villeneuve d'Ascq cedex - France
   {talbi,melab}@lifl.fr

Summary. Solving exactly Combinatorial Optimization Problems (COPs) using a Branch-and-Bound algorithm (B&B) requires a huge amount of computational resources. The efficiency of such algorithm can be improved by its hybridization with meta-heuristics such as Genetic Algorithms (GA) which proved their effectiveness, since they generate acceptable solutions in a reasonable time. Moreover, distributing at large scale the computation, using for instance Peer-to-Peer (P2P) Computing, provides an efficient way to reach high computing performance. In this chapter, we propose ParallelBB and ParallelGA, which are P2P-based parallelization of the B&B and GA algorithms for the computational Grid. The two algorithms have been implemented using the ProActive distributed object Grid middleware. The algorithms have been applied to a mono-criterion permutation flow-shop scheduling problem and promisingly experimented on the Grid5000 computational Grid.

Keywords: P2P Computing, Branch and Bound, Genetic Algorithms, Grid Middleware, Flow-Shop Scheduling.

11.1 Introduction

In practice, many problems can be modelled as combinatorial optimization problems. These problems are often large and classed NP-hard [20] such as scheduling and quadratic assignment problems. To solve these problems, various methods were proposed in the literature. Meta-heuristics proved their effectiveness, since they generate acceptable solutions, in a reasonable time. Searching an exact solution for this kind of problem remains unpractical when the problem size grows, because the execution time increases in an exponential way. To mitigate this constraint, hybridization between exact and heuristic methods and their parallelization are two effective ways in terms of improving the computing performances, in particular the use of large scale parallelism based on *Grid Computing* [19] or *Peer-to-Peer Computing* [34, 36].

Grid and P2P Computing are emerging technologies allowing to share various resources at a large scale. Grid Computing uses an infrastructure for globally sharing compute-intensive resources such as supercomputers or computational clusters. P2P Computing, using for instance *XtremWeb* [18] or *ProActive* [1], is based on the exploitation of non used CPU cycles or completely idles. Nowadays, these two technologies provide effective tools to achieve high performance in solving large-scale problems. Particularly, solving exactly complex combinatorial optimization problems is a good challenge for GRID/P2P Computing.

The Branch-and-Bound algorithm (*B&B*) is the most known method for exact resolution of combinatorial optimization problems (*COP*). B&B explores the search space by implicitly enumerating subtrees. The whole exploration of this space is impossible considering the exponential increase in the number of solutions when the size of the problem increases. The use of good lower and upper bounds reduces the number of subtrees to enumerate. Meta-heuristics provide sub-optimal solutions in a reasonable time (they allow us to reach an acceptable solution in a short time). Genetic Algorithms (GAs) belongs to Evolutionary Algorithms (EAs) which make use of a randomly generated population of solutions. The initial population is enhanced through a natural evolution process. At each generation of the process, the whole population or a part of the population is replaced by newly generated individuals. Several parallel versions of B&B [11, 13, 35, 39, 40, 41] and GA [3, 22, 30] are studied in the literature. The B&B algorithm is suitable to be parallelized given that the subtrees can be explored independently. The only shared information in the algorithm is the value of the best known solution (upper bound). Likewise, the parallelism is necessary to not only reduce the resolution time of GAs, but also to improve the quality of the provided solutions. The hybridization of these two categories permits to improve the performances of the total execution time.

Recently, some approaches [5, 6, 8, 12, 33] aiming at exploiting P2P/GRID computing and at deploying scientific applications requiring a great computing power, have been developed. [5, 6] are based on the *Master/Worker* paradigm [23, 38]. The main drawback of this approach is bottlenecks created on the master process because the inter-worker communications transit throw the master. Our work presents two parallel B&B and GA Algorithms and their hybridization based on master/worker paradigm with direct communications between workers. Therefore, bottlenecks are eliminated. We develop the peer-to-peer version using ProActive middleware which enables direct communications between the various peers (*workers*) of the network without flowing throw an intermediary (*master*). We applied the two algorithms and their combination version to mono-criterion permutation flow-shop problem *PFSP*. PFSP consists to find a schedule of a set of jobs on a set of machines that minimizes the completion time (makespan).

Jobs are scheduled in the same order on all machines and each machine can not be simultaneously assigned to two jobs.

The remainder of this chapter is structured as follows : Section 11.2 highlights the major points for the parallelization of a Branch-and-Bound algorithm and a brief description of parallel genetic algorithms. The concept of P2P Computing, ProActive middleware and the various tools that it offers to develop a distributed application on a peer-to-peer system are presented in Section 11.3. In Section 11.4, we present our parallelization of the Branch and Bound algorithm *ParallelBB* and genetic algorithm *ParallelGA* intended to be deployed on a large scale computing. In Section 11.5, its peer-to-peer implementation on top of ProActive middlware (namely *PHyGABaB*). Preliminary large scale deployment and performance evaluation on a P2P computing network formed and managed by ProActive showed in Section 11.6. We conclude this chapter in Section 11.7.

11.2 Parallel Combinatorial Optimization

Combinatorial optimization problems are often NP-hard, complex and CPU time-consuming. Exact methods and meta-heuristics are two major traditionally used approaches [7]. Exact techniques may be useful for solving small problem instances, but in realistic cases they are inefficient as they are extremely time-consuming. Conversely, meta-heuristics provide near-optimal solutions and allow to meet the resolution delays often imposed in the industrial field. The parallelization of these two categories is an efficient way to solve larger instances of problems in a reasonable time. In the following, we present parallelization of theses two categories as described in the literature.

11.2.1 Parallel Branch-and-Bound Algorithms

Branch-and-Bound algorithms are the most known techniques for an exact resolution of COPs. They make an implicit enumeration of the whole search space, because of the impossibility of a complete enumeration of all solutions of the search space due to the exponential growing of the potential solutions. B&B algorithms are characterized by four operations: *branching, bounding, selection and elimination*. In the first operation, the solution space of a given problem is partitioned into a number of smaller subsets on which the same optimization problem is defined. The bounding rule is used to compute the lower bound of the optimal solution of the considered problem. When a new solution (*upper bound*) is identified, it is compared to the actual lower bound in order to decide whether it is necessary to decompose the subproblem or not. The elimination rule uses these bounds to determine when further decomposition of a subproblem is unnecessary, so it identifies nodes which do not lead to the optimal solution and eliminates them. The subproblems are explored according to the selection rule. We can find the following exploration methods: *depth first search, breath first search, best bound,... etc*. A serial implementation of the algorithm consists of a sequential execution of these four operations.

The subtrees generated when executing a B&B algorithm can be explored independently, this makes the parallelization of these algorithms easier. The only global information in the algorithm is the value of the upper bound. Its parallelization may be attached to the architecture of the calculator machine, synchronization, granularity of generated tasks, communication between different processes and the number of computing processors. In the literature, several works on parallelization of B&B had been conducted [11, 13, 35, 39, 40, 41]. Geondron and Crainic [11] classified the parallelization strategies into three classes according to the degree of parallelization: *parallelism of type 1* introduces parallelism when performing the operations (generally the bounding operation) on generated subproblems (e.g. executing the bounding operation in parallel for each subproblem). This type of parallelism depends on the problem to be solved. In *parallelism of type 2* the search tree is built in parallel (e.g. processes work on several subproblems simultaneously). The *parallelism of type 3* also implies the building of several trees in parallel. The information generated when building one tree can be used for the construction of another. Thus the tree is explored concurrently.

The processes which participate in the computation of the parallel algorithm select their tasks from a *work pool*. A work pool is a memory where the processes select and store their work units (generated and not yet explored subproblems). Two types of work pool can be distinguished: *single work pool* and *multiple work pool*. Generally the first type is implemented on shared memory systems [13] and the second type uses several allocation memories. The first type is more adequate for the applications based on the *master/worker* [5] paradigm. Indeed, *master* process distributes part of computing (*tasks*) on a set of *workers* processes. When workers finishes their execution, the main process collects the obtained results. This paradigm is very used in scientific applications dedicated to be deployed on massively parallel systems (cluster, Grid computing). However, this paradigm presents a major drawback, it creates bottlenecks on the *master* process [4, 5].

11.2.2 Parallel Genetic Algorithms

Evolutionary Algorithms (EAs) [7] are stochastic search techniques and population-based algorithms. Genetic Algorithms (GAs), proposed by Holland [25] are the most known algorithms in this field [17]. They are powerful search techniques that are used successfully to solve problems in different disciplines. They are based on principles of natural selection and recombination. They attempt to find the optimal solution to the problem at hand by manipulating a population of candidate solutions. The population is evaluated and the best solutions are selected to reproduce and mate to form the next generation. Over a number of generations, good traits dominate the population, resulting in the improvement of the quality of the solutions.

Starting with a randomly generated population, a new generation is produced with three genetic operators: selection, reproduction and mutation. With the Selection operator we decide which individuals to survive. In the reproduction operator, two individuals are selected to produce a child which inherits its two parents. The mutation operator alters the genetic code of an individual to promote diversity (see [17] for more details on GAs).

In the literature, several works have been dedicated to parallel GAs (see [14, 21, 24, 32]) and classified them into two categories: parallelization of computation (fine grained) and parallelization of population (coarse-grained). In the first model, the operations commonly applied to each of the individuals are performed in parallel. The coarse-grained type is the most popular and used category. In this type, an initial population is divided into sub-populations which will evolve separately (in parallel) and exchange individuals following a migration protocol. They are usually implemented on distributed memory computers (MIMD) and based on the island model. The most recent example is the work of Mezmaz *et al* [33]. The authors proposed a P2P hybrid Genetic-Mimetic Algorithm based on the island model aiming at exploiting P2P/GRID-Computing.

11.3 P2P Computing and the ProActive Middleware

Distributed systems and applications are called Peer-to-Peer (P2P) if they employ distributed resources to perform a function in a decentralized manner. Here, resources includes (computer power, data storage and network bandwidth), the function concerns (distributed computing, data sharing, communication and collaboration) and decentralization (algorithms, data or both of them). That is the definition given in [34]. In distributed computing area, the idea is to exploit sparse computing resources (idle CPU cycles) and high performance can be obtained by using a large number of standard machines. *XtremWeb* [18], *SETI@Home* [8] and *ProActive* [1] are some examples of Peer-to-Peer middlewares dedicated to distributed computing. In this chapter we are interested in ProActive. It is a Java library which proposes an API, a graphical interface and parallel, distributed and concurrent programming tools [1]. A distributed application built with ProActive is composed of active objects *AO* [15]. An active object is a remote object having its own thread and receives calls on its public methods. Each *AO* has its own activity and the capability to decide in which order it will serve the method calls. The AOs are created on a support called Virtual Nodes VN. Association between a JVM and a VN is made by an XML deployment descriptor. ProActive is a SPMD (*Single Program Multiple Data*) middleware where a great number of interconnected nodes execute the same application operating on several distributed data. As the majority of P2P middlewares, designed to a distributed computing, the ProActive motivation is to use idle CPU cycles. A P2P network formed by ProActive, is a set of dynamic JVMs or VN which operates as a network of computing nodes. The concept of resource in ProActive is reduced to JVMs. Each JVM which wants to take part in calculation, launches a P2P Service which is a "*daemon*" executing on each node [16].

With ProActive, communications are done by remote method calls between AOs. It includes three types of communications: (1) *Synchronous calls:* the method call is blocking, the execution is suspended until the arrival of the called method result; (2) *Asynchronous calls:* calls are not blocking and the execution of the program can continue without waiting for the result. An appointment ensures that the request arrives well at the called before continuing the activity. A future object is created waiting for any result. A future object represents the result of one of method call of this object which did not arrive yet; (3) *Single direction calls:* calls are not blocking (the appointment is always present), no result is awaited and no future is created.

The groups of communication[9] are another power tool provided in ProActive for distributed programming. A group of communication is the local representant of a set of objects distributed on interconnected machines. When a method is called upon a group, the execution environment sends an invocation request of the method on the group members, awaits one or more answers of the members according to the defined policy, and returns back the result to the caller. For more details on ProActive middleware see [9]).

11.4 Parallel B&B and GA for P2P Environment

11.4.1 ParallelBB

The Parallel B&B algorithm *"ParallelBB"* we developed is a high level parallelization algorithm, and belongs to *type 2* of the Gendron and Crainic classification. ParallelBB is developed with the *(Master/Worker)* paradigm with direct communication worker/worker and worker/master, to avoid the bottlenecks created on the master process. The master divides the initial problem into a set of subproblems *(tasks)*. Indeed, it builds reduced, independent and fine grained subproblems which can be treated in parallel by mono-processors. A single work pool is available on the master process which distributes the tasks among the workers. After this blocks waiting for the results of each one of them. In the following, we present principal operations of parallelBB.

Branching

The branching operation is performed serially by the master process. The master prepares an initial tree (Fig. 11.1) with a depth equal to K. Let n be the number tasks explored by the workers, N the initial size of the problem: $n \leq \prod_{0 \leq i \leq k} (N - i)$.

$T_1, T_2 \ldots T_n$, in the figure represent subtrees, each one contains a partial solution having a size equal to the current level in the tree. K is a parameter of ParallelBB which depends on two important factors: initially, it depends on the size of the considered problem, for example, the number of jobs in the case of a permutation Flow Shop Problem. The depth of the tree increases with the increasing of the number of jobs. Therefore, K must be sufficiently great to generate a large number of subtrees which will be treated in parallel by the workers. Thus, we allow to generate subproblems of a reasonable granularity to be performed by each worker. K also depends on (the size of the computing network).

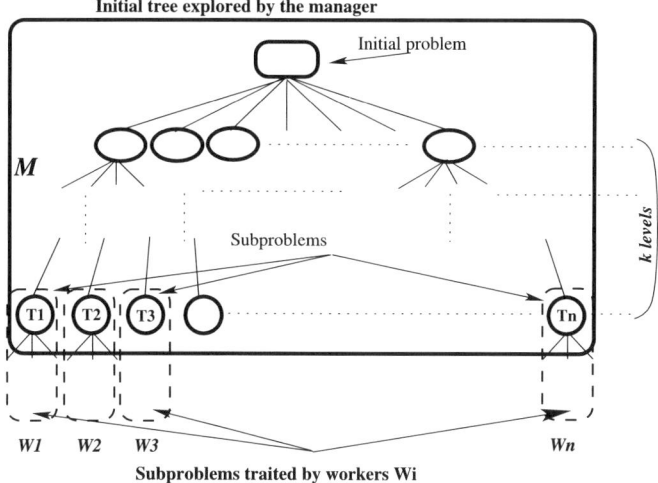

Fig. 11.1. General scheme of ParallelBB

In our case, K depends on the number of workers available in the network. If there is a reduced number of peers in the network, it is more interesting to have a reduced number of parallel tasks. Otherwise, we will loose more time in the communication between the workers and the distribution of the tasks. Among the roles of the master, the attribution of tasks (subtrees) to the workers. If the number of workers is greater than the number of tasks, the master considers only the workers which it needs. On the contrary, the master will make a redistribution of tasks to each new available[1] worker.

Selection and Elimination

The elimination operation is used only to eliminate subtrees having a lower bound greater than or equal to the upper bound. The policy of the tree exploration used by the master and the workers is different. The master explores the initial tree in width to build subtrees which will be explored in parallel. The master explores nodes by priority to the most promising nodes i.e. nodes having a lower bound less than or equal to the upper bound found until now, by all other workers. These subtrees are qtored in a priority-based queue. The workers explore their subtrees in depth and use *Best First Search* policy. They use a stack with opposite priority stacking of the subtrees nodes according to least promising ,i.e. at the top of the stack we find the most promising nodes. Thus, the workers start initially with promising nodes.

---

[1] A worker is said available if it accomplished its task or it finished the calculation which was assigned to it.

Communication and knowledge sharing

The global knowledge (all processes knowledge) related to the upper bound is increased and updated each time a given worker finds a new upper bound. This operation is performed by broadcasting the upper bound to all workers. The collaborative work between the workers, using the communication of the upper bound, allows us to gain much in computation time. Several branches can be eliminated, more quickly than in a traditional B&B (sequential B&B) before their exploration, quite simply by consulting the solution found so far. Unlike traditional B&B, where this same upper bound is known only when the exploration process reaches into the current node. By using this algorithm, a significant number of branches can be eliminated. These branches can't be cut in a sequential B&B because the upper bound making it possible can be found only in the future. This solution is situated in a search space which will be explored only later.

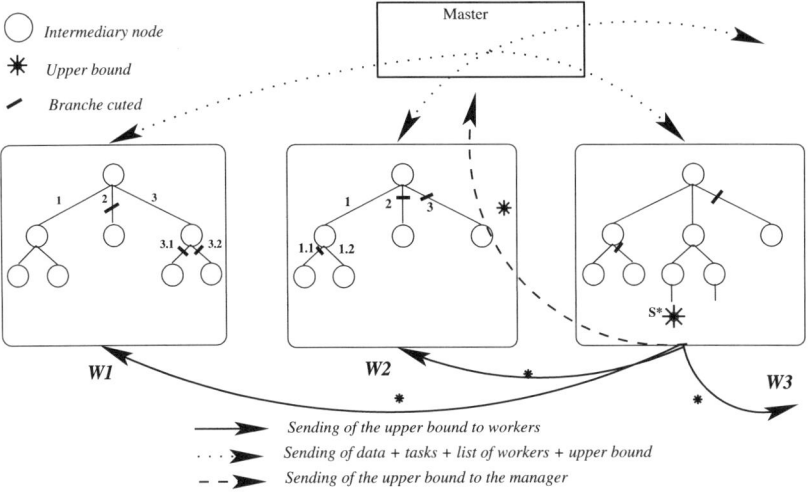

Fig. 11.2. Communications between processes of ParallelBB

In Fig. 11.2, the upper bound (solution S^*) was found by the worker W_3. This solution is in a future search space[2] compared to the search spaces of W_1 and W_2. When W_3 sends the upper bound S^* to W_1 and W_2, it allows then to eliminate the branches: (2 and 3.1) in the subtree of W_1 and (1.1, 2 and 3) in the subtree of W_2.

The master increases also the workers knowledge concerning all other workers executing in the system (dynamic management). The various types of communication can be summarized as follows (see Fig. 11.2) :

---

[2] In a serial execution, S^* will not be found before exploring all subtrees belonging to the search space of W_1 and W_2.

- *The Master to a Worker:* (1) Sending of the task to perform (data of the problem and the subtree to explore). (2) Sending of the pool of executing workers. This knowledge allows each worker to know its environment concerning other workers in progress for a collaborative work. (3) Initialization of the global upper bound. This knowledge allows each worker to eliminate branches from the beginning in its search space.
- *A Worker to the Master:* (1) Sending of the final solution obtained by the worker (if the worker finishes the exploration of the subtree). (2) Sending of the upper bound which is better than the global current knowledge of the algorithm. this allows the master to improve the knowledge of the future workers with this upper bound.
- *A Worker to a Worker:* Each worker sends the upper bound to all the workers in its communication window (workers in progress) so that these workers will be able to reduce the search space by eliminating a great number of branches. The communication window of a worker is reduced to its neighbors i.e. a worker communicates only with the workers in progress (its neighbors).

11.4.2 ParallelGA

In this section, we present briefly the parallel genetic algorithm we developed *ParallelGA*. As ParallelBB, ParallelGA is a master/worker-based algorithm with direct communications between workers. The master process divides the initial population into subpopulations with a reduced size. The exploration of these subpopulations will be considered as parallel tasks and can be handled by a single processor. All sub-populations will evolve in parallel by the available workers, each worker executes its instance of the algorithm. The master redistributes not handled subpopulation each time a worker terminates its part of calculation. The size of initial generated population must be sufficiently great to increase the search space and then increase chances to reach acceptable solution. The master fixes the size of sub-populations according to the number of available workers and the power of processors.

Like in ParallelBB, communications are very important in the case of a ParallelGA. The workers communicate their best individuals to their neighbors (see Fig. 11.3). This migration of individuals allows us to prevent convergence and to prevent workers to turn in locals minimum. After a fixed number of generations (or after a fixed time interval), the migration of elite individuals occurs.

The different types of communications of ParallelGA can be summarized as follows:

- *Master to Workers:* The master communicates the task to perform (data of the problem, subpopulation and different parameters of the GA) as well as the pool of executing workers.
- *A worker to the master:* the only information that a worker sends to the master is the final result when it terminates its part of computation.
- *a worker to a worker:* The communication between workers consists in the exchange of individuals (see Fig. 11.3).

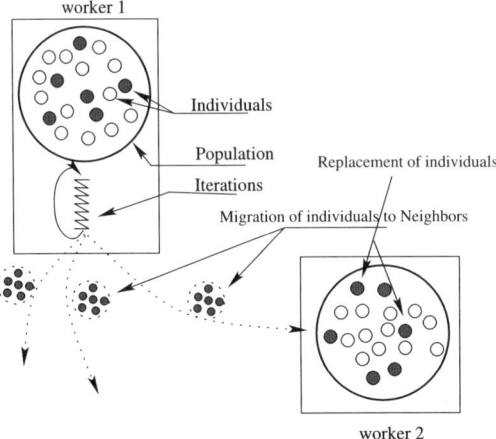

Fig. 11.3. Migration of individuals

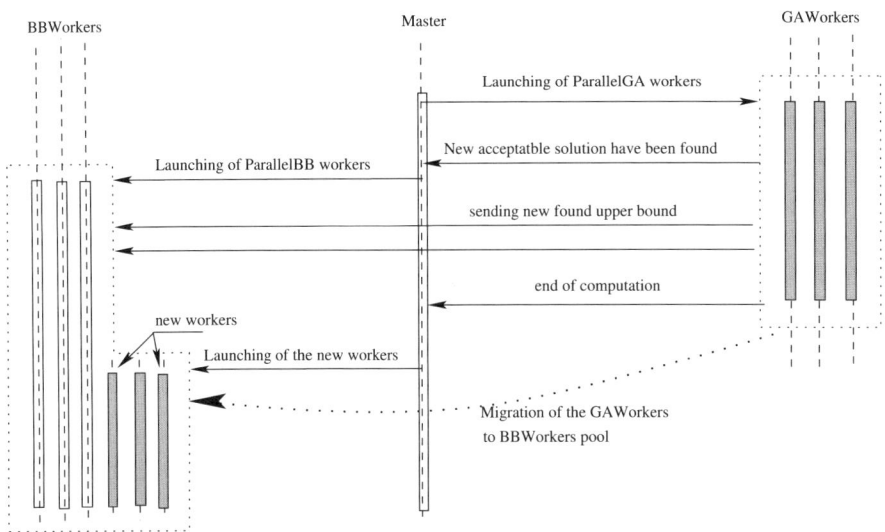

Fig. 11.4. Behavior of the different workers

11.4.3 Hybridization

A quick execution of an exact algorithm like *ParallelBB* needs to start computation with a near-optimal upper bound that we can obtain by *ParallelGA*. As shown in (Fig. 11.4), we launch first *ParallelGA* to obtain an acceptable initial value of the upper bound which is passed to *ParallelBB*. We use two types of workers: workers participating in the exploration of the B&B search space

BBWorkers and those participating in the exploration of the GA population *GAWorkers*. *ParallelBB* starts computation before the termination of *ParallelGA* and run in parallel. *BBWorkers* receive the value of the upper bounds from *GAWorkers* each time new ones have been found. At the end of exploration of all *ParallelGA*'s subpopulations, the concerned *GAWorkers* migrate and join the exploration of *ParallelBB* search tree process, thus they take a *BBWorkers* behavior.

11.5 Peer-to-Peer Implementation on Top of ProActive

The implementation of *ParallelBB* and *ParallelGA* on *ProActive* gave rise to our Active Application[3] *PHyGABaB*. This application is based on two entities (active objects *AOs*): *P2PWorker* and *P2PMaster*. ProActive provides active nodes *ANs* (JVMs), recovered on the whole of the network, which are ready to receive calculation. These ANs are *P2PWorkers* receiving tasks (subtrees or subpopulation to explore). In the case of a static grid managed by ProActive, the P2P services *P2PService* are already launched, i.e., each host shares its JVM and at least one *P2PWorker* which can receive AOs. An *AN* can receive one or more *AOs*. When the *P2PMaster* is created on the local JVM, it consults an XML deployment descriptor where *P2PMaster* will find ANs. At the end of this stage, *P2PMaster* will have a list of *P2PWorkers* ready to receive calculation. When such nodes are ready, they can directly receive computational work coming from the *P2PMaster*, new active objects will be launched otherwise. In this case, the XML descriptor must be modified so that it can deal with the dynamic nature of a P2P network and that receive new peers which arrive into the initial network (see Section. 11.5.3 for the handling of new arrivals).

11.5.1 Distribution of the Computation among Workers

After initializing the workers, *P2PMaster* generates a set of independent tasks (subtrees and/or subpopulations). These tasks are represented by passive objects (see Fig. 11.5). Before the *P2PMaster* sends a task to a *P2PWorker*, it increases the knowledge of each *P2PWorker* concerning its environment. This knowledge concerns the set of workers executing other tasks, the best solution found so far (when the worker participates in the branch and bound tree exploration) or only the initial subpopulation (when the worker participates in the exploration of the GA population).

Each time a task is assigned to a *P2PWorker*, a future object is created and added to the future list (*futureList*)of *P2PMaster*. *P2PMaster* waits all the future objects coming from the *P2PWorkers* appearing in its list. The future *P2PWorker* represents the result of the calculation of its task that the *P2PMaster* assigned to it. This is accomplished by listening its response (the

---

[3] An Active Application is an application based on active objects. Any application developed using the ProActive middleware must be Active so that it can be deployed on the Computing Network.

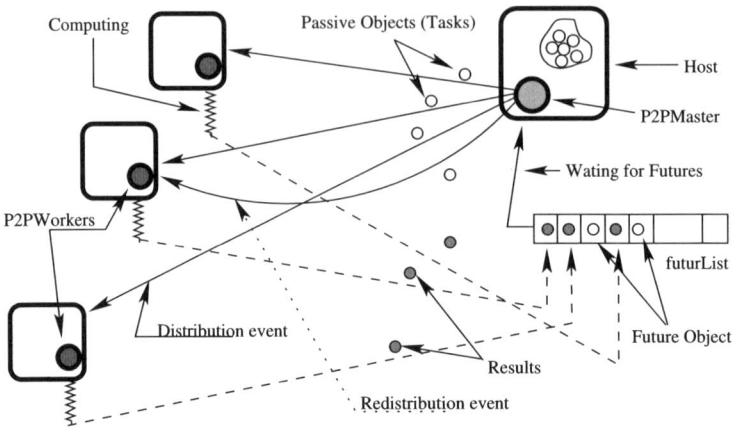

Fig. 11.5. Tasks distribution on workers

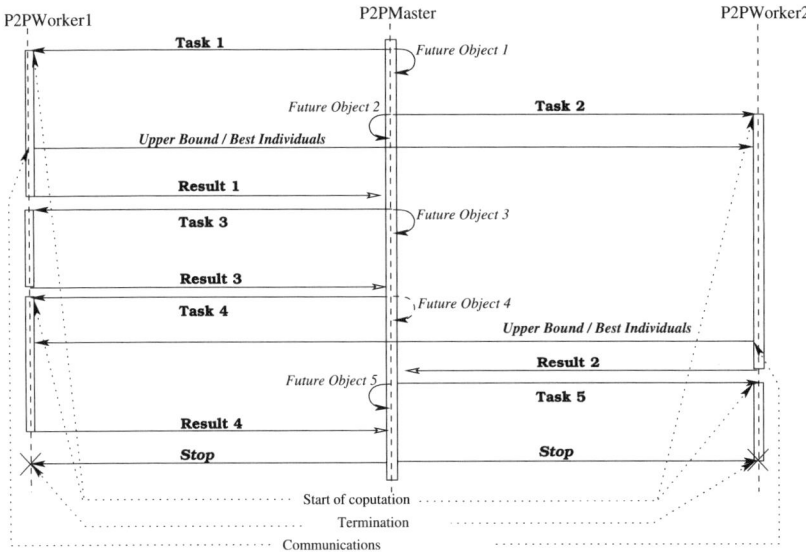

Fig. 11.6. Sequence diagram of tasks distribution

computation result) in the future. The listening is of type *wait for any event* made by the method *waitForAny(futureList)* and is accomplished by waiting for any event coming from *P2PWorkers* appearing in the list *futureList*. The event is started with each termination of a task treatment. After that the *P2PMaster* recovers the result produced by the future *P2PWorker*, it creates and reallocates a new passive object to it.

The chronology of events presented in (Fig. 11.5) is not really true because all operations are in asynchronous mode i.e. an event of type redistribution can arrive before other events of type distribution when one of *P2PWorkers* returns back result before *P2PMaster* finishes distributing of all tasks. We can see well on (Fig. 11.6) the sequence diagram of the chronology of all events of the tasks distribution.

11.5.2 Communications

Communication between different components (*P2PMaster* and *P2PWorkers*) is very important for its good functioning. We have seen previously that the *workers* communicate frequently to ensure the freshness of the upper bound and the migration of individuals. The use of classic communication between *P2PWorkers*, i.e., by sending one message for each *P2PWorker* is not efficient in this type of application where the communication cost is very high. If we use the classical communication we will need to broadcast the same message to all *P2PWorkers* in our computing pool. This procedure requires the traversal of the whole *P2PWorkers* list (thousands or millions), in other words, each *P2PWorker* must have one copy of all *P2PWorkers* taking part in the computation. This solution is not interesting because of huge amount of time required by the traversal of the entire list and the memory space necessary to store requiring when saving the set of *P2PWorkers* while this space is to be minimized.

With ProActive, we opted for the communication groups with single direction, non blocking and asynchronous methods invocation. We created *BBWorkerGroup* and *GAWorkerGroup* which are the two local representants of a set of *P2PWorker* recipients of a message. *BBWorkerGroup* represents all *P2PWorkers* participating in ParallelBB computation and *GAWorkerGroup* represents *P2Pworkers* participating in the exploration of subpopulations in the ParallelGA. When a *P2PWorker* wants to send a message to its colleagues, it passes by these two groups, which implement the same communication method which is even implemented on all the *P2PWorkers*. We developed two communication methods: *shareBestValue* and *shareSubPopulation* allowing the workers to share respectively the best value of the current solution (upper bound) and their selected individuals (elite). A *P2PWorker* calls these two communication methods in order to share their upper bound or subpopulation. Thus *BBWorkerGroup* and *GAWorkerGroup* call this same method on the set of *P2PWorkers* they represent.

11.5.3 New Arrivals (New Peers)

The dynamic availability of peers is one of the P2P networks characteristics where the resources (here JVMs) join and leave frequently the system. Each JVM is at the same time client and server for other JVMs. The ProActive middleware manages the new coming peers in the system by a listener implemented with the method (*nodeCreated*). *P2PMaster* implements this interface which listens for eventually peers connections in the network. A P2P daemon is launched on each machine participating in the computation. When a new node is connected, an AN is created there and one *P2PWorker* is established to receive the computation units. *P2PMaster* decides the affiliation of this new peer, indeed, it will behave as a *BBWorker* or as *GAWorker*. *P2PMaster* adds this new *P2PWorker* to the *P2PWorkers* set list and to the corresponding group of communication (*BBWorkerGroup* or *GAWorkerGroup*). After that, if this new *P2PWorker* belongs to *BBWorkerGroup* it could be informed of the global upper bound. Other *P2PWorkers* will be able to have an idea on the progress and the solutions obtained by this new *P2PWorker*. Whatever the affiliation of this new peer, the *P2PMaster* sends to it its task and will behave as other *P2PWorkers*. A new peer, arriving at the computational network, adheres to the group of communication. The peers forming the old group of communication update their group by adding this new peer. This operation is managed by *P2PMaster* which sends to all the group of communication members the new configuration of the group, i.e., the adding of this new peer. This operation allows new *BBWorkerGroup* members to avoid the exploration of subtrees unnecessarily, this allows reduce execution time. To obtain the real global upper bound, the new peer selects randomly a peer and then sends its initial upper bound. If the upper bound of this selected peer is inferior to the received one, it proceeds to its correction by broadcasting the real global upper bound.

11.5.4 Fault Tolerance

The peers failure is taken into account by both ProActive (middleware-level) and our application (application-level). With ProActive we create tow types of servers: *Resource server* and *Fault tolerance servers*.

The *resource server* returns a free node that can host the recovered AO, this server can store free nodes by two different ways:

- At deployment time: the user can specify in the deployment descriptor a resource virtual node. Each node mapped on this virtual node will automatically register itself as free node at the specified resource server.
- At execution time: the resource server can use an underlying P2P network (see [1]) to reclaim free nodes when a hosting node is needed.

The *fault tolerance servers* are used for checkpointing operations, the localization of AOs, and the failure detection.

In our application, when a peer disconnects, the *P2PMaster* sends its part of calculation to one or more other available peer(s) and recover(s) only the first returned solution of the same task and ignores other results representing the same task. This process is performed at the end of the computation of all tasks.

11.6 Large Scale Deployment and Performance Evaluation

In the following, we present the different experiments and obtained results of the exact algorithm *ParallelBB* hybridized with the heuristic *ParallelGA* applied to the permutation flow-shop problem (PFSP) which is a reference problem in the given its importance in many industrial areas.

11.6.1 PFSP Formulation

A Permutation Flow-Shop Problem (PFSP) is a scheduling in which all tasks of all jobs are scheduled on all machines in the same order. The execution of a job J_i on the machine M_k is called operation $O_{i,k}$ and its execution time will be noted $p_{i,k}$, $t_{i,k}$ represents its starting date and $c_{i,k}$ its release time. We designate also by $r_{i,k} = \sum_{l<k} p_{i,l}$ the early instant in which the job J_i can start its operation on M_k and by $q_{i,k} = \sum_{l>k} p_{i,l}$ the latency duration (minimal time selling between the end of J_i on M_k and the end of the total scheduling).

11.6.2 Modeling and Lower Bound Calculation

We applied our algorithm to the PFSP and we considered the total completion time *Makespan* (C_{Max}) cost function. In ParallelGA, an individual (permutation) is considered as a vector of jobs. The root of the tree generated by ParallelBB represents a configuration where no task is assigned to any machine. A node with depth n will have a configuration with n assigned tasks.

The effectiveness of B&B algorithms resides in the use of a good estimation of the optimal solution. M. R. Garey, D. S. Johnson and R. Sethi (Garey and al., 1976) proved that the PFSP problem becomes NP-hard beyond 3 machines. The calculation of the lower bound for a PFS problem is based on two results. The first one is found by Johnson [27] (*rule of Johnson*). A transitive rule $\preceq$ is defined as follows:

$$J_i \preceq J_j \Leftrightarrow min(p_{i,1}; p_{j,2}) \leq min(p_{i,2}; p_{j,1}) \tag{11.1}$$

If $J_i \preceq J_j$, then there exists an optimal scheduling for a FSP (P) in which the job J_i precedes the job J_j[27]. Thus, the PFS problem with two machines $F2||C_{max}$ can be solved in $O(nlogn)$ [31]. The optimal solution is obtained by sorting the jobs having the execution times on the first machine shorter than the second in the ascending order. Then, sort jobs having their execution time on the second machine shorter than on the first one in the descending order. This result was extended by Jackson [28] and Mitten [29] for the resolution of a two machines PFS problem with lags $F2|l_j, permut|C_{max}$ where each job has a *lag* l_j which represents the minimal duration between $t_{j,2}$ and $c_{j,1}$. They demonstrated that the optimal solution of this problem is obtained using the Johnson to the values $p_{i,1} + l_i$ for the jobs on the 1st machine and $l_i + p_{i,2}$ on the 2nd one.

$$J_i \preceq J_j \Leftrightarrow min(p_{i,1} + l_i; l_j + p_{j,2}) \leq min(l_i + p_{i,2}; p_{j,1} + l_j) \tag{11.2}$$

The most known lower bound for the PFS problem with m machines is the bound proposed by Lageweg et al. [26]. They used the optimal solutions values for all 2 machines subproblems with lags. Given two machines M_k and M_l (with c $k < l$), it is indeed possible to extract such problem posing:

$$p_{j,1} = p_{j,k}; l_j = \sum_{k<\mu<l} p_{j,\mu}; p_{j,2} = p_{j,l} \qquad (11.3)$$

In practice, we consider all couples of machines M_k et M_l (with $k < l$) and we extract for each couple a PFS with two machines lags substituting the values $p_{i,1}$ by $p_{i,1} + l_i$ and $p_{i,2}$ by $l_i + p_{i,2}$. We notate $P^*_{Ja}(\jmath, M_k, M_l)$ the Jackson-Mitten optimal solution of the obtained subproblem considering the set of jobs $\jmath$ and machines M_k and M_l. B.J. Lageweg et al obtained thus the lower bound (with $O(m^2 n log n)$ complexity) which we used in our work :

$$LB(\jmath) = \max_{1 \leq k < l \leq m} \{ P^*_{Ja}(\jmath, M_k, M_l) + \min_{(i,j) \in \jmath^2, i \neq j} (r_{i,k} + q_{j,l}) \} \qquad (11.4)$$

11.6.3 Experiments

The studied problem instances are those of E. Taillard [37]. We treated the benchmarks: $ta\_20\_5\_2$, $ta\_20\_5\_3$, $ta\_20\_10\_1$, $ta\_20\_10\_2$ and $ta\_100\_5\_1$[4]. Parameters of *ParallelGA* are fixed as follows: 500 individuals in the population, the size of each subpopulation is fixed to 20, migrations occur every 10 generations with 10 migrants.

Table 11.1. Experimentation hardware platform

Site	CPU characteristic x number of CPU / node	Number of nodes	Number CPUs
Lille	AMD Opteron 248, 2.2 GHz x 2	70	140
Lyon	AMD Opteron 246, 2.0 GHz x 2	55	110
Nancy	AMD Opteron 246, 2.0 GHz x 2	35	70
Orsay	AMD Opteron 246, 2.0 GHz x 2	290	580
Rennes	Intel Xeon 5148 LV, 2.33 GHz x 2	60	120
	AMD Opteron 246, 2.0 GHz x 2	90	180
	AMD Opteron 248, 2.2 GHz x 2	50	100
Nice	AMD Opteron 246, 2.0 GHz x 2	55	110
	AMD Opteron 275, 2.2 GHz x 2	50	100
Total		775	**1510**

We made large scale deployment of the application (more than 1500 processors) gathered on six geographically distributed sites located at (*Lille, Rennes, Orsay, Nice, Lyon and Nancy*) belonging to the French grid GRID'5000 [2]. The experimentation hardware platform characteristics are presented in Table 11.1. As we said previously, our objective is the development of hybrid algorithm

---
[4] $ta\_i\_j\_k$: a Taillard benchmark with i: number of jobs, j: number of machines and k: the instance number.

Table 11.2. Some obtained execution times

Number of Processors	Deployment Time	$ta\_20\_5\_2$	$ta\_20\_5\_3$	$ta\_20\_10\_1$	$ta\_20\_10\_2$	$ta\_100\_5\_1$
06 (1/5)	15	(3) 4	(492) 401	(1815) 1287	(277) 234	-
20 (5/15)	46	(10) 8	(409) 391	(170) 129	(111) 98	-
50 (15/35)	112	(16) 11	(277) 263	(100) 97	(59) 51	-
100(20/80)	234	17	193	81	50	-
200(40/160)	504	-	151	77	-	-
300(60/240)	713	-	152	-	-	7h [5572]
600(100/400)	1949	-	-	-	-	6h 57min [5571]
1500(300/1200)	4186	-	-	-	-	6h 57min [5571]

to solve exactly complex instances of benchmarks with large scale deployment. Here, we made this preliminary deployment just to show that the application is scalable and can be used in this sense.

We made other deployments of our application on 6, 20, 100, 200 and 300 processors on GRID'5000. In this experiments a portion of workers are assigned to *ParallelGA* computation and the rest to *ParallelBB*. As shown in Section 11.4.3, when *GAWorkers* terminate their calculation parts they join *BBWorkers*, so after they terminate, *ParallelBB* exploits the whole pool of workers. The application was launched in P2P mode where all processes run with the lowest priority to reach one of P2P Computing characteristic which is the exploitation of idle CPU cycles. Table 11.2 shows the execution times obtained by our hybrid application compared to an older version of P2P non hybrid parallel branch and bound algorithm [10]. The old times are presented in the table in parentesis. In the first column we have the number of used processors (i/j): i number of *GAWorkers* and j number of *BBWorkers*.

The first point we notice is that the hybridization improves efficiency. Comparing with the non hybrid method, practically, all benchmarks are solved more efficiently when using hybridization. For example, the exact resolution of the benchmark $ta\_20\_10\_01$ on 20 machines took 129 seconds using the hybrid algorithm whereas it was solved in 170 using non hybrid version. This same benchmark was solved in 97 seconds whereas it took 100 sec. The only exception is when solving $ta\_20\_05\_02$ it was solved three times more quickly on 6 machines than on 20 and 5 times more efficient more than on 50. This can be explained if we take a look on the situation of the solution regarding the space of solutions. It was found in the 3rd node of the solutions tree, this means that the six machines was sufficient to find the solution in short time, and the 50 machines take an additional time to manage tasks and free all workers deployed.

In the second column we have deployment times (deployment time includes, detection and handling of nodes and distribution of tasks). For small instances of benchmarks, we don't need a large scale deployment, $ta_{\_20\_5\_2}$ was exactly solved on 50 machines in 11 seconds whereas the deployment time is 112 seconds. This is not the case for large instances where the deployment time is insignificant compared to time of resolution. Though, instance $ta_{100\_5\_1}$ wasn't exactly solved, but we remark easily that this time is negligible. The calculation of these two instances wasn't terminated, values in the table represent the time of calculation and reached upper bound between square brackets.

11.7 Conclusions

Using exact methods for the resolution of COPs, such as B&B which is the most used for an exact resolution of these problems, is very important. However, their use on applications of industrial size is possible only by the use of a great computational power. Hybridization and large scale parallelism based on the use of Grid Computing or P2P Computing proves today a potential tool which offers such power. Several factors have to be taken into account for a better parallelization of these methods, for their implementations on a Peer-to-Peer systems such as ProActive and for a better exploitation of the computing power. (1) A study and a good choice of a suitable model of parallelism; (2) A good management of the knowledge generated by these algorithms; (3) Exploitation of all the tools that P2P middlewares offers for controlling of the computational network.

In this chapter, we developed a parallel branch-and-bound algorithm hybridized with a parallel genetic algorithm for resolution of COPs on a Peer-to-Peer system. We applied it to the Permutation Flow-Shop problem which is a reference problem in this area. We chose a high level parallelism of branch-and-bound and a coarse grained parallel genetic algorithm. In this direction, we developed *ParallelBB*, *ParallelGA* and a hybrid version based on *master/worker* paradigm which is a most appropriate technique for the development of scientific applications dedicated to an intensive computing on large scale systems. The performances of the algorithm were improved with the knowledge sharing between the workers. This was realized by the use of the *master/worker* paradigm with direct communications between workers.

We implemented the peer-to-peer version of our algorithms on top of ProActive and we took benefits from the maximum of its functionalities. We took advantage of the communication groups and the asynchronous methods invocation in single direction for the knowledge sharing between workers and master. We used the listeners and daemons in order to take into account the new arrivals, their detection and the management of their connections. Finally, we used the future active objects for collecting of computation results. The experiments made on a P2P network managed by ProActive showed the interest of collaborative work between nodes of the computation network as well as the importance of hybridization. We have shown the ability of our application to support scalability and dynamic availability of peers in the network.

As a perspective, we project to extend our work to other type of COPs (Quadratic Assignment Problems QAP and Quadratic three dimensional Assignment Problems Q3AP). In addition, we plan to improve the performances of ParallelBB with: (1) The load balancing of the tasks generated by the algorithm so that they become more equitable; (2) The production of several forms of granularity of tasks and their distribution to the corresponding station (Calculator, PC, laptop...).

References

1. http://proactive.objectweb.org
2. http://www.grid5000.fr
3. Adamidis, P.: Review of parallel genetic algorithms bibliography. Technical report, Aristotle University of Thessaloniki, Thessaloniki, Greece (1994)
4. Aida, K., Futakata, Y., Hara, S.: High-performance parallel and distributed computing for the bmi eigenvalue problem. In: Proc. the 16th IEEE International Parallel and Distributed Processing Symposium, pp. 71–78 (2002)
5. Aida, K., Natsume, W., Futakata, Y.: Distributed Computing with Hierarchical Mast-Worker Paradigm for Parallel Branch-and-Bound Algorithm. In: IEEE/ACM International Symposium on Cluster Computing and the Grid (CCGrid 2003), p. 156 (2003)
6. Aida, K., Osumi, T.: A case Study in Running a Parallel Branch and Bound Application on the Grid. In: IEEE, Symposium on Applications and the Internet (SAINT 2005), January 2005, pp. 164–173 (2005)
7. Alba, E., Luque, G., Talbi, E.-G., Melab, N.: Meta-heuristics and parallelism. In: Parallel Meta-heuristics, pp. 79–104. John Wiley & Sons, Chichester (2005)
8. Anderson, D.P., Cobb, J., Corpela, E., Lepofsky, M., Werthimer, D.: SETI@Home: An Experiment in Public-Resource Computing. Communications of the ACM 45(11), 56–61 (2002)
9. Baduel, L., Baude, F., Caromel, D., Contes, A., Huet, F., Morel, M., Quilici, R.: Grid Computing: Software Environments and Tools. In: Programming, Deploying, Composing, for the Grid. Springer, Heidelberg (2006)
10. Bendjoudi, A., Melab, N., Talbi, E.-G.: A Parallel P2P Branch-and-Bound Algorithm for Computational Grids. In: Seventh International Workshop on Global and Peer-to-Peer Computing. IEEE/ACM International Symposium on Cluster Computing and the Grid (IEEE/ACM CCGRID), May 14-17, pp. 749–754 (2007)
11. Gendron, B., Crainic, T.G.: Parallel Branch-and-Bound Algorithms: Survey and Synthesis. Operations Research 42(06), 1042–1066 (1994)
12. Bolze, R., Cappello, F., Caron, E., Daydé, M., Desprez, F., Jeannot, E., Jégou, Y., Lantéri, S., Leduc, J., Melab, N., Mornet, G., Namyst, R., Primet, P., Quetier, B., Richard, O., Talbi, E.-G., Touche, I.: Grid 5000: a large scale and highly reconfigurable experimental Grid testbed. Intl. Journal of High Performance Computing Applications 20(4), 481–494 (2004)
13. Bourbeaua, B., Crainica, T.G., Gendron, B.: Branch-and-bound parallelization strategies applied to a depot location and container fleet management problem. Parallel Computing (26), 27–46 (2000)
14. Cahon, S., Melab, N., Talbi, E.-G.: ParadisEO: A Framework for the reusable Design of Parallel and Distributed Meta-heuristics. Journal of Heuristics 10(3), 357–380 (2004)

15. Caromel, D., Klauser, W., Vayssière, J.: Towards seamless computing and metacomputing in Java. In: Fox, G.C. (ed.) Concurrency Practice and Experience, vol. 10, pp. 1034–1061. Wiley & Sons, Ltd., Chichester (1998)
16. Mathieu, C., Caromel, D., di-Costanzo, A.: Peer-to-peer for computational grids: Mixing clusters and desktop machines. Parallel Computing Journal on Large Scale Grid (to appear, 2007)
17. Davis, L.: Handbook of Genetic Algorithms. Van Nostrand Reinhold, New York (1991) BU: 511.6 HAN
18. Fedak, G., Germain, C., Néri, V., Cappello, F.: XtremWeb: A Generic Global Computing System. In: Proceedings of the first International Symposium on Cluster Computing and the Grid (CCGRID 2001), p. 582. IEEE, Los Alamitos (2001)
19. Foster, I., Kesselman, C.: The Grid: Blueprint for a New Computing Infrastructure. Morgan Kaufmann, San Francisco (1999)
20. Garey, M.R., Johnson, D.S.: Computers and Intractability: A Guide to the Theory of NP-Completeness. W.H. Freeman & Co., New York (1979)
21. Goodman, E.D.: An Introduction to GALOPPS v3.2. Technical report, 96-07-01, CARAGE, I.S. Lab. Dpt. of C.S and C.C.C.A.E.M., Michigan State University, East Lansing, MI (1996)
22. Gordon, V.S., Whitley, D.: Serial and Parallel Genetic Algorithms as Function optimizers. In: Forrest, S. (ed.) Proceedings of the Fifth International Conference on Gentic Algorithms, pp. 177–183. Morgan Kaufmann, San Francisco (1993)
23. Goux, J., Kulkami, S., Linderoth, J., Yoder, M.: An enabling framework for masterworker applications on the computational grid. In: IEEE Symposium and High Performance Distributed Computing (HPDC 2000), vol. 9, p. 43 (August 2000)
24. Herrera, F., Lozano, M.: Gradual distributed real-codeed genetic algorithm. IEEE Transaction in Evolutionary computation 4, 43–63 (2000)
25. Holland, J.H.: Adapation in natural and artificial systems. The university of Michigan Press, Ann Arbor (1975)
26. Lenstra, J.K., Lageweg, B.J., Rinnooy Kan, A.H.G.: A General boundind scheme for the permutation flow-shop problem. Operations Research 26(1), 53–67 (1978)
27. Johnson, S.M.: Optimal two and three-stage production schedules with setup times included. Naval Research Logistis Quarterly 1, 61–68 (1954)
28. Jackson, J.R.: An Extension of Johnson's results on Job-Lot Scheduling. aval Research Logistis Quarterly 3(3) (1956)
29. Mitten, L.G.: Sequencing n jobs on two machines with arbitrary time lags. Management Science (1959)
30. Lin, S.-C., Punch, W., Goodman, E.: Coarse-Grain Parallel Genetic Algorithms: Categorization and New Approach. In: IEEE Symposium on Parallel and distributed Processing, pp. 28–37. IEEE computer Society Press, Los Alamitos (1994)
31. Péridy, L., Pinson, E., Rivreau, D.: Elimination Rules for the Flow-Shop and Permutation Flow-Shop Problems. Technical report, Institut de Mathématiques Appliquées, Université Catholique de l'Ouest (May 27, 2002)
32. Mejia-Olvera, M., Cantu-Paz, E.: DGENESIS-software for the execution of destributed genetic algorithms. In: Proceedings XX Conference Latinoamerica de Informatica, pp. 935–946 (1994)
33. Melab, N., Talbi, E.-G., Mezmaz, M., Wei, B.: Parallel Hybrid Multi-objective Meta-heuristics on P2P Systems. In: Olaru, S., Zomaya, A.Y. (eds.) Handbook of Bioinspired Computing, pp. 649–663. CRC Press, Boca Raton (2006)

34. Milojicic, D., Kalogeraki, V., Lukose, R., Nagaraja, K., Pruyne, J., Richard, B., Rollins, S., Xu, Z.: HP Laboratories Palo Alto. Peer-to-Peer Computing. Technical Report HPL-2002-57, HP (March 8, 2002)
35. Mitten, L.G.: Branch-and-bound Methods: General formulation and properties. Operations Research 18, 24–34 (1970)
36. Schollmeier, R.: A definition of Peer-to-Peer networking for the classification of Peer-to-Peer architectures and applications. In: 2001 International Conference on Peer-to-Peer Computing (P2P 2001), Linko pings Universitet, Sweden. IEEE, Los Alamitos (2001)
37. Taillard, E.: Benchmarks for basic scheduling problems. European Journal of Operations Research 64, 278–285 (1993)
38. Tanaka, Y., Sato, M., Hirano, M., Nakada, H., Sekiguchi, S.: Performance evaluation of firewall compliant globus-based wide-area cluster system. In: IEEE Symposium on High-Performace Distributed Computing (HPDC), vol. 9, p. 121 (2000)
39. Trienekens, H.W.J.M.: Parallel Branch and Bound on an MIMD System. Report 8640/A, Econometric Institute, Erasmus University Rotterdam (1989)
40. Trienekens, H.W.J.M., Bruin, A.: Towards a Taxonomy of Parallel Branch and Bound Algorithms. Report EUR-CS-92-01, Dept. of Computer Science, Erasmus University, Rotterdam (1992)
41. Yang, M.K., Das, C.R.: A Parallel Branch-and-Bound Algorithm for MIN-Based Multiprocessors, pp. 222–223. ACM, New York (1991)

12

Peer-to-Peer Neighbor Selection Using Single and Multi-objective Population-Based Meta-heuristics

Hongbo Liu[1,2], Ajith Abraham[3], and Fatos Xhafa[4]

[1] School of Computer Science and Engineering, Dalian Maritime University,
    116026 Dalian, China
[2] Department of Computer, Dalian University of Technology,
    116023 Dalian, China
    lhb@dlut.edu.cn
[3] Centre for Quantifiable Quality of Service in Communication Systems,
    Norwegian University of Science and Technology, NO-7491 Trondheim, Norway
    ajith.abraham@ieee.org
    http://www.softcomputing.net
[4] Dept. de Llenguatges i Sistemes Informàtics
    Universitat Politècnica de Catalunya
    C/Jordi Girona 1-3, 08034 Barcelona, Spain
    fatos@lsi.upc.edu

Summary. Peer-to-peer (P2P) topology has significant influence on the performance, search efficiency and functionality, and scalability of the application. In this Chapter, we introduce the problem of neighbor selection in peer-to-peer networks using two population based meta-heuristics: Particle Swarm Optimization (PSO) algorithms and Genetic Algorithms (GAs). Both a single objective and a multi-objective problem are formulated, and then the P2P neighbor selection problem is defined. We present the neighbor selection strategy based on PSO and GA algorithm. Each particle encodes the upper half of the peer-connection matrix through the undirected graph, which reduces the search space dimension. We also discuss the characteristics of ergodicity during particle swarm searching process. We also illustrate the algorithm performance and trace its feasibility and effectiveness with the help of some examples.

Keywords: P2P computing, Neighbor selection, Multi-objective optimization, Population-based meta-heuristics, Genetic Algorithms, Particle Swarm Optimization.

12.1 Introduction

Peer-to-peer computing has attracted great interest and attention of the computing industry and gained popularity among computer users and their networked virtual communities [1]. It is no longer just used for sharing music files over the Internet. Many P2P systems have already been built for some new purposes and are being used. An increasing number of P2P systems are used in corporate networks or for public welfare (e.g. providing processing power to fight cancer) [2].

P2P comprises peers and the connections between these peers. These connections may be directed, may have different weights and are comparable to a graph with nodes and vertices connecting these nodes. Defining how these nodes are connected affects many properties of an architecture that is based on a P2P topology, which significantly influences the performance, search efficiency and functionality, and scalability of a system. A common difficulty in the current P2P systems is caused by the dynamic membership of peer hosts. This results in a constant reorganization of the topology [3, 4, 5, 6, 7].

Kurmanowytsch et al. [8] developed the P2P middleware systems to provide an abstraction between the P2P topology and the applications that are built on top of it. These middleware systems offer higher-level services such as distributed P2P searches and support for direct communication among peers. The systems often provide a pre-defined topology that is suitable for a certain task (e.g., for exchanging files). Koulouris et al. [9] presented a framework and an implementation technique for a flexible management of peer-to-peer overlays. The framework provides means for self-organization to yield an enhanced flexibility in instantiating control architectures in dynamic environments, which is regarded as being essential for P2P services to access, routing, topology forming, and application layer resource management. In these P2P applications, a central tracker decides about which peer becomes a neighbor to which other peers.

Genetic Algorithms (GAs) are adaptive heuristic search algorithm premised on the evolutionary ideas of natural selection. GAs have been widely studied, experimented and applied in many fields in engineering worlds. Finding optimal parameters for many real world problems prove difficult for traditional methods but is suitable for GAs [10]. PSO (PSO) algorithm is inspired by social behavior patterns of organisms that live and interact within large groups. In particular, PSO incorporates swarming behaviors observed in flocks of birds, schools of fish, or swarms of bees, and even human social behavior, from which the Swarm Intelligence(SI) paradigm has emerged [11, 12]. It could be implemented and applied easily to solve various function optimization problems, or the problems that can be transformed to function optimization problems. As an algorithm, the main strength of PSO is its fast convergence, which compares favorably with many global optimization algorithms [13, 14]. In this chapter, we introduce the P2P neighbor-selection problem based GA and PSO for P2P networks.

This chapter is organized as follows. We formulate the problem in Section 12.2. The considered approaches based on GAs and PSO algorithms are presented in Section 12.3. In Section 12.4, experiment results and discussions are provided in detail, followed by some conclusions in Section 12.5.

12.2 Neighbor-Selection Problem in P2P Networks

Based on existing research [15, 16, 17, 18, 19, 20], we formulate the neighbor-selection problem for P2P networks. We introduce first the model of P2P networks, and then discuss metrics for measuring neighbor selection.

12.2.1 Modelling P2P Networks

P2P networks can be modelled by an undirected graph $G = (V, E)$ where the vertex set V represents units such as hosts and routers, and the edge set E represents physical links connecting pairs of communicating units. Further, $f : V \to \{1, \cdots, n\}$ is a labelling of its nodes, where $n = |V|$. For instance, G could model the whole or part of the Internet. Given an undirected graph $G = (V, E)$ modelling an interconnection network, and a subset $X \subseteq V(G)$ of communicating units (peers), we can construct a corresponding weighted graph $D = (V, E)$, where $V(D) = X$, and the weight of each $uv \in E(D)$ is equal to the length of a shortest path between peer u and peer v in G. Usually we start with a physical network G (perhaps representing the Internet), and then choose a set of communicating peers X. The resulting distance graph D is the basis for constructing a P2P graph $H = (V, E)$, which is done as follows. The vertex set $V(H)$ will be the same as $V(D)$, and edge set $E(H) \subseteq D(G)$. The key issue here is how to select $E(H)$. If $E = [e_{ij}]_{n \times n}$ is such that $e_{ij} = 1$ if $(i, j) \in E$, and 0 otherwise, i.e., E is the incidence matrix of G, then the neighbor-selection problem is to find a permutation of rows and columns which brings all non-zero elements of E into the optimal possible interconnection around the diagonal.

12.2.2 Metrics

In P2P networks, specially for file sharing, an interested file is divided into many fragments. The size of each fragment ranges from several hundred kilobytes to several megabytes. When a new peer joins the network, it begins to download fragments from other peers. As long as it obtains one fragment of the file, the new peer can start to serve other peers by uploading fragments. Since peers are downloading and uploading at the same time, when the network becomes large, although the demands increase, the service provided by the network also increases [21]. Given N peers, a graph $G = (V, E)$ can be used to denote a network, where the set of vertices $V = \{v_1, \cdots, v_N\}$ represents the N peers and the set of edges $E = \{e_{ij} \in \{0, 1\}, i, j = 1, \cdots, N\}$ represents their connectivity : $e_{ij} = 1$ if peers i and j are connected, and $e_{ij} = 0$ otherwise. For an undirected graph, it is required that $e_{ij} = e_{ji}$ for all $i \neq j$, and $e_{ij} = 0$ when $i = j$. Let C be the entire collection of content fragments, and $\{c_i \subseteq C, i = 1, \cdots, N\}$ denotes the collection of the content fragments each peer i shares. The disjointness of contents from peer i to peer j is denoted by $c_i \setminus c_j$, which can be calculated as:

$$c_i \setminus c_j = c_i - (c_i \cap c_j). \tag{12.1}$$

This disjointness can be interpreted as the collection of content fragments that peer i has but peer j does not. In other words, it denotes the fragments that peer i can upload to peer j. Note that the disjointness operation is not commutative, i.e., $c_i \setminus c_j \neq c_j \setminus c_i$. Let $|c_i \setminus c_j|$ denote the cardinality of $c_i \setminus c_j$, which is the number of content fragments peer i can contribute to peer j. In order to maximize the disjointness of content, we maximize the number of content fragments each peer can contribute to its neighbors by determining the connections e_{ij}'s. Let us define ϵ_{ij}'s to be sets such that $\epsilon_{ij} = C$ if $e_{ij} = 1$, and $\epsilon_{ij} = \emptyset$ (null set), otherwise. Therefore we have the following optimization problem:

$$f(x) = \max_E \sum_{j=1}^{N} \left| \bigcup_{i=1}^{N} (c_i \setminus c_j) \cap \epsilon_{ij} \right| \qquad (12.2)$$

It is desirable to select peers with the most mutually disjoint collection of content fragments as neighbors. However, downloading the file fragments between each peer pair would consume a lot of bandwidth and connection cost, etc. Let τ_{ij} denote the cost coefficient between peers i and j. The performance of the whole system can be expressed as follows. The neighbor selection strategy is expected not only to assure maximum content availability but also to minimize the downloading cost to improve the overall throughput of the system. Therefore we have the following multi-objective optimization problem:

$$f_1(x) = \max_E \sum_{j=1}^{N} \left| \bigcup_{i=1}^{N} (c_i \setminus c_j) \cap \epsilon_{ij} \right| \qquad (12.3)$$

$$f_2(x) = \min_E \sum_{j=1}^{N} \sum_{i=1}^{N} \tau_{ij} |(c_i \setminus c_j)| |\epsilon_{ij}| \qquad (12.4)$$

In the network, every node is a potential neighbor of each other node since the network's topology is a logical one. So the full connection is an ideal solution for the peer's connectivity. For the networks, however, we have to consider some constraints [3, 20]:

- based on the underlying network characteristics, i.e., delay or capacity of actual links;
- based on location of data and services;
- based on the nodes's capabilities of managing peers, e.g., the number of direct neighbors a node can maintain; some peers are tied down since they could possess relatively more content fragments. Note that this resource constraint can be independent of the underlying network.

In the environment, the maximum number of each peer needs to be considered, i.e., each peer i will be connected to a maximum of d_i neighbors, where $d_i < N$. Therefore there are two constraints for each peer:

$$\begin{aligned} \sum_{j=1}^{N} e_{ij} &\leq d_i, \quad \text{for all } i \\ \sum_{i=1}^{N} e_{ij} &\leq d_j, \quad \text{for all } j \end{aligned} \qquad (12.5)$$

Definition 1. *A neighbor-selection problem in P2P networks problem can be defined as* $\prod = (N, C, M, F, s)$, *in which N is the number of peers, C is the entire* collection of content fragments, M is the maximum number of the peers, which each peer can connect steadily in the session, F is a single objective to optimize the number of swap fragments, or multi-objective to optimize the number of swap fragments, and to minimize the downloading cost; s denotes the environment constraints. The key components are operations, machines and data-hosts. A P2P state is determined by N, C and M, i.e. $S = (N, C, M)$. For the sake of simplify, the neighbor-selection problem can be also represented in triple $\prod = (S, F, s)$.

12.3 P2P Neighbor-Selection Strategy

GA and PSO algorithms share many similarities [22]. In GA, a population of candidate solutions (for the optimization task to be solved) is initialized. New solutions are created by applying reproduction operators (mutation and crossover). The fitness (how good the solutions are) of the resulting solutions are evaluated and suitable selection strategy is then applied to determine which solutions will be maintained to the next generation. PSO algorithm is inspired by social behavior patterns of organisms that live and interact within large groups. It incorporates swarming behaviors observed in flocks of birds, schools of fish, or swarms of bees, and even human social behavior. In this section, we discuss P2P neighbor selection strategy based on PSO and GA algorithms.

12.3.1 Particle Swarm Algorithm for Single Objective Neighbor Selection

To apply the particle swarm algorithm successfully for the NS problem, one of the key issues is the mapping of the problem solution into the particle space, which directly affects its feasibility and performance. Usually, the particle's position is encoded to map each dimension to one directed connection between peers, i.e. the dimension is $N * N$. But the neighbor topology in P2P networks is an undirected graph, i.e. $e_{ij} = e_{ji}$ for all $i \neq j$. We set up a search space of D dimension as $N * (N-1)/2$. Accordingly, each particle's position is represented as a binary bit string of length D. Each dimension of the particle's position maps one undirected connection. The domain for each dimension is limited to 0 or 1.

The particle swarm model consists of a swarm of particles, which are initialized with a population of random candidate solutions. They move iteratively through the D-dimension problem space to search the new solutions, where the fitness f can be measured by calculating the number of swap fragments in the potential solution. Each particle has a position represented by a position-vector $\mathbf{p}_i$ (i is the index of the particle), and a velocity represented by a velocity-vector $\mathbf{v}_i$. Each particle remembers its own best position so far in a vector $\mathbf{p}_i^{\#}$, and its j-th dimensional value is $p_{ij}^{\#}$. The best position-vector among the swarm so far is then stored in a vector $\mathbf{p}^*$, and its j-th dimensional value is p_j^*. When the particle moves in a state space restricted to zero and one on each dimension, the change of probability with time steps is defined as follows:

$$P(p_{ij}(t) = 1) = f(p_{ij}(t-1), v_{ij}(t-1), p_{ij}^{\#}(t-1), p_j^*(t-1)), \quad (12.6)$$

where the probability function is

$$sig(v_{ij}(t)) = \frac{1}{1 + e^{-v_{ij}(t)}}. \quad (12.7)$$

At each time step, each particle updates its velocity and moves to a new position according to Eqs. (12.8) and (12.9):

$$v_{ij}(t) = wv_{ij}(t-1) + c_1 r_1(p_{ij}^{\#}(t-1) - p_{ij}(t-1)) \\ + c_2 r_2(p_j^*(t-1) - p_{ij}(t-1)) \quad (12.8)$$

$$p_{ij}(t) = \begin{cases} 1 & \text{if } \rho < sig(v_{ij}(t)); \\ 0 & \text{otherwise.} \end{cases} \quad (12.9)$$

where c_1 is a positive constant, called coefficient of the self-recognition component, c_2 is a positive constant, called coefficient of the social component; r_1 and r_2 are random numbers in the interval [0,1]. The variable w is called as the inertia factor, whose value is typically setup to vary linearly from 1 to near 0 during the iterated processing and ρ is a random number in the closed interval [0, 1]. From Eq. (12.8), a particle decides where to move next, considering its current state, its own experience, which is the memory of its best past position, and the experience of its most successful particle in the swarm. The particle has a priority levels according to the order of peers. The sequence of the peers will be not changed during the iteration. Each particle's position indicates the potential connection state. The pseudo-code for the particle swarm search method is illustrated in Algorithm 12.1.

In multi-dimensional search space, the characteristics of ergodicity is of vital importance to an algorithm. We discuss them for the particle swarm optimization. Clerc and Kennedy have stripped the particle swarm model down to a most simple form [23, 24]. If the self-recognition component c_1 and the coefficient of the social-recognition component c_2 in the particle swarm model are combined into a single term c, i.e. $c = c_1 + c_2$, the best position $\mathbf{p}_i$ can be redefined as follows:

$$\mathbf{p}_i \leftarrow \frac{(c_1 \mathbf{p}_i + c_2 \mathbf{p}_g)}{(c_1 + c_2)} \quad (12.10)$$

Then, the update of the particle's velocity is defined by:

$$\mathbf{v}_i(t) = \mathbf{v}_i(t-1) + c(\mathbf{p}_i - \mathbf{x}_i(t-1)) \quad (12.11)$$

Algorithm 12.1. Neighbor Selection Algorithm Based on Particle Swarm

01. Initialize the size of the particle swarm n, and other parameters.
02. Initialize the positions and the velocities for all the particles randomly.
03. While (the stopping criterion is not met) do
04. $t = t + 1$;
05.   For $s = 1$ to n
06.   For $i = 1$ to N
07.   For $j = 1$ to N
08.   If $j == i$, $e_{ij} = 0$;
09.   If $j < i$, $a = j; b = i$;
10.   If $j > i$, $a = i; b = j$;
11. $e_{ij} = p_{[a*N+b-(a+1)*(a+2)/2]}$;
12.   if $e_{ij} = 1$, calculate $c_i \setminus c_j$;
13.   Calculate $f = f + \left|\bigcup_{i=1}^{N}(c_i \setminus c_j) \cap \epsilon_{ij}\right|$;
14. $\mathbf{p}^* = argmin_{i=1}^{n}(f(\mathbf{p}^*(t-1)), f(\mathbf{p}_1(t)),$
15. $f(\mathbf{p}_2(t)), \cdots, f(\mathbf{p}_i(t)), \cdots, f(\mathbf{p}_n(t)))$;
16.   For $s = 1$ to n
17. $\mathbf{p}_i^{\#}(t) = argmin_{i=1}^{n}(f(\mathbf{p}_i^{\#}(t-1)), f(\mathbf{p}_i(t)))$;
18.   For $d = 1$ to D
19.   Update the d-th dimension value of $\mathbf{p}_i$ and $\mathbf{v}_i$
20.   according to Eqs. (12.8) and (12.9);
21. End While

The system can be simplified even further by using $\mathbf{y}_i(t-1)$ instead of $\mathbf{p}_i - \mathbf{x}_i(t-1)$. Thus, the reduced system is then:

$$\begin{cases} \mathbf{v}(t) = \mathbf{v}(t-1) + c\mathbf{y}(t-1) \\ \mathbf{y}(t) = -\mathbf{v}(t-1) + (1-c)\mathbf{y}(t-1) \end{cases}$$

This recurrence relation can be written as a matrix-vector product, so that

$$\begin{bmatrix} \mathbf{v}(t) \\ \mathbf{y}(t) \end{bmatrix} = \begin{bmatrix} 1 & c \\ -1 & 1-c \end{bmatrix} \cdot \begin{bmatrix} \mathbf{v}(t-1) \\ \mathbf{y}(t-1) \end{bmatrix}$$

Let

$$\mathbf{P}_t = \begin{bmatrix} \mathbf{v}_t \\ \mathbf{y}_t \end{bmatrix}$$

and

$$A = \begin{bmatrix} 1 & c \\ -1 & 1-c \end{bmatrix}$$

we have an iterated function system for the particle swarm model:

$$\mathbf{P}_t = A \cdot \mathbf{P}_{t-1} \qquad (12.12)$$

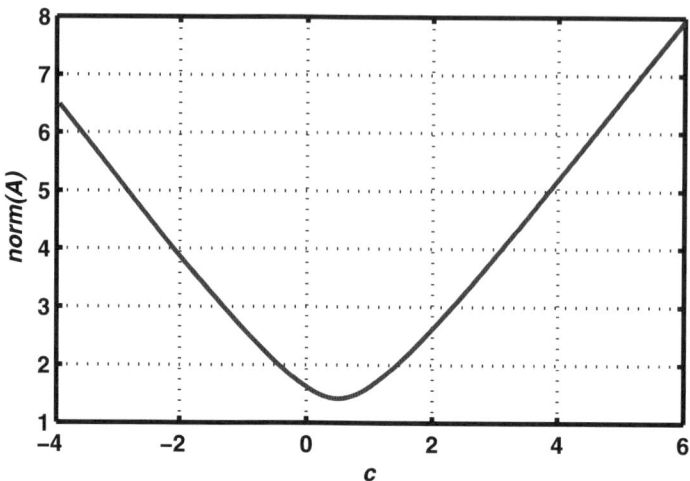

Fig. 12.1. Norm of A

Thus, the system is completely defined by A. Its norm $\|A\|_2$ (also written $\|A\|$) is determined by c. The relationship of A and its dependence on c is illustrated in Fig. 12.1.

Note that it is possible to find different trajectories of the particle for various values of c. Fig. 12.2(a) illustrates the system for a torus when $c=2.9$; Fig. 12.2(b), a hexagon with spindle sides when $c=2.99$; Fig. 12.2(c), a triangle with spindle sides when $c=2.999$; Fig. 12.2(d) and a simple triangle when $c=2.9999$. As depicted in Fig. 12.2, the iteration time step used is 100 for all the cases. Another system sensitivity instance is illustrated in Fig. 12.3. It is to be noted that Figs. 12.2 and 12.3 illustrate only some 2-dimensional representations of the iterated process. A comparison between 2D and 3D is illustrated in Fig. 12.4.

12.3.2 Genetic Algorithm for Multi-objective Neighbor Selection

Multi-objective GAs are very popular multi-objective techniques, which normally exhibit good overall performance. Many multi-objective optimization techniques using evolutionary algorithms have been proposed in recent years [22, 25, 26]. Given a P2P state S, the multi-objective neighbor selection is not only to maximize Eq. (12.3) but also to minimize Eq. (12.4) with the constraint in Eq. (12.5).

Similarly, we adopt the upper-half-triangle encoding representation in our genetic algorithm for the NS problem. We also set up a search space of D dimension as $N*(N-1)/2$. Accordingly, each individual is represented as a binary bit string of length D. The pseudo-code for our P2P neighbor selection method is illustrated in Algorithm 12.2.

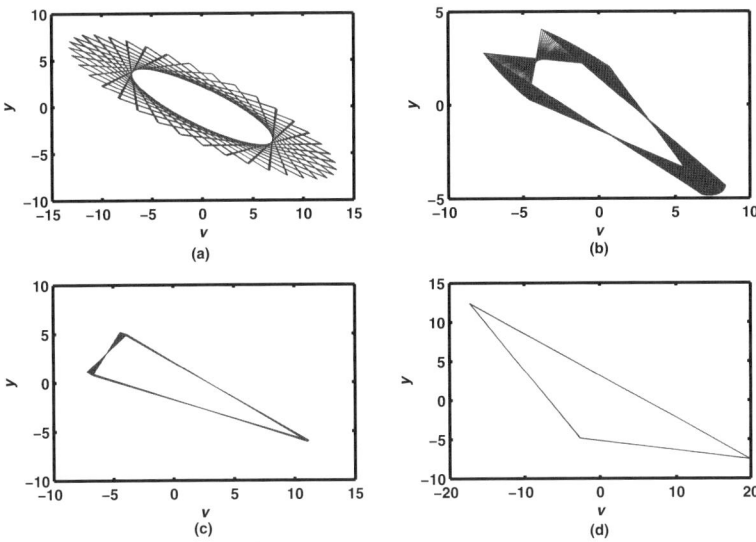

Fig. 12.2. Trajectory of the particle (a) $c = 2.9$, (b) $c = 2.999$, (c) $c = 2.999$, (d) $c = 2.9999$

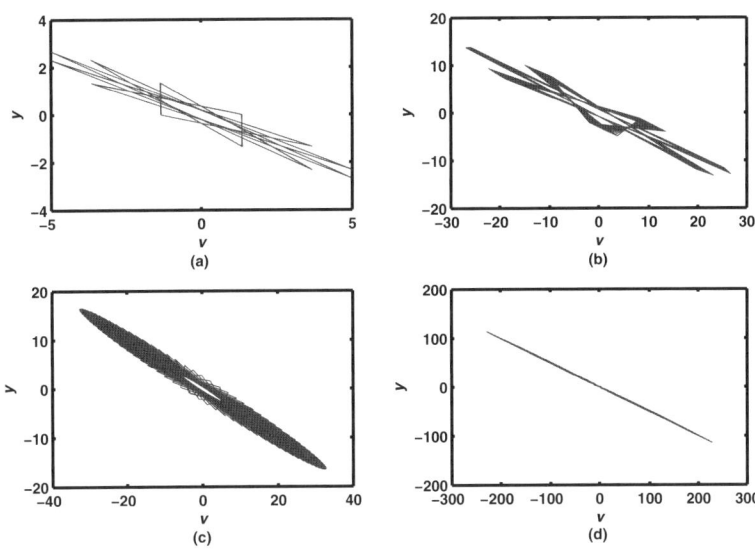

Fig. 12.3. Trajectory of the particle (a) $c = 3.7321$, (b) $c = 3.8$, (c) $c = 3.9$, (d) $c = 3.999$

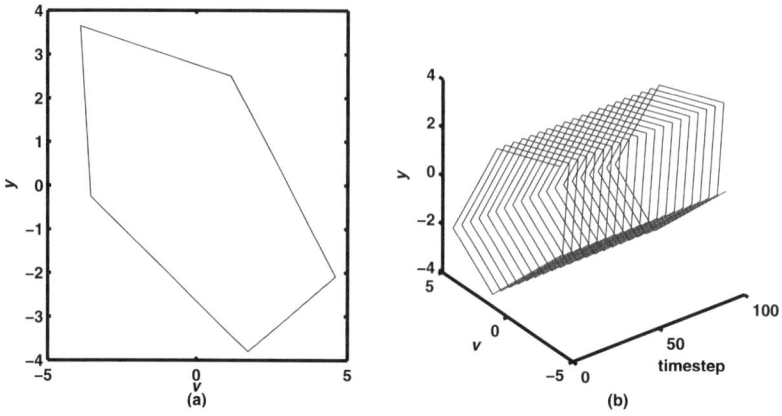

Fig. 12.4. 2D versus 3D (a) 2D: $c = 1.3820$, (b) 3D: $c = 1.3820$

Algorithm 12.2. Neighbor Selection Algorithm Based on Genetic Algorithm
---
01. Initialize the population, and other parameters.
02. While (the stopping criterion is not met) do
03.   Evaluate();
04.   for $i = 1$ to N
05.   for $j = 1$ to N
06.   if $j == i$, $e_{ij} = 0$;
07.   else if $j < i$, $a = j; b = i$;
08.   else if $j > i$, $a = i; b = j$;
09. $e_{ij} = p_{[a*N+b-(a+1)*(a+2)/2]}$;
10.   If $e_{ij} = 1$, calculate $c_i \setminus c_j$;
11.   Calculate $f_2 = f_2 + \tau_{ij}|(c_i \setminus c_j)|$;
12.   Next j
13.   calculate $f_1 = f_1 + \left|\bigcup_{i=1}^{N}(c_i \setminus c_j) \cap \epsilon_{ij}\right|$;
14.   Rank();
15.   If nondomCtr> MaxArchiveSize, maintenance-archive();
16.   Generate-new-pop();
17.   Crossover();
18.   Mutation();
19. $t++$;
20.   If rank $== 1$ output the fitness;
21. End While

12.4 Algorithm Performance Evaluation

To illustrate the effectiveness and performance of the considered algorithms, we illustrate the neighbor-selection process and results through some test problems. The specific parameter settings of the algorithms are described in Table 12.1.

Table 12.1. Parameter settings for the algorithms

Algorithm	Parameter name	Value
	Size of the population	$int(10 + 2sqrt(D))$
GA	Probability of crossover	0.8
	Probability of mutation	0.08
	Swarm size	$int(10 + 2sqrt(D))$
	Self coefficient c_1	2
PSO	Social coefficient c_2	2
	Inertia weight w	0.9
	Clamping Coefficient ρ	0.5

12.4.1 Single Objective Neighbor Selection

We first illustrate an execution trace of the algorithm for the NS problem. A file of size 7 MB is divided into 14 fragments (512 KB each) to distribute, 6 peers download from the P2P networks, and the connecting maximum number of each peer is 3, which is represented as (6, 14, 3) problem. In some session, the state of distributed file fragments is as follows:

$$\begin{bmatrix} 1 & 0 & 0 & 4 & 0 & 6 & 7 & 8 & 0 & 10 & 0 & 12 & 0 & 14 \\ 0 & 0 & 0 & 4 & 5 & 0 & 7 & 0 & 9 & 0 & 11 & 0 & 13 & 0 \\ 0 & 2 & 0 & 0 & 0 & 6 & 0 & 0 & 0 & 0 & 11 & 12 & 0 & 14 \\ 0 & 2 & 3 & 4 & 0 & 6 & 0 & 0 & 0 & 0 & 11 & 0 & 0 & 0 \\ 0 & 2 & 0 & 0 & 0 & 0 & 7 & 8 & 0 & 10 & 0 & 12 & 0 & 14 \\ 1 & 2 & 0 & 0 & 5 & 0 & 0 & 0 & 9 & 10 & 11 & 0 & 13 & 14 \end{bmatrix}$$

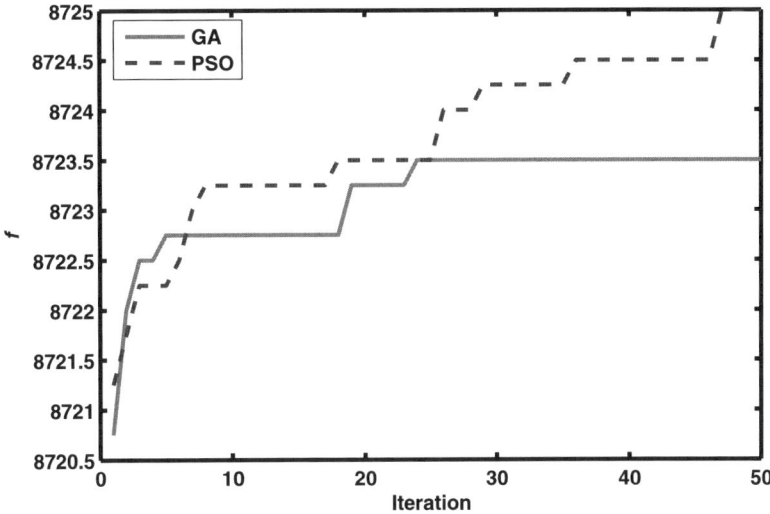

Fig. 12.5. Performance for the NS (25, 1400, 12)

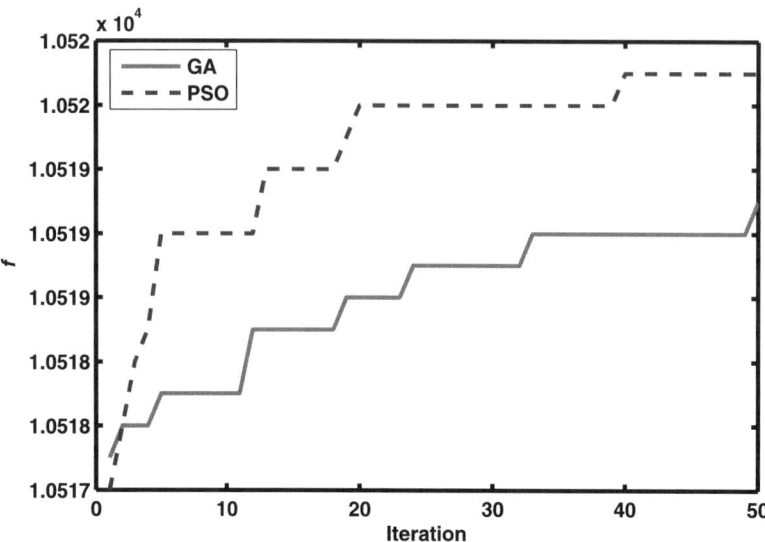

Fig. 12.6. Performance for the NS (30, 1400, 15)

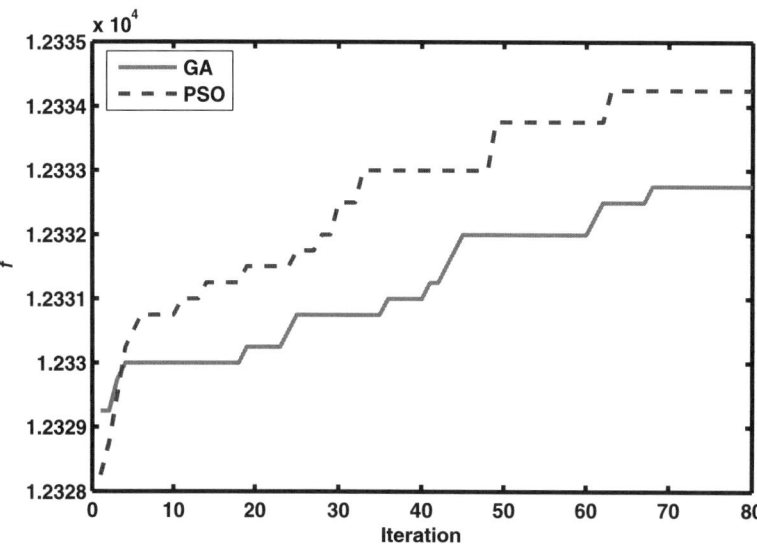

Fig. 12.7. Performance for the NS (35, 1400, 17)

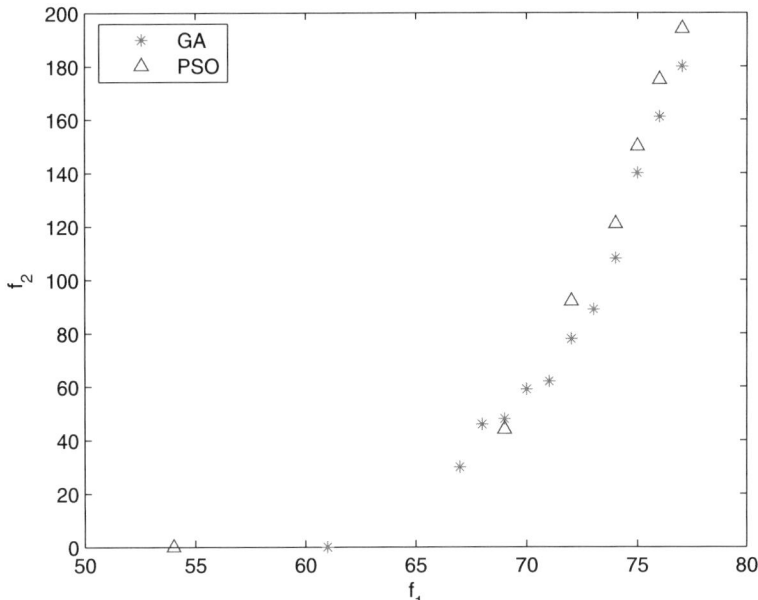

Fig. 12.8. Performance for the NS $(6, 60, 3)$

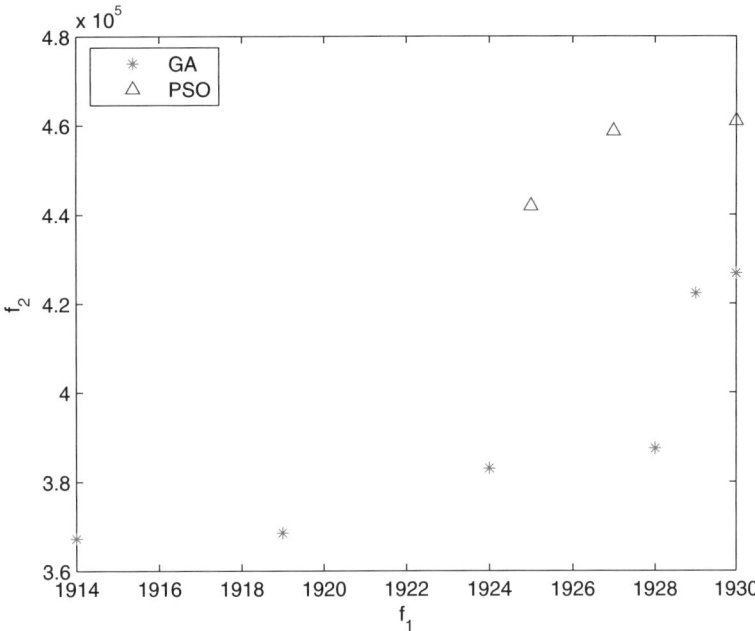

Fig. 12.9. Performance for the NS $(25, 300, 12)$

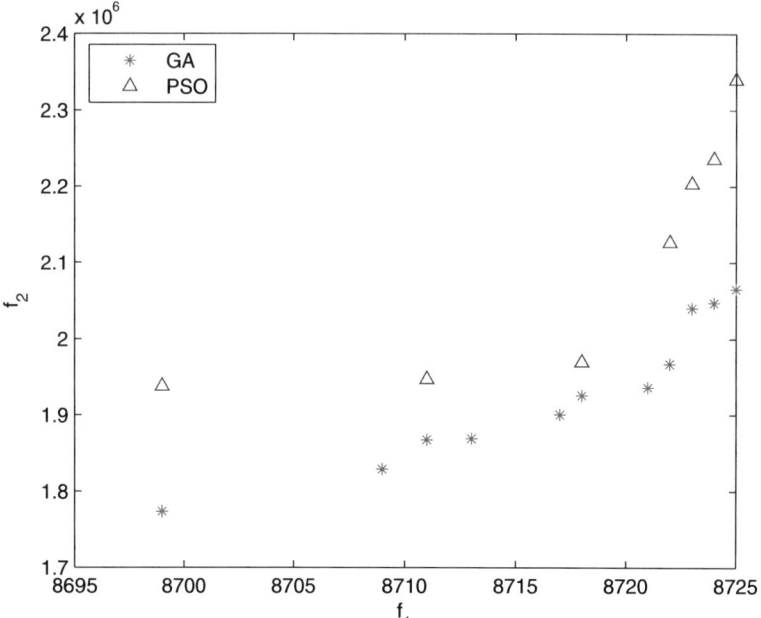

Fig. 12.10. Performance for the NS (25, 1400, 12)

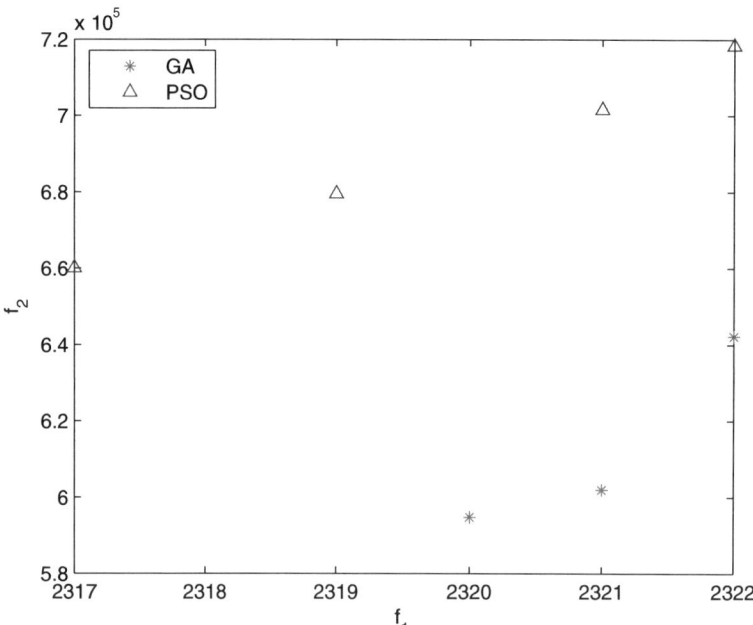

Fig. 12.11. Performance for the NS (30, 300, 15)

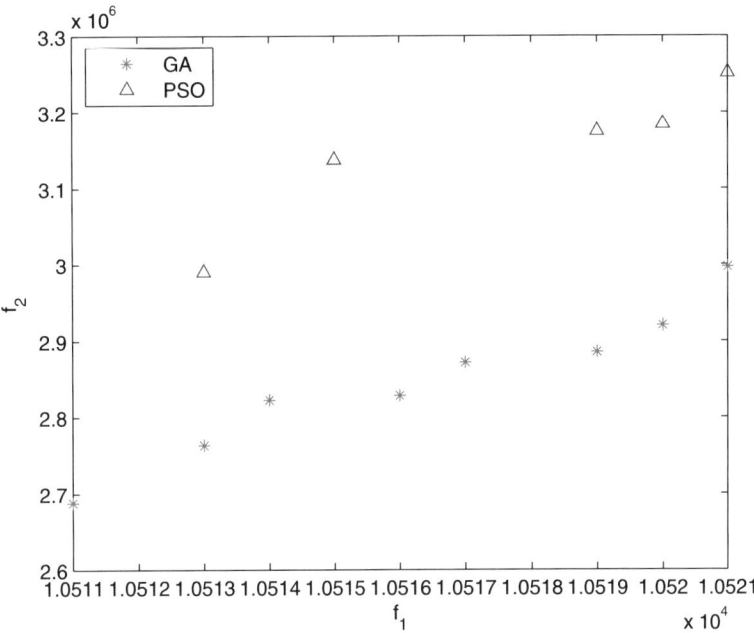

Fig. 12.12. Performance for the NS $(30, 1400, 15)$

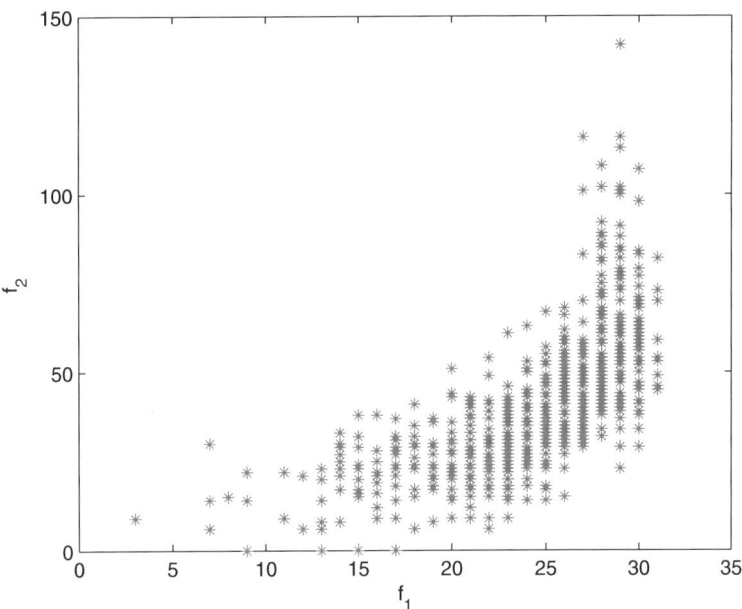

Fig. 12.13. Performance for the NS $(6, 60, 3)$

The optimal result search by the proposed algorithm is 31, and the neighbor selection solution is shown in the matrix below:

$$\begin{array}{c c} & \begin{array}{c c c c c c} 1 & 2 & 3 & 4 & 5 & 6 \end{array} \\ \begin{array}{c} 1 \\ 2 \\ 3 \\ 4 \\ 5 \\ 6 \end{array} & \left(\begin{array}{c c c c c c} 0 & 0 & 0 & 1 & 1 & 1 \\ 0 & 0 & 0 & 0 & 1 & 1 \\ 0 & 0 & 0 & 1 & 1 & 1 \\ 1 & 0 & 1 & 0 & 0 & 0 \\ 1 & 1 & 1 & 0 & 0 & 0 \\ 1 & 1 & 1 & 0 & 0 & 0 \end{array} \right) \end{array}$$

We also tested three other representative instances (problem (25,1400,12), problem (30,1400,15) and problem (35,1400,17)). In our experiments, the algorithms used for comparison were GA and PSO.

The PSO/GA algorithms were repeated 4 times with different random seeds. Each trial had a fixed number of 50 / 80 iterations. Other specific parameter settings of the algorithms are described in Table 12.1. The average fitness values of the best solutions throughout the optimization run were recorded. The average and the standard deviation were calculated from the 4 different trials. Figs. 12.5, 12.6 and 12.7 illustrate the PSO/GA performance during the search processes for the NS problem. As evident, PSO obtained better results much faster than GA, especially for large scale problems.

12.4.2 Multi-objective Neighbor Selection

We demonstrate an execution trace of the algorithm for the first NS problem in last subsection, i.e., $(6, 14, 3)$ problem. In this problem, the network cost is considered; the corresponding cost matrix is as follows:

$$\begin{bmatrix} 0 & 5 & 2 & 4 & 1 & 0 \\ 5 & 0 & 3 & 0 & 2 & 2 \\ 2 & 3 & 0 & 0 & 0 & 0 \\ 4 & 0 & 0 & 0 & 5 & 2 \\ 1 & 2 & 0 & 5 & 0 & 10 \\ 0 & 2 & 0 & 2 & 10 & 0 \end{bmatrix}$$

The PSO/GA algorithms were repeated 3 times with different random seeds. Each trial had a fixed number of 200 iterations. The average fitness values of the best (rank = 1) solutions throughout the optimization run were recorded. The performance output is illustrated in Fig. 12.13 by the proposed algorithm. We also tested other five representative instances (problem (6,60,3), problem (25,300,12), problem (25,1400,12), problem (30,300,15), problem (30,1400,15)) further. Figs. 12.8, 12.9, 12.10, 12.11 and 12.12 illustrate the GA/PSO performance during the search processes for the NS problem. As evident, GA usually obtained better results than PSO.

12.5 Conclusions

In this chapter, we introduced the problem of neighbor selection in peer-to-peer networks using a Particle Swarm Optimization and Genetic Algorithms. We first introduced the model of Peer-to-Peer networks and discussed measuring metrics for P2P neighbor selection. Both a single and a multi-objective formulations are given, and then the P2P neighbor selection problem is defined. In the considered approaches, we presented an upper-half-triangle encoding representation method. The particle/individual was encoded by the upper half matrix of the peer connection through the undirected graph, which reduces the dimension of the search space. We evaluated the performance of the algorithms. The results indicate that PSO usually required shorter time than GA, specially for large scale problems. PSO could be an ideal approach for solving the single objective NS problem, while GA usually obtain better results than PSO in the multi-objective NS problems.

Acknowledgements

This work was partly supported by NSFC (60373095), DLMU (DLMU-ZL-200709). F. Xhafa acknowledges partial support by Projects ASCE TIN2005-09198-C02-02, FP6-2004-ISO-FETPI (AEOLUS) and MEC TIN2005-25859-E and FORMALISM TIN2007-66523.

References

1. Kwok, S.: P2P searching trends: 2002-2004. Information Processing and Management 42, 237–247 (2006)
2. Idris, T., Altmann, J.: A Market-managed topology formation algorithm for peer-to-peer file sharing networks. In: Stiller, B., Reichl, P., Tuffin, B. (eds.) ICQT 2006. LNCS, vol. 4033, pp. 61–77. Springer, Heidelberg (2006)
3. Surana, S., Godfrey, B., Lakshminarayanan, K., Karp, R., Stoica, I.: Load balancing in dynamic structured peer-to-peer systems. Performance Evaluation 63, 217–240 (2006)
4. Duan, H., Lu, X., Tang, H., Zhou, X., Zhao, Z.: Proximity neighbor selection in structured P2P network. In: Proceedings of Sixth IEEE International Conference on Computer and Information Technology, p. 52 (2006)
5. Koo, S., Kannan, K., Lee, C.: A genetic-algorithm-based neighbor-selection strategy for hybrid peer-to-peer networks. In: Proceedings of the 13th IEEE International Conference on Computer Communications and Networks, pp. 469–474 (2004)
6. Schollmeier, R.: A definition of peer-to-peer networking for the classification of peer-to-peer architectures and applications. In: Proceedings of the First International August Conference on Peer-to-Peer Computing, pp. 101–102 (2001)
7. Ghosal, D., Poon, B.K., Kong, K.: P2P contracts: a framework for resource and service exchange. Future Generation Computer Systems 21, 333–347 (2005)
8. Kurmanowytsch, R., Kirda, E., Kerer, C., Dustdar, S.: OMNIX: A topology-independent P2P middleware. In: Proceedings of the 15th Conference on Advanced Information Systems Engineering (2003)

9. Koulouris, T., Henjes, R., Tutschku, K., de Meer, H.: Implementation of adaptive control for P2P overlays. In: Wakamiya, N., Solarski, M., Sterbenz, J.P.G. (eds.) IWAN 2003. LNCS, vol. 2982, pp. 292–306. Springer, Heidelberg (2004)
10. Quagliarella, D., Périaux, J., Poloni, C., Winter, G. (eds.): Genetic Algorithms in Engineering and Computer Science. John Wiley & Sons Ltd., Chichester (1997)
11. Kennedy, J., Eberhart, R.: Swarm Intelligence. Morgan Kaufmann, San Francisco (2001)
12. Clerc, M.: Particle Swarm Optimization. ISTE Publishing Company, London (2006)
13. Abraham, A., Guo, H., Liu, H.: Swarm intelligence: foundations, perspectives and applications. In: Nedjah, N., Mourelle, L. (eds.) Swarm Intelligent Systems. Studies in Computational Intelligence, pp. 3–25. Springer, Heidelberg (2006)
14. Liu, H., Sun, S., Abraham: A Particle swarm approach to scheduling work-flow applications in distributed data-intensive computing environments. In: Proceedings of The Sixth International Conference on Intelligent Systems Design and Applications, pp. 661–666 (2006)
15. Sen, S., Wang, J.: Analyzing Peer-to-Peer Traffic Across Large Networks. IEEE/ACM Transactions on Networking 12(2), 219–232 (2004)
16. Liu, Y., Xiao, L., Esfahanian, A., Ni, L.M.: Approaching Optimal Peer-to-Peer Overlays. In: Proceedings of the 13th IEEE International Symposium on Modeling, Analysis, and Simulation of Computer and Telecommunication Systems, pp. 407–414 (2005)
17. Belmonte, M.V., Conejo, R., Díaz, M., Pérez-de-la-Cruz, J.L.: Coalition Formation in P2P File Sharing Systems. In: Marín, R., Onaindía, E., Bugarín, A., Santos, J. (eds.) CAEPIA 2005. LNCS (LNAI), vol. 4177, pp. 153–162. Springer, Heidelberg (2006)
18. Koo, S., Kannan, K., Lee, C.: On neighbor-selection strategy in hybrid peer-to-peer networks. Future Generation Computer Systems 22, 732–741 (2006)
19. Ghanea-Hercock, R.A., Wang, F., Sun, Y.: Self-Organizing and Adaptive Peer-to-Peer Network. IEEE Transactions on Systems, Man, and Cybernetics - Part B: Cybernetics 36(6), 1230–1236 (2006)
20. Wang, S., Chou, H., Wei, D., Kuo, S.: On the Fundamental Performance Limits of Peer-to-Peer Data Replication in Wireless Ad hoc Networks. IEEE Journal on Selected Areas in Communications 25(1), 211–221 (2007)
21. Qiu, D., Sang, W.: Global Stability of Peer-to-Peer File Sharing Systems. Computer Communications (2007) doi:10.1016/j.comcom.2007.08.012
22. Abraham, A.: Evolutionary computation. In: Sydenham, P., Thorn, R. (eds.) Handbook for Measurement Systems Design, pp. 920–931. John Wiley and Sons Ltd., London (2005)
23. Clerc, M., Kennedy, J.: The Particle Swarm-explosion, Stability, and Convergence in A Multidimensional Complex Space. IEEE Transactions on Evolutionary Computation 6, 58–73 (2002)
24. Liu, H., Abraham, A., Clerc, M.: Chaotic Dynamic Characteristics in Swarm Intelligence. Applied Soft Computing 7, 1019–1026 (2007)
25. Abraham, A., Jain, L.: Evolutionary Multi-objective Optimization. In: Abraham, A., Jain, L.C., Goldberg, R. (eds.) Evolutionary Multi-objective Optimization: Theoretical Advances and Applications, ch. 1, pp. 1–9. Springer, London (2005)
26. Srinivas, N., Deb, K.: Multi-objective optimization using nondominated sorting genetic algorithms. Evolutionary Computation 2(3), 221–248 (1994)

13

An Adaptive Co-ordinate Based Scheduling Mechanism for Grid Resource Management with Resource Availabilities

B.T. Benjamin Khoo and Bharadwaj Veeravalli[*]

Computer Networks and Distributed Systems (CNDS) Laboratory, Department of Electrical and Computer Engineering, National University of Singapore, 4 Engineering Drive 3, Singapore 117576

Summary. In this chapter, we propose a novel resource-scheduling strategy capable of handling multiple resource requirements for jobs that arrive in a Grid Computing Environment. In our proposed algorithm, referred to as Multi-Resource Scheduling (MRS) algorithm, we take into account both the site capabilities and the resource requirements of jobs. The main objective of the algorithm is to obtain a *minimal execution schedule* through efficient management of available Grid resources. We introduce the concept of a 2-dimensional virtual map and resource potential using a co-ordinate based system. To further develop this concept, a third dimension was added to include resource availabilities in the Grid environment. Based on the proposed model, rigorous simulation experiments shows that the strategy provides excellent allocation schedules as well as superior avoidance of job failures by at least 55%. The aggregated considerations is shown to render high-performance in the Grid Computing Environment. The strategy is also capable of scaling to address additional requirements and considerations without sacrificing performance. Our experimental results clearly show that MRS outperforms other strategies and we highlight the impact and importance of our strategy.

Keywords: Multiple resource scheduling, Resource requirements, Minimal execution schedule, 2-dimensional virtual map.

13.1 Introduction

As computing systems has evolved from single monolithic systems to the networked multi-core systems of today, distributed computing has began to make an impact beyond research and has grown into the commercial space. This growth has accelerated the usage of distributed computing systems either as dedicated clusters, or as workstations that share its computing resources with an interactive user. The usage of Grids [3] adds further complexity into these two classes of systems by considering geographical separation, multi-organizational identity management as well as resource co-ordination.

Vendors such as IBM, HP and Sun Microsystems have all introduced hardware solutions that aims to effectively lower the cost-per-gigaflop of processing

---
[*] Corresponding author.

while maintaining high performance using locally distributed systems. Additionally, solution vendors such as Sybase, DataSynpase and United Devices have also further pushed the envelope of distributed computing beyond research and academia, moving traditionally local resources such as memory, disk and CPUs to a wide area distributed computing platform sharing these very same resources for commercial workloads. Consequently, what had used to be optimal in performance for a local environment has suddenly become a serious problem when high latency networks, uneven resource distributions, and low node reliability guarantees, are added into the system.

Allocation strategies for such distributed environments are also affected as more resources and requirements have to be addressed in a Grid system. Coupled with un-reliable information availability and possibilities of failures, such environments has also resulted in failure of traditional scheduling algorithms where changes adversely affects the robustness of scheduling algorithms that are available for Grids.

In this chapter, we propose a novel scheme that considers various resource requirements of jobs while taking into consideration the distributed computation environment where the job resides in. The technique we propose shall then devise an allocation, which can be used to provide what it believes as the most efficient job execution sequence to handle the jobs. Below we summarize our contributions in this chapter.

13.1.1 Our Contributions

We propose a novel methodology referred to as Multiple-Resource-Scheduling (MRS) strategy that would enable jobs with multiple resource requirements to be run effectively in a Grid Computing Environment (GCE). A job's resource dependencies in computational, data requirements and communication overheads will be considered. A parameter called Resource Potential is also introduced to ease in situations where in inter-resource communication relations need to be addressed. An n-dimensional resource aggregation and allocation mechanism is also proposed. The resource aggregation index, derived from the n-dimensional resource aggregation method, and the Resource Potential sufficiently allows us to mathematically describe the relationship of resources that affects general job executions in a specific dimension into a single index. Each dimension is then put together to form an n-dimensional virtual map that allows us to identify the best allocation of resources for the job. The performance of such a scheduling algorithm promises respectable waiting times, response times, as well as an improved level of utilization across the entire GCE. The number of dimensions considered depends on the number of job related attributes we wish to schedule for.

We evaluate the performance of our proposed strategy firstly in 2 dimensions, namely computation and data, while addressing requirements of resources such as, FLOPS, RAM, Disk space, and data. We study our strategy with respect to several influencing factors that quantify the performance. We then further extend MRS into a third dimension to accommodate availability considerations in the Grid environment. Our study shows that MRS out performs most of the commonly available schemes in place for a GCE.

The organization of this chapter is as follows. In Section 13.2, we describe the Grid Computing Environment and in Section 13.3 we introduce our MRS strategy and algorithm. Section 13.4 evaluates the performance of both MRS in 2 and 3 dimensions. Section 13.5 concludes the chapter.

13.2 Grid Environment Model

In this section, we define the GCE in which the MRS strategy was designed. We first clearly identify certain key characteristics of resources as well as the nature of jobs. A GCE comprises many diverse machine types, disks/storage, and networks. In our resource environment, we consider the following.

1. Resources can be made up of individual desktops, servers, clusters or large multi-processor systems. They can provide varying amounts of CPU computing power, RAM, Hard disk space and bandwidth. Communication to individual nodes in the cluster will be done through a Local Resource Manager (LRM). We assume that the LRM will dispatch a job immediately when instructed by the Grid Meta-Scheduler (GMS). The GMS thus treats all resources exposed under a single LRM as a single resource. We find this assumption to be reasonable as GMS usually does not have the ability to directly contact resources controlled by the LRM.
2. Negligible propagation delay of information is assumed in the GCE. We also assume that every node in the GCE is able to execute all jobs when evaluating the performance of the MRS strategy.
3. Each computation resource is connected to each other through networks which are possibly asymmetrical in bandwidth.
4. All resources have prior agreement to participate on the Grid. From this, we safely assume an environment whereby all resources shared by sites are accessible by every other participating node in the Grid if required to do so.
5. In our simulations, we assume that the importance of the resources with respect to each other is identical.
6. The capacity for computation in a CPU resource is provided in the form of GFlops. While we are aware that this is not completely representative of a processor's computational capabilities, it is at current one of the most basic measure of performance on a CPU. Therefore, this is used as a gauge to standardize the performance of different CPU architectures in different sites. However, the actual units used in the MRS strategy does not require actual performance measures, rather, it depends on relative measures to the job requirements. We will show how it is done in later sections.

The creation of the job environment is done through the investigation of the workload models available in the Parallel Workload Archive Models [14] and the Grid workload model available in [16]. The job characteristics are thus defined by the set of parameters available in these models and complemented with additional resource requirements that are not otherwise available in these two models. Examples of these resources includes information such as job submission locations and data size required for successful execution of the task. In our job execution environment, we assume the following.

1. Resource requirement for a job does not change during execution and are only of (a) Single CPU types, or (b) massively parallel types written in either MPI such as MPICH[1] or PVM[2].
2. The job resource estimates provided are the upper bound of the resource usage of a given job.
3. Every job submitted can have its data-source located anywhere within the GCE.
4. A job submitted can be scheduled for execution anywhere within the GCE as applications are assumed to be available in all sites.
5. Jobs resource requirements are divisible into any size prior to execution.
6. Every job also has a data requirement where-by the main data source and size is stated.
7. The effective run time of a job is computed from the time the job is submitted, till the end of its result file stage-out procedure. This includes the time required for the data to be staged in for execution and the time taken for inter-process communication of parallel applications.
8. Resources are locked for a job execution once the distribution of resources start and will be reclaimed after use.

A physical illustration of the resource environment that we consider is shown in Fig. 13.1, and the resource view of how the Grid Meta-Scheduler will access all resources through the LRM is shown in the Fig. 13.2.

In such an environment, we consider *Failure* to be the breakdown of network communication between computing resources, thereby leading to a loss in status updates in the progress of an executing job. This failure can be due to a variety of reasons such as hardware or software failures. We do not specifically identify the cause of the failure, but generalize it for any possible kind. We also assume that a failed resource will be restarted and all history of past executions will be cleared.

We take the view of the resource by an external agent in order to classify if a resource has entered a state of a general *failure* or has *recovered* from its unavailable failed state. Thus, under these assumptions, we are able to break down the participation of a resource in a GCE into the following stages:-

1. Resource becomes available to the GCE
2. Resource continues to be available pending that none of the components within itself has failed

---

[1] MPICH: http://www-unix.mcs.anl.gov/mpi/mpich/
[2] Parallel Virtual Machines: http://www.csm.ornl.gov/pvm/pvm_home.html

13 An Adaptive Co-ordinate Based Scheduling Mechanism   345

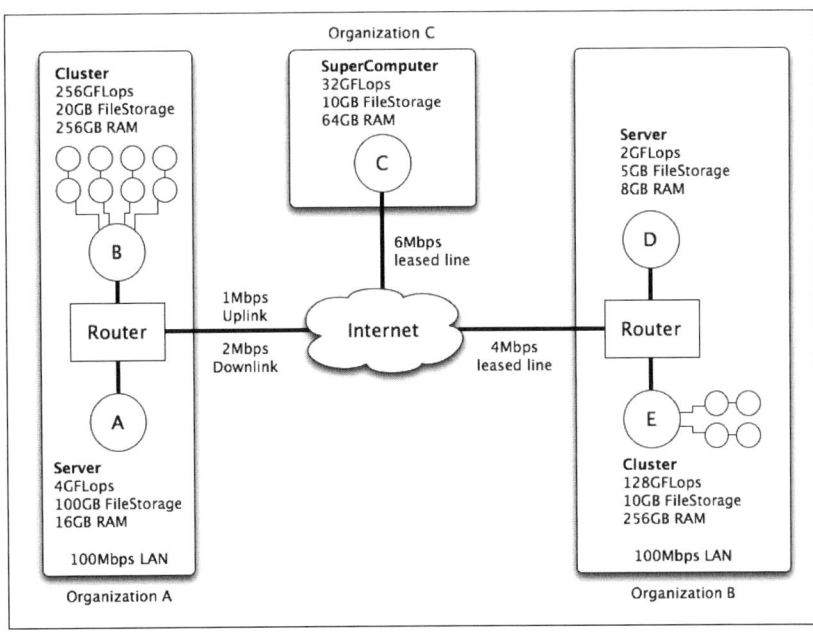

Fig. 13.1. Illustration of a physical network layout of a GCE

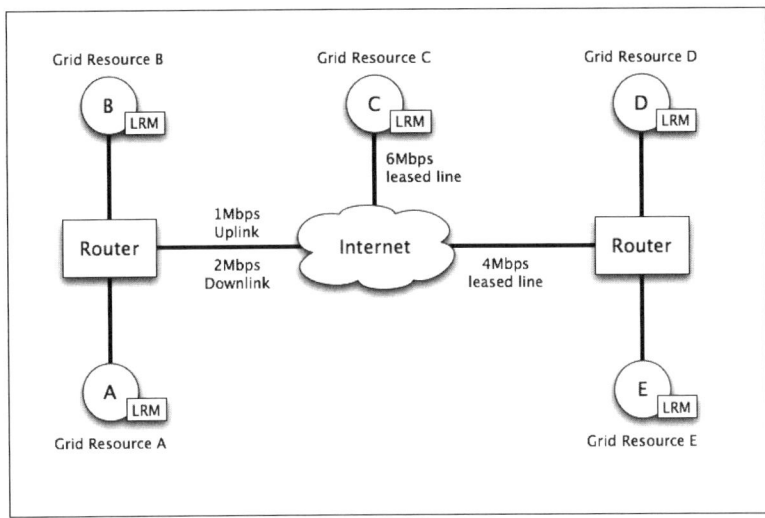

Fig. 13.2. Resource view of physical environment with access considerations

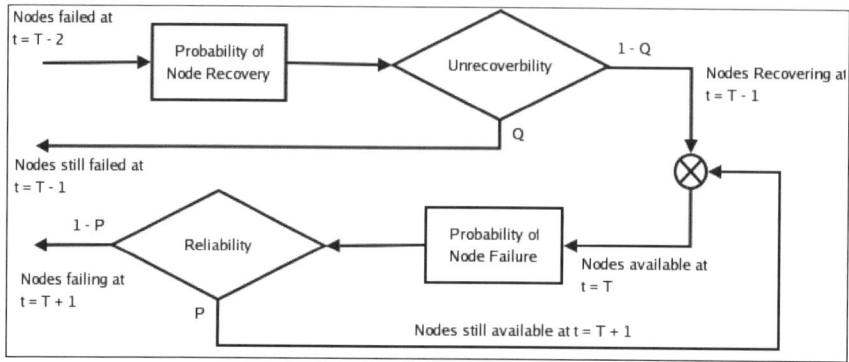

Fig. 13.3. Resource Life Cycle Model for resources in the GCE

3. Resource encounters a failure in one of its components and goes offline for maintenance and fix
4. Resource goes through a series of checks, replacements or restarts to see if it is capable to re-join the GCE
5. Resource comes on-line once it is capable and becomes available to the GCE (return to first stage)

From the above, it was observed that in Stages (2) and (4), the resource undergoes a period of uncertainty. This uncertainty stems from the fact that the resource probably might not fail or recover for a certain period of time. Based on these stages the model presented in [6] was constructed. The Resource Life Cycle (RLC) Model shown in Fig. 13.3 identifies the stages where by Grid resources under-go cycles of failures and recovery, and also accounts for the probabilities of *each* resource being able to recover or fail in the next epoch of time. Thus using this model, we are able to describe a general form of resource failure that would cause a loss of job control or connectivity to the said resource. This in turn affects the capacity of the GCE.

13.3 Scheduling Strategy

From the Grid Environment Model, we note that the system environment of the Grid consists of heterogeneous nodes. This results in an environment whereby a wide range of resources are available. These resources may or may not be well connected to each other depending on network connectivity and thus require proper allocation and grouping before jobs can be executed efficiently.

MRS addresses the various job requirements and resource capabilities by dividing decision factors into separate dimensions. By using some performance metrics, it then decides which resources the jobs would be best dispatched to. It combines several inter-dependent factors within each selected dimension and simplifies it into a single index which is then used to decide how a job is to be sent or distributed to the resources. MRS also treats each submitted job as an independent entity and does not address work-flow requirements of any application. We feel that this is done without any loss of generality as work-flow requirements should be addressed at an orchestration layer independent of the scheduling middle-ware. With MRS always allocating every job to sites that best provides its resources, it ensures that the job execution environment will remain optimal for both serial as well as parallel jobs.

The jobs request and site representations of CPU resources is done in terms of GFLOPs as an indication of performance. Future changes in unit representations will not affect the strategy as the aggregation algorithm will result in dimensionless indexes as long as the request and site resource representation units are the same. This applies to all other resources shared within MRS.

In this chapter, we first consider 2 dimensions (2D) within MRS and then extend it to a third dimension (3D). The two basic dimensions (1) Computation, and (2) Data are used in our design. These two dimensions are chosen due to the general requirement to achieve faster computation through proper resource allocation such as GFLOPs, RAM and disk, and better data resource allocation to achieve higher I/O throughput. It is to be noted that these two components are highly related to each other in the scheduling process. Each of them on its own, would be unable to provide optimality in resource allocation. Aggregation of the various available resources are then combined into two major indices based on these two basic dimensions. We refer to these indices as the Computational and Data Index respectively.

The third dimension of capacity is subsequently added to MRS as another component that affects the optimality of the allocation strategy. While in an ideal GCE, this dimension can be ignored, the inclusion of this dimension would allow better representation of how an allocation strategy can adapt dynamically to changing GCEs. This makes the allocation strategy much more versatile compared to traditional algorithms.

The Computational and Data Index allows us to create a 2D plot which describes the virtual topology, which we call a *Virtual Map*, of the GCE. The distance to the origin will describe the matching proximity of the resource to the job. Similarly, the extension of the third dimension to include the availability of resources extends the *Virtual Map* into a *Virtual Space*. The most suited resource providers will continue to be the sites whereby it is located nearest to the origin. The sections below will demonstrate how we construct the two basic dimensions and the process of aggregation that leads to the final aggregated Indexes used in the Virtual Map. A description of the simplicity of extending this to a Virtual Space is then described.

13.3.1 Computation Dimension

Resources in the computation dimension consist of entities that would impact the efficient computation of a job. Each resource is in turn represented by a capability value and a requirement value. In our simulations, we make use of the following allocatable resources as basis for scheduling in the computation dimension:

- GFLOP (C)
- RAM (M)
- Disk space (F)

However, we note that this is insufficient to represent a collection of sites and how they can possibly inter-operate with each other. A job submitted to a poorly connected site will be penalized when job fragmentation occurs or when the data required for processing is located in another location.

In order to minimize the detrimental effects in such cases, we introduce a parameter referred to as the *Resource Potential*. This is to assist in the evaluation of the Computation Index. We denote m as the total number of sites in a GCE. The potential, denoted as P_i, of a resource R_i quantifies the level of network connectivity between itself and its neighboring sites. For simplicity, we assume that the network latencies as well as the communication overhead of a resource is inversely proportional to its bandwidth. We refer to the Resource Potential, P_i of a resource R_i, as a form of "Virtual Distance", where $1 \leq i \leq m$. This is computed as $P_i = \sum B_{ij}$ where, B is the upload bandwidth, expressed in bits per sec, from R_i to R_j for $i \neq j$ and $B_{ij} = 0$ if $i = j$. This effectively eliminates all network complexities and "flattens" the bandwidth view of all the resources to the maximum achievable bandwidth between resources. This also inherently includes all sub-net routing overheads and communication overheads when a bandwidth monitoring system such as NWS [19] is employed. We illustrate this "flattening" process in Fig. 13.4. The values C, M, F and P_i dynamically change with resource availability over time t, and is constantly monitored for changes in our simulation. Thus, in a GCE where we characterize the resource environment as a set $S = \{R_1, ..., R_m\}$, we can represent the allocatable computational resources within a site i as a set $S_c = \{R_i, t\}$ where $S_c \subseteq S$. R_i is further represented by 4-tuple of $f_i(<C, M, F, P_i>, t)$ denoting the four resources considered in our allocation strategy.

In order to ascertain an aggregated Computation Index of a site to a job, resources are also requested based on the same GFLOPs, RAM and Harddisk space required. Similar to a node's Resource Potential described earlier, jobs are also additionally characterized by a potential value. However, this potential value is not obtained from the location where the job is submitted from, rather, it is obtained from the location of the source file required for the job to execute efficiently. In our simulations, we assume that each job only requires data from one data resource. This data resource can be either local to the job submission site or remote. As MRS is expected to operate in a GCE, we also simulate scenarios wherein users can submit jobs from different locations[3].

We characterize the job environment by $J = \{A_i, ..., A_j\}$, and the computational requirement of each job A_j in the set of J jobs is represented by $g_j(<C, M, F, P_{src}>, t)$.

---

[3] In our simulations, we have assumed that applications are pre-staged at the sites.

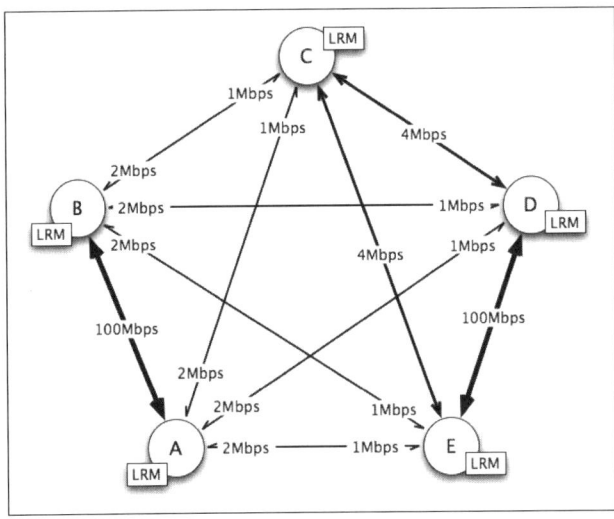

Fig. 13.4. Flattened network view of resources for computation of Potential

13.3.2 Computational Index through Aggregation

Evaluation of various resource requirements of sites and jobs allows us to aggregate and encode inter-resource relationships in order to arrive at a single index which can be used to obtain the allocation score. This is done by obtaining a ratio of provision (R_{ij}), for site i and job j, between what is requested and what is possibly provided. For computational resources, it is given by, $R_{ij}\{C\} = 1 - \frac{f_i\{C\}}{g_j\{C\}}$. Only the positive values of $R_{ij}\{C\}$ are considered, such that and $R_{ij}\{C\} = 0$ if the above evaluates to be less than zero. $f_i\{C\}$ and $g_j\{C\}$ are the GFLOP resource provided at site i and GFLOP resource required by job j. We only consider positive values in the Virtual Map, and therefore truncate the values at zero.

We apply the same ratio of provision to all resource and requirements which also includes RAM (M) and Harddisk (F) requirements. Additionally we also include the ratio of provision between the potential value of the site (P_i) and the source file potential (P_{src}). This allows us to evaluate if a site connectivity is equal or better to where the source data file is located. This ensures that the possible target job submission site will not be penalized more than required if job fragmentation is to occur, when compared to executing the job in place at the data source location.

These ratios are then aggregated into a dimensionless computation index (x_{ij}) for site i on job j using the following equation. Constants K_C, K_M, K_F and K_P represents weights that provide modification to the importance of the respective provisioning ratios in terms of importance to each other. An increasing value of $K > 0$ signifies an increasing importance of a specific resource requirement relative to the other resources. This steers the strategy away from the default allocation to one that is weighted towards the more important resource.

After the sites providing resources are indexed to obtain x_{ij}, the site i with the lowest computation index, x_{ij}^* is deemed to provide the best resources suited for a job j. In our simulations, we set the K constants such that $K = 1$. This provides equal importance to all components making up the computational index. Detailed derivation and formulations can be found in [5]. Essentially, such a strategy does not restrict itself to specific units of measure for C, M or F. Potentially, any arbitrary unit is suitable for this approach, as long as the entire GCE is in agreement.

$$x_{ij} = \sqrt{(K_C R_{ij} \{C\})^2 + (K_M R_{ij} \{M\})^2 + (K_F R_{ij} \{F\})^2 + (K_P R_{ij} \{P\})^2} \quad (13.1)$$

13.3.3 Data Dimension and Indexing through Resource Inter-relation

In the data dimension, we wish to determine the best resource that would execute a job considering its I/O requirements. The expected time for I/O is determined based on the estimated data communications required and the bandwidth between the source file location and the target job allocation site. The ratio between the I/O communication time to the estimated local job run-time is then taken. This ratio allows us to evaluate the level of advantage a job has in dispatching that job to a remote site. This is because a site capable of executing a job locally would incur a minimal (non-zero) I/O time as compared to any other remote location. Thus, allocation of a job to the intended target resource should be one whereby this ratio is as low as possible.

The I/O time is time dependent resource which is based on the instantaneous bandwidth availability at a resource. We annotate bandwidth B between two sites i and j as $B_{ij} = min\{B_{ij}^{download}, B_{ji}^{upload}\}$ which changes over time t as data capabilities of a resource $S_d\{R_i, t\}$. Where each item in the set is represented by $d_i\{<B>, t\}$. The data requirement of a job j is thus represented by $e_j\{<F, A^{runtime}>, t\}$ where $A^{runtime}$ is the estimated run-time of the job.

We make use of this ratio to create the Data Index. This evaluation is an example of aggregation based on resource inter-relation. I/O time is affected by the amount of data for a job and the actual bandwidth resource available. In the worst case scenario, the amount of data required for the job would also be the amount of hard-disk resource required at the site to store the data to be processed. This, therefore inter-relates the data resources to the bandwidth resources available. The ratio is written as follows.

$$y_{ij} = \frac{e_j\{F\}}{d_i\{B_{ij}\}} \cdot \frac{1}{A^{runtime}} \quad (13.2)$$

13.3.4 Dimension Merging

From the individual Computation and Data Indices described above, we observe that the best allocated resources are represented by those with low index values. Each of the individual indices are also encoded with resource requirements considerations in its evaluation through aggregation. These points when plotted on a 2-dimensional axis creates what we termed as the Virtual Map. As we have observed, sites that position themselves closest to the origin are those that deviate from the resource requirements by the least amount. An illustration of the virtual map is shown in Fig. 13.5. The euclidean distance from the origin therefore denotes the best possible resources that matches the resource requirements of a job for an instance in time.

In Fig. 13.5, the computation and data index is computed by Eq. 13.1 and (13.2 for each job in the queue. As job requirements differs for each job, the Virtual Map is essentially different for each job submitted. This has to be computed each time a job is to be submitted or re-submitted to the GCE.

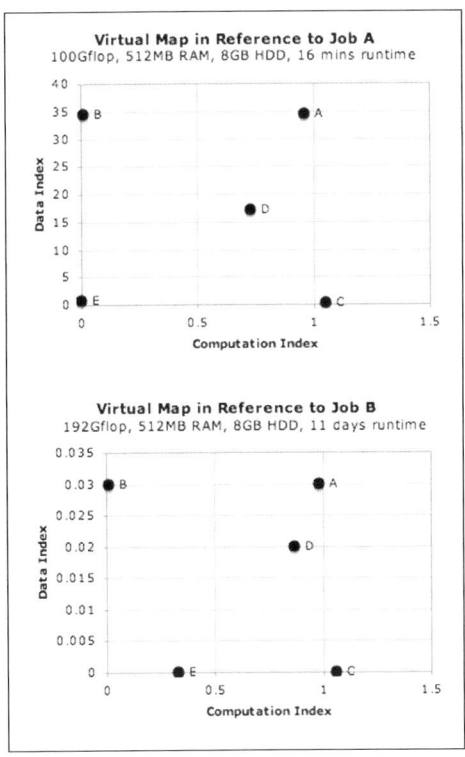

Fig. 13.5. A Virtual Map is created for each job to determine allocation

13.3.5 Availability Index

We first define the following notations used in this section, and the definitions associated with them:-

- $MTTF_j$ and λ_j^F : The Mean Time to Failure represents the average amount of time a resource is available to the GCE before going offline. We also term the average rate of failure to be $\lambda_j^F = \frac{1}{MTTF_j}$. Where j denotes the node index in the GCE.
- τ, τ_j^U : τ represents a specific time instance after the time period T, while τ_j^U is defined as the duration of the of the jth node in the UP state.
- P_j : Denotes the resource reliability is a single value representing the likelihood of a resource staying on-line at any given time. This value is influenced by information such as the resource availability pattern to the GCE, the reliability of the various components in the resource and the reliability value provided by the creators of this resource.
- Pr_j^{UP}: The probability of a resource j remaining in its UP, or on-line, state.

Using the Poisson Distribution to model the event of a single change in state, with the assumption that the resources remain in constant state within τ period of time, it is possible to estimate the probability of Pr^{UP} at time $(T + \tau)$. This is captured by Eq. 13.3.

$$Pr^{UP}\{n_{T+\tau}\} = 1 - \prod_{j=1}^{n} \lambda_j^F (1 - P_j) \sum_{t=0}^{\tau_j^U + \tau - 1} e^{-(\lambda_j^F t)} (\lambda_j^F t) \qquad (13.3)$$

We make use of Eq. 13.3 to obtain an index of Availability represented by $(1 - Pr^{UP})$. This is coupled as a third dimension in addition to the Compute and Data Index for each resource in consideration. One would notice that a resource that is more likely to stay up will have a value closer to zero than one that is likely to fail. The inclusion of this index allows one to pro-actively estimate which resources will be available during a jobs run-time such that the likelihood of job failure is reduced. This is different from other proposed failure handling strategies where-by the failed job is trapped and then restarted.

The merging process is similar to that described in section 13.3.4, where the shortest Euclidean distance from the origin denotes the best possible resources that matches the resource requirements of a job for that instance in time.

13.4 Performance Evaluation

13.4.1 Computation and Data Index in 2D MRS

We compare our basic 2-dimensional MRS with the Backfilling strategy (BACK-FILL) [4, 13] and a job Replication (REP) strategy [8], which is similar to that used in SETI@Home [7]. The workload model provided by [16] was used as the workload input. The following metrics where used as the performance measure of the algorithm.

1. Average Wait-Time (AWT)
   This is defined as the time duration for which a job waits in the queue before being executed. The wait time of a single job instance is obtained by taking the difference between the time the job begins execution (e_j) and the time the job is submitted (s_j). This is computed for all jobs in the simulation environment. The average job waiting time is then obtained. If there are a total of J jobs submitted to a GCE, the AWT of a job is given by,

$$AWT = \frac{\sum_{j=0}^{J-1}(e_j - s_j)}{J}$$

   This quantity is a measure of responsiveness of the scheduling mechanism. A low wait time suggests that the algorithm can potentially be used to schedule increasingly interactive applications due to reduced latency before a job begins execution.

2. Queue Completion Time (QCT)
   This is defined as the amount of time it takes for the scheduling algorithm to be able to process all the jobs in the queue. This is computed by tracking the time when the first job enters the scheduler until the time the last job exits the scheduler. In our experiments, the number of jobs entering the system is fixed, to make the simulation more trackable. This allows us a quantitative measure of throughput, where the smaller the time value, the better. The queue completion time is given by,

$$QCT = e_{J-1} + E_{J-1} - s_0$$

   where, E_{J-1} is the execution time of the last job. This includes the I/O and communication overheads that occurs during job execution.
   This metric, when coupled with the average waiting time of a job, allows us to deduce the maximum amount of time a typical job will spend in the system for a given workload.

3. Average Grid Utilization (AGU)
   This quantity investigates how well the algorithm is capable of organizing the workload and the GCE resources so as to optimize the performance. Thus, the higher the utilization, the better optimized the environment is. The utilization of the GCE at each execution time step is captured and represented as $U(t) = \frac{M_u}{M}$, where M is the total computational resources available. M_u is the number of computational resources utilized. The average grid utilization is thus given by the following equation.

$$AGU = \frac{\sum_{t=s_0}^{QCT} U(t)}{QCT}$$

The results of our experiments is summarized in Fig. 13.6. The significance of these results are discussed below.

It was noted that in terms of AWT, both REP and MRS significantly outperforms BACKFILL by 40% and 50% respectively. This is due to the fact that

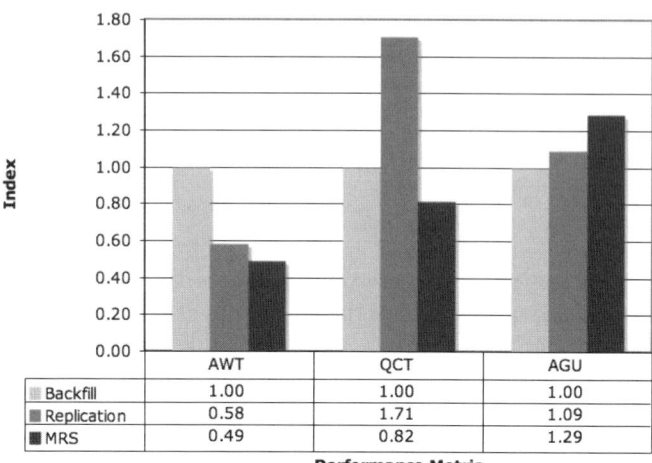

Fig. 13.6. Normalized comparison of MRS and REP simulation to Backfill Algorithm

the backfill algorithm does not allocate jobs in consideration of the data distribution time. The improved performance of REP compared to BACKFILL on AWT can be attributed to the fact that as a job gets replicated, the likelihood of being allocated to a faster resource or bandwidth increases. This is however non-optimal as it was achieved without making full use of the information available in the execution environment. This non-optimality is verified by the fact that MRS is able to achieve an even better AWT by making use of inter-resource relationships defined within its indices.

From the figure, we can also clearly see that the utilization for BACKFILL is the lowest in all the experiments. REP and MRS exhibits increasing levels of utilization which accounts for a shorter AWT. However, it may be noted that in the replication algorithm, every job is essentially submitted twice in order to achieve better performance. This replication potentially hinders the execution of other jobs that might require more CPUs in the GCE. This is clear from the fact that an increase in utilization using the REP strategy does not translate to an improvement in QCT. It has, instead, induced a detriment to the GCE by almost 70% when compared to BACKFILL. This could be attributed to inefficient allocation of resources. In contrast, we can see that an improvement in utilization of 29% between MRS and BACKFILL is also directly reflected by a 18% improvement in QCT.

As discussed earlier, we have ascertained that replication can lead to a degradation of performance when the entire queue is considered. In contrast to BACKFILL and REP, our simulations have shown that MRS has been able to achieve a 50% improvement AWT, an 18% improvement over QCT and a 29%

improvement in AGU. This is due to the fact that MRS makes use of comparative measures on the benefits of allocation to each node. This is inherent to the algorithm during the process of Virtual Map creation. A lower AWT is very much due to a good allocation decision of the resources when MRS is presented with a queue of jobs. This allows for more jobs to be allocated per unit time, which is clearly reflected in MRS's improvement in QCT over BACKFILL. This continues to be achieved when compared to REP, indicating that MRS is able to allocate resources more effectively when compared to REP. This is clearly shown when comparing the results in Fig. 13.6. The matching of resources using the computation and data indexes, also results in a much higher utilization, dispatching jobs to nodes that are able to satisfy the jobs while intelligently deciding which jobs to keep local and which jobs to dispatch.

13.4.2 Inclusion of Availability Index in 3D MRS

From the positive results in the implementation of 2D MRS in section 13.4.1, we extended our GCE to exhibit failures in resources and also included the Availability Index as the third dimension in MRS. This is to further investigate if the extension of the MRS methodology will continue to exhibit positive results even in higher dimensions. It is also to apply the strategy in a more realistic GCE environment where failures do affect how a job should be scheduled. We used the following metrics as a measure of performance.

1. Job Processing Rate (JPR):

$$JPR = \frac{NumberOfJobsSuccessfullyCompleted}{TotalQueueCompletionTime} = \frac{J_{Success}}{T_Q}$$

A higher JPR will indicate larger number of successfully completed jobs or a lower queue completion time. A high JPR will therefore indicate that an algorithm is capable of high throughput.

2. Job Failure Rate (JFR):

$$JFR = \frac{NumberOfJobsFailedAtRuntime}{TotalQueueCompletionTime} = \frac{J_{Fail}}{T_Q}$$

A low JFR is desired as it signifies the number of jobs failing during the course of its queue completion is low. This thus indicates that a strategy is able to allocate resources will to reduce the number of jobs failing in its course of execution.

3. Job Rejection Rate (JRR):

$$JRR = \frac{NumbeOfJobsRejected}{TotalQueueCompletionTime} = \frac{J_{Rej}}{T_Q}$$

A low JRR indicates the ability of an algorithm to handle all types of jobs submitted to the queue based on the workload model used. A high JRR will therefore mean that the algorithm is unable to execute jobs due to insufficient capacity. A low JRR is thus desired to indicate that an allocation strategy is able to handle the workload presented using the workload model.

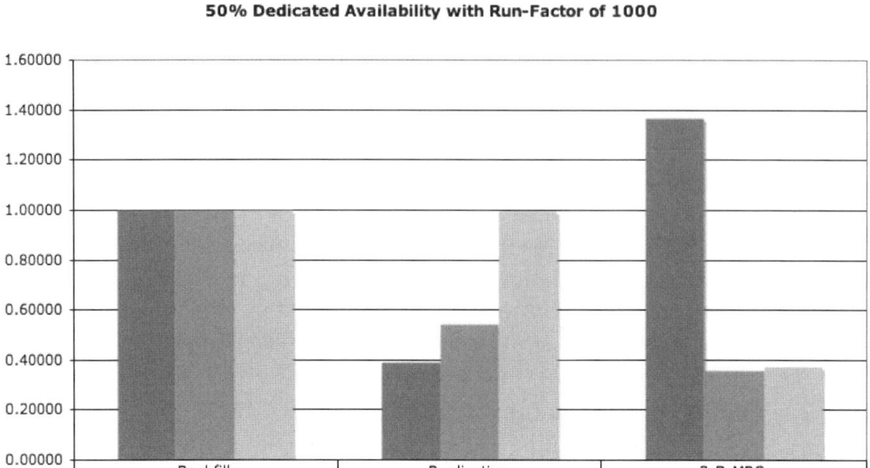

50% Dedicated Availability with Run-Factor of 1000

	Backfill	Replication	3-D MRS
■ NJPR	1.00000	0.39011	1.36816
■ NJFR	1.00000	0.54348	0.35870
■ NJRR	1.00000	0.99494	0.37421

Fig. 13.7. Normalized comparison of 3-D MRS and REP simulation to Backfill Algorithml

The GCE was simulated to include 50% of dedicated resources while the remaining are volatile resources that goes on and offline periodically based on a set of random generated normally distributed $MTTF$ value. The workload environment was also modified to address the problem where workload models tends to generates much lesser jobs with long run-times. It is done such that the longest job run-time is 1000 times that of the average $MTTF$ in the GCE. This would induce a much larger number of failures in the volatile resources, providing a better view into the effectiveness of the algorithm. The normalized result of the simulation is shown in the Fig. 13.7.

From the results, we can clearly see an approximately 30% increase in the JPR of 3-D MRS when compared to BACKFILL and an even larger increase when compared to REP. This is likely due to the more effective job to resource matching strategy employed in MRS, allowing more jobs to complete within each period of high resource availability. The approximately 64% and 61% improvements in JFR and JRR also clearly demonstrates the effectiveness of being able to include availability information as part of the allocation strategy. It is clear from the simulations that the job failures resulting in a REP strategy is approximately half that of the replication factor (which was 2 in the simulation). It can be reasoned then that for every 2 jobs replicated, there is a 50% chance that one of them will fail due to the lack of knowledge in which half of the resources is likely to fail. One would also notice that the JRR of REP is similar to that of

BACKFILL further pointing to limited improvement in resource utilization in the GCE.

In general, it is observed that MRS is able to render a performance that is much suited for scheduling resources over a GCE, and is able to be extended to include availability factors of the GCE into the strategy while continuing to provide superior performance to traditional algorithms.

13.5 Conclusions

In this chapter, we have proposed a novel distributed resource scheduling algorithm capable of handling several resources to be catered among jobs that arrive at a Grid system. Our proposed algorithm, referred to as Multi-Resource Scheduling (MRS) algorithm, takes into account the different resource requirements of different tasks and shown to obtain a minimal execution schedule through efficient management of available Grid resources. We have proposed a model in which the job and resource relations are captured and are used to create an aggregated index. This allows us to introduce the concept of virtual map that can be used by the scheduler to efficiently determine a best fit of resources for jobs prior to execution. We also introduced the concept of Resource Potential to identify inter-relations between resources such as bandwidth and data. This allows us to identify sites that has least execution overheads with respect to a job. A third dimension was also introduced to extend the idea of a virtual map into a virtual space. This new dimension proposes the use of availability of each resource such as to provide a more accurate resource allocation strategy.

In order to quantify the performance, we have used various measures to ascertain the performance of our strategy in both the 2D and 3D aspects. We considered practical workload models that are used in real-life systems to quantify the performance of MRS. Performance of MRS has been compared with conventional backfill and replication algorithms that are commonly used in a GCE. Our experiments have also conclusively elicited several key performance features of MRS with respect to the backfill and replication algorithms, yielding performances improvements up to 50% on some performance measures. The strategy presented continues to exhibit performance gains even when extended to a GCE that exhibits failures. Presenting more than 60% improvements in job failure rates while improving throughput by up to 30%.

The strategy discussed in this chapter introduces a mechanism where individual resources can be compounded into dimensions. Within these dimensions, the inter-relations of resources are addressed. By placing multiple dimensions together, it creates a virtual map, which can be extended into a virtual space when more than 2 dimensions are used. The ability for this strategy to be able to extend itself through additional dimensions, provides a mechanism where other forms of specialized optimization heuristics can be used. These individual dimensions, which are optimal in itself, could then be aggregated in a strategy similar to MRS where each can be weighted and inter-related for further optimization over multiple resources. The idea of a virtual map or space also potentially provides a starting point where more complex optimizations can be applied based on virtual "spatial" considerations in a GCE.

Some possible immediate extensions to the strategy we have proposed in this chapter could look at providing a computationally less intensive mechanism to compute the predicted availabilities in the GCE such as to provide a gauge of capacity in the GCE. It could also be possible that other parameters such as, Quality-of-Service, economic considerations as well as real-time applications can be included into the model by simply extending the number of dimensions of consideration.

Related Work

There have been other strategies introduced to handle resource optimization for jobs submitted over Grids. However, while some investigated strategies to obtain optimizations in the computational time domain, others looked at optimizations in data or I/O domain. Very few works address failure on Grids. We classify the current available work on Grid failures into pro-active and post-active mechanisms. By pro-active mechanisms, we mean algorithms or heuristics where the failure consideration for the Grid is made *before* the scheduling of a job, and dispatched with hopes that the job does not fail. Post-active mechanisms identifies algorithms that handles the job failures *after* it has occurred.

In [17], job optimization is handled by redundantly allocating jobs to multiple sites instead of sending it only to the least loaded site. The rationale in this scheme was that the fastest queue will allow a job to execute before its replicas and this provides low wait times and improves turn-around time. Job allocation failures due site availabilities would also be better handled due to this redundancy. However, this strategy leads to problems where queue lengths of different sites are unnecessarily loaded handle the same job. The frequent changes in queue length can also potentially hamper on-site scheduling algorithms to work effectively as schedules are typically built by looking ahead in the queue. In addition, the method proposed does not investigate the problems that can arise when the data required for the job is not available at the execution site and needs to be transported for a successful execution.

In [20], Zhang has highlighted that the execution profiles of many applications are only known in real-time, which makes it difficult for an "acceptance test" to be carried out. The study also broke down the various scheduling models into Centralized, Decentralized and Hierarchical models where jobs are submitted to a meta-scheduler but are dispatched to low-level schedulers for dispatch. Effective virtualization of resources was also proposed in order to abstract the resource environment and hide the physical boundaries defined. A buddy set as in [15] was also proposed, and its effectiveness also highlighted in [1], where it was shown that an establishment of relationships in resources can lead to better performance.

In the work presented in [10], the ability to schedule a job in accordance to multiple (K) resources is explored. This approach shows clearly the benefits where scheduling with multiple resources is concerned. Effective resources-awareness in the scheduling algorithm provided performance gains of up to 50%. Similar resource awareness and multi-objective based optimizations where studied in [18]. In both cases, the limitations of conventional methods was also identified as there was have no mechanism for utilizing additional information known about the system and its environment. However, in [10], there was no data resources identified, while in [18], we believe that the over simplicity of resource aggregation was in-adequate in capturing resource relationships.

Of works that look into failures in the GCE, many works are primarily post-active in nature and deal with failures through Grid monitoring as mentioned in [12]. These methods mainly do so by either checkpoint-resume or terminate-restart [9,11]. Two pro-active failure mechanisms are introduced in [8,17] and [2]. While [8,17] operates by replicating jobs on Grid resources, [2] only looks at volunteer Grids. The former can possibly lead to an over allocation of resources, which will be reflected as an opportunity cost on other jobs in the execution queue. While the latter only addresses independent task executing on the resources.

The formulation of 2D MRS in [5] has allowed us to build on a effective mechanism, providing an alternative solution to the problem of resource allocation. This also highlights how the strategy can be extended to consider more complex environmental requirements, which is presented in this chapter.

Acknowledgment

Parts of this work addressed certain objectives of a funded project in Grid Computing under the Grant 052 015 0024/R- 263-000-350-592 from National Grid Office via A*Star, Singapore. Bharadwaj Veeravalli would like to thank the funding agency in supporting this research.

References

1. Azzedin, F., Maheswaran, M.: Integrating Trust into Grid Resource Management Systems. In: Proc. ICPP, pp. 47–52 (2002)
2. Choi, S., Baik, M., Hwang, C.S.: Volunteer Availability based Fault Tolerant Scheduling Mechanism in Desktop Grid Computing Environment. In: The Proceedings of the 3rd IEEE International Symposium on Network Computing and Applications, Boston, Massachusetts, August 30th - September 1st, pp. 366–371 (2004)
3. Foster, I., Kesselman, C.: The Grid: Blueprint for a new Computing Infrastructure, 2nd edn. Morgan Kaufmann, San Francisco (2004)
4. Hamscher, V., Schwiegelshohn, U., Streit, A.: Evaluation of Job-Scheduling Strategies for Grid Computing. In: The Proceedings of 1st IEEE/ACM International Workshop on Grid Computing, Brisbane Australia, pp. 191–202 (2000)
5. Khoo, B., Boon, T., Veeravalli, B., Hung, T., See, S.: A Multi-Dimensional Scheduling Scheme in a Grid Computing Environment. Journal of Parallel and Distributed Computing (JPDC) 67(6), 659–673 (2007)

6. Khoo, B., Veeravalli, B.: Cluster Computing and Grid 2005 Works in Progress: A Dynamic Estimation Scheme for Fault-Free Scheduling in Grid Systems. IEEE Distributed Systems 6(9) (2005)
7. Korpela, E., Werthimer, D., Anderson, D., Cobb, J., Lebofsky, M.: SETI@home-Massively distributed computing for SETI. Computing in Science and Engineering 3(1), 78–83 (2001)
8. Li, Y., Mascagni, M.: Improving Performance via Computational Replication on a Large-Scale Compuational Grid. In: IEEE/ACM CCGRID 2003, Tokyo, p. 442 (2003)
9. Lee, H.M., Chin, S.H., Lee, J.H., Lee, D.W., Chung, K.S., Jung, S.Y., Yu, H.C.: A Resource Manager for Optimal Resource Selection and Fault Tolerance Service in Grids. In: The Proceedings of 4th IEEE International Symposium on Cluster Computing and the Grid, Chicago, Illinois, USA, pp. 572–579 (2004)
10. Leinberger, W., Karypis, G., Kumar, V.: Job Scheduling in the presence of Multiple Resource Requirements. In: Proceedings of the IEEE/ACM SC 1999 Conference, Portland, Oregon, USA, November 13-18, pp. 47–48 (1999)
11. Litzkow, M., Livny, M., Mutka, M.: Condor - A hunter of Idle Workstations. In: The Proceedings of the 8th International Conference of Distributed Computing Systems, pp. 104–111 (June 1988)
12. Medeiros, R., Cirne, W., Brasileiro, F., Sauve, J.: Faults in Grids: Why are they so bad and What can be done about it? In: The proceedings of the Fourth international Workship on Grid Computing (GRID 2003), pp. 18–24 (2003)
13. Mu'alem, A.W., Feitelson, D.G.: Utilization, Predictability, Workloads, and User Runtime Estimates in Scheduling the IBM SP2 with Backfilling. IEEE Transactions on Parallel & Distributed Systems 12(6), 529–543 (2001)
14. Parallel Workload Archive: Models, http://www.cs.huji.ac.il/labs/parallel/workload/models.html
15. Shin, K.G., Chang, Y.: Load sharing in distributed real-time systems with state change broadcasts. IEEE Transactions on Computers 38(8), 1124–1142 (1989)
16. Song, B., Ernemann, C., Yahyapour, R.: User Group-based Workload analysis and Modelling. In: Cluster and Computing Grid Workshop 2005, Cardiff United kingdom, pp. 953–961 (2005)
17. Subramani, V., Kettimuthu, R., Srinivasan, S., Sadayappan, P.: Distributed Job Scheduling on Computational Grids Using Multiple Simultaneous Requests. In: The Proceedings of 11th IEEE International Symposium on High Performance Distributed Computing HPDC-11 20002 (HPDC 2002), Edinburgh, Scotland, July 24-26, pp. 359–368 (2002)
18. Vijay, K.N., Chuang, L., Yang, L., Wagner, J.: On-line Resource Matching for Heterogeneous Grid Environments. In: Cluster and Computing Grid, Cardiff, United Kingdom, pp. 607–614 (2005)
19. Wolski, R., Obertelli, G.: Network Weather Service (2003), http://nws.cs.ucsb.edu
20. Zhang, L.: Scheduling algorithm for Real-Time Applications in Grid Environment. In: The Proceedings on IEEE International Conference on Systems, Man and Cybernetics, USA, vol. 5 (2002)

Index

Advanced Job Scheduler, Markov Availability Model, Resource Selection, Desktop Grid Computing, Stochastic scheduling   153
Agent   222, 224–226, 228, 233, 236–238
ApMon   223, 224, 239

Backlog   40, 45
Batch mode scheduling   178
Batch of Task   234
Best-effort based workflow scheduling   176
Branch-and-Bound algorithms   303
  subtrees   303
Broker   225, 226, 233
Budget constrained scheduling   198

Cellular Memetic Algorithms   274
  Cell updating   280
  Local search methods   284
  Mutation operator   283
  Population initialization   280
  Population shape   286
  Population topology   278
  Recombination operator   282
  Replacement Policy   286
Chromosome   218, 225–230, 232–234
Cluster based scheduling   185
Cluster Scheduling
  Non-combinatorial policies   101
communication model
  One-port   122, 124
Computational Grid   3

Grid scenario   6
  Resource utilization   19
Computational models
  Cluster Grids model   16
  ETC model   14
  Grid Information System model   16
  Multi-Cluster Grids model   16
  TPCC model   15
Condor   224
Crossover, operator   218, 228, 229, 235
Crossover, single-point   228
Crossover, two-point   228
Crossover, uniform   228

Data Grids   5
Deadline constrained scheduling   198
Decentralized Grid Scheduling, Genetic Algorithms, Task assignment, Lookup services   215
Decentralized Scheduler Architecture   217
Dependency mode scheduling   182
Desktop Grids   5
DIOGENES   219, 222, 223, 226
Direct Acyclic Graph,DAG   96
Distributed computing   215
Distributed systems   216
Duplication based scheduling   185
Dynamic programming   48
Dynamic real-time systems   68

Encoding representation   330
Enterprise grid   39
Enterprise Grids   5

Index

ETC, *see* expected-time-to-compute matrix
Evolutionary Algorithms 103, 304
  Elitism 105
  Estimation of Distribution Algorithms 104
  Genetic Algorithms 103
  Genotypes 104
  Phenotypes 104
  Steady State Algorithms 104
Expected-time-to-compute matrix
  Consistent 125
  Inconsistent 125

Fitness function 188
Flow-shop Scheduling 97
Flowtime 274
Fork-graph 124

Genetic Algorithm
  Initialization methods 29
Genetic algorithm 49
Genetic Algorithms 188, 218, 219, 222, 224–227, 233–236, 240, 242, 324
GRASP 186
Greedy optimization 48
Grid 215–220, 223–225, 234, 240, 244
Grid Computing 273
  eScience applications 4
Grid computing 1
Grid middleware 3
Grid monitoring 216, 222–224, 233, 244
Grid scheduler 2
Grid Scheduling 6, 220, 222, 226
  Average Weighted Response Time 21
  Matching proximity 20
  Particle Swarm Algorithm 254
  Total weighted completion time 20
  Turnaround time 20
Grid scheduling
  Adaptive scheduling 13
  Ant Colony Optimization 253
  Batch scheduling 13
  Centralized scheduling 12
  Computational models 14
  Decentralized scheduling 12
  Dynamic benchmark 293
  Dynamic environment 292
  Dynamic scheduling 12
  Economy-based scheduling 31
  Evolutionary Algorithms 250
  Evolutionary Multi-objective Optimization 251
  Fuzzy scheme 257
  Grid security 31
  Grid services scheduling 31
  Hierarchical scheduling 12
  Immeadiate scheduling 13
  Nature Inspired Meta-heuristics 249
  Optimization criteria 17
  Performance requirements 17
  Phases 10
  Scheduling in data grids 14
  Simulated Annealing 251
  Simulator 292
  Static benchmark 288
  Static scheduling 12
Grid system 215
Grid workflows 11

HEFT 183
Heterogeneous systems 215
Heuristics
  Ad hoc methods 29
  Ant Colony Optimization 29
  Hill Climbing 25
  Hyper-heuristic method 30
  Local search 24
  Memetic Algorithms 28
  Population-nased 28
  Simulated Annealing 25
  Tabu Search 25
  Variable Neighborhood Search 26
Hybrid approaches
  Fuzzy Logic 30
  QoS approach 30
  Reinforced learning 30
Hypergraph 123, 130
  partitioning problem 130

Independent job scheduling 11
Individual task scheduling 178

Job flows 39
Job scheduling 6
  Characteristics 6
  Completion time 18
  Definition 8

Flowtime 18
Makespan 18
Performance requirements 17
Terminology 8
Job scheduling
  Completion time 18
Job-shop Scheduling 98

Knapsack Problem 99

LAN 224
List scheduling 178
LoadLeveler 102
Local search
  emptiest resource rebalance 26
  Flowtime rebalance 26
  Local move 25
  Local rebalance 26
  Local short hop 26
  Local swap 26
  Resource flowtime rebalance 26
  Steepest local move 25
  Steepest local swap 26
  Variable Neighborhood Search 26
Lookup services 215
LSF 224

Makespan 154, 274
Markov modelling 154
Master process,Worker process 309
Maui 102
Max-Min 180
Memetic Algorithm 28
  Initialization methods 29
Memetic Algorithms 277
Meta-scheduler 40
Min-Min 180
Moab 102
MonALISA 215, 222–224, 234, 239
Multi-dimensional robustness metric 71
Multi-objective genetic algorithm 330
Multi-objective optimization
  Hierarchic approach 21
  Pareto set 251
  Simultaneous approach 21
Multiple Offspring Sampling 95, 102–111
  Algorithm-based MOS 106
  Coding-based MOS 107

Fitness landscape 107
Genotype encodings 108
Hybrid MOS 107
Multiple Codings 95
Operator-based MOS 107
Parameter-based MOS 107
Participation Function 109
Technique 106
Transformation Functions 108
Multiple Resource Management 344
Multiple resource scheduling 341
Multiprocessor Scheduling 98
Mutation, Additive 230
Mutation, operator 218, 228, 229, 235
Mutation, Order-based (Swap) 230
Mutation, Partial-gene 229

Neighbor Selection
  Multi-objective 338
  Single objective 332
Neighbor-selection problem 326

OpenPBS 102

P2P Computing 305
P2P Computing, Branch and Bound, Genetic Algorithms, Grid Middleware, Flow-Shop Scheduling 301
Packing Problem 99
Parallel architecture 215
Parallel GA 309
Particle Swarm Optimization 324
PBS 102 224
PBS Resource Manager 101
Peer-to-Peer
  Middleware 305
Peer-to-peer computing 323
Performance Metrics 220
Policy
  Backfilling 101
  Backfilling with Reservations 101
  Conservative Backfilling 101
  First-come-first-serve 101
  Shortest-job-first 101
Predictive Failure Handling 344
ProActive
  Fault tolerance 314
Proxy server 224

QoS   197
QoS guided min-min   181

RCL   187
Real Time Computing, Real Time Allocation, Robust Allocation, Scheduling, Heuristics, Distributed Real-time Systems   61
Recovery Service   244
Resource requirements   344
Robust allocation   76
  Multi-dimensional problem   76
Routing, Backlogs, Distributed System, Job Flows, Cluster, Genetic Algorithms, Dynamic Programming   39
Routing constraints   45
Routing policy   45

Scaling   55
Scavenging Grid   4
Scheduling   95, 174, 216–220, 223
  Flow-shop Scheduling (FSS)   95
  Job-shop Scheduling (JSS)   95
  Multiprocessor Scheduling (MPS)   95
  Supercomputer Scheduling   95
Scheduling, File-Sharing Tasks, Iterative-Improvement Heuristics, Heterogeneous Platforms, Neighborhood exploration   121

Scheduling, Supercomputing, Evolutionary Algorithms, Multiple Offspring Sampling, Genetic Algorithms   95
SGE   224
Simulated Annealing   192
SLURM   102
Smooth objective function   126, 131
Sufferage   180
Supercomputer Scheduling   99

TANH   185
Task   225, 227, 228, 230, 231, 233, 235, 240–242
Task allocation   240
Task assignment   242
Task Queue   225
Task Scheduling   232, 233
Torque   102

Variable Neighbourhood Search   107
Virtual Maps   347

WAN   224
Workflow   173
Workflow scheduling, Inter-dependent tasks, Distributed resources, Heuristics   173

XML   221
XSufferage   181
XtremWeb,ProActive   305

Author Index

Abraham, Ajith 1, 247, 273, 323
Alba, Enrique 273
Aykanat, Cevdet 121

Bendjoudi, A. 301
Boboila, Marcela 215
Buyya, Rajkumar 173
Byun, EunJoung 153

Choi, SungJin 153
Cristea, Valentin 215

Dorronsoro, Bernabé 273
Duran, Bernat 273

Grosan, Crina 247
Gu, Dazhang 61
Guerdah, S. 301

Hwang, ChongSun 153

Iordache, George 215

Kaya, Kamer 121
Khoo, B.T. Benjamin 341
Kim, HongSoo 153

LaTorre, A. 95
Lee, SangKeun 153
Liu, Hongbo 247, 323

Mansoura, M. 301
Melab, N. 301
Miguel, P. de 95
Montana, David 39

Peña, J.M. 95
Pop, Florin 215

Ramamohanarao, Kotagiri 173
Robles, V. 95

Stratan, Corina 215

Talbi, E-G. 301

Uçar, Bora 121

Veeravalli, Bharadwaj 341

Welch, Lonnie 61

Xhafa, Fatos 1, 247, 273, 323

Yu, Jia 173

Zinky, John 39